BIBLIOGRAPHIC RECORD TARGET

ACM2310
UL FMT B RT a BL m T/C DT 07/18/88 R/DT 08/04/94 CC STAT mm E/L 1
035/1: : |a (RLIN)MIUG86-B26506
035/2: : |a (CaOTULAS)160431889
040: : |a MiU |c MiU
100:1 : |a Bass, Edgar W. |q (Edgar Wales), |d 1843-1918.
245:00: |a Elements of differential calculus. |c By Edgar W. Bass ...
250: : |a 2d ed. 1st thousand.
260: : |a New York, |b J. Wiley & sons; |a London, |b Chapman & Hall, ltd.,
|c 1903.
300/1: : |a viii, 354 p. |b diagr. |c 20 cm.
650/1: 0: |a Differential calculus.
998: : |c DPJ |s 9124

ELEMENTS

OF

DIFFERENTIAL CALCULUS.

BY

EDGAR W. BASS,

Colonel United States Army, Retired. Professor of Mathematics in the U. S. Military Academy April 17, 1878, *to October* 7, 1898.

SECOND EDITION.

FIRST THOUSAND.

NEW YORK:

JOHN WILEY & SONS.

LONDON: CHAPMAN & HALL, LIMITED.

1993.

ROBERT DRUMMOND, ELECTROTYPER AND PRINTER, NEW YORK.

PREFACE.

THIS text-book has been prepared for the use of the Cadets of the U. S. Military Academy who begin the subject with a knowledge of the elements of Algebra, Geometry, and Trigonometry which ranges from fair to excellent. The time allotted to the subject (ten and one half weeks), and the requirements of the subsequent courses, especially Mechanics, Ordnance and Gunnery, and Engineering, limit and determine the scope of the work.

My experience leads me to the belief that the more rigorous and comprehensive method of infinitesimals is suitable only for a treatise, and not for a text-book intended for beginners.

At the same time I believe that any presentation of the subject, no matter how elementary, should in no manner prejudice the student against any established method. On the contrary, it should, I think, endeavor to lead him to an understanding of the relations between those in general use, and, above all, it should aim to construct the best possible ground work for the subsequent study of the subject treated in the most rigorous and extended form.

The principle of interchange of infinitesimals, which consists in replacing one infinitesimal by another when unity is the limit of their ratio, has been used to overcome the difficulty encountered by beginners in the determination of the limits of ratios of infinitesimals.

To the Officers of the U. S. Army who have taught the subject with me, and in many cases to my pupils, I am greatly indebted for much valuable assistance.

To Captain Wm. Crozier, Lieut. J. A. Lundeen, Lieut. H. H. Ludlow, and Lieut. F. McIntyre I am under obligations for many demonstrations and solutions.

Associate Professor W. P. Edgerton has been my collaborator throughout the work, and to him much credit is due for numerous demonstrations, improvements, and suggestions.

I have added a list of the works of authors which I have freely consulted in the preparation of this book, for the purpose of acknowledging my indebtedness to them, and for the benefit of students who may desire to extend their knowledge of the subject.

EDGAR W. BASS.

WEST POINT, N. Y., June 15, 1896.

PREFACE TO THE SECOND EDITION.

FOR the corrections and changes made in this edition I am indebted to the Department of Mathematics, U. S. M. A. My thanks are especially due to Professor W. P. Edgerton, Associate Professor Chas. P. Echols, Lieut. George B. Blakely, and Lieut. F. W. Coe.

EDGAR W. BASS.

524 FIFTH AVENUE, NEW YORK CITY,
June 1, 1901.

LIST OF AUTHORS WHOSE WORKS HAVE BEEN CONSULTED IN THE PREPARATION OF THIS BOOK.

AMERICAN.

Church,
Rice and Johnson,
Byerly,
Taylor,
Newcomb,
Bowser,
Hardy,
Osborne.

ENGLISH.

Todhunter,
Williamson,
Edwards,*
Greenhill,*
Price,*
Haddon, Examples.

FRENCH.

Bertrand,*
Jordan,*
Duhamel,*
Hoüel,*
Haag,*
Serret.*

Harnack's Introduction to the Calculus, translated from the German by Cathcart. is a rigorous treatment of the subject.

---

* Treatises

CONTENTS.

DIFFERENTIAL CALCULUS.

INTRODUCTION.

DEFINITIONS. NOTATION AND FUNDAMENTAL PRINCIPLES.

PART I.

DIFFERENTIALS AND DIFFERENTIATION.

PART II.

ANALYTIC APPLICATIONS.

PART III.

GEOMETRIC APPLICATIONS.

DIFFERENTIAL CALCULUS.

INTRODUCTION.

DEFINITIONS, NOTATION AND FUNDAMENTAL PRINCIPLES.

CHAPTER I.

CONSTANTS, VARIABLES AND FUNCTIONS.

1. In the Calculus quantities are divided into two general classes, *constants* and *variables*.

A Constant is a quantity that has, or is supposed to have, an absolute or relative fixed value.

A Variable is a quantity that is, or is supposed to be, continually changing in value.

In general, constants are represented by the first letters of the alphabet, and variables by the last; but they should not, therefore, be confused with the known and unknown quantities of Algebra.

The same quantity may sometimes be either a variable or a constant, depending upon the circumstances under which it is considered. Thus, in the equation of a curve, the coördinates of its points are variables; but in the equation of a tangent to the curve, the coördinates of the point of tangency are generally treated as constants. It is, therefore, necessary to determine from the circumstances, or object in view, which quantities are to be regarded as variables, and which as constants, in each discussion.

In general, any or all of the quantities represented by letters in any mathematical expression or equation may have definite values assigned to them, and be regarded as constants; or they may be considered as changing in value, and treated as variables. Thus, in the expression $4\pi r^2$, r is a constant if we suppose it to represent the radius of a particular sphere; but if r is considered as changing in value, it will be a variable. In the first case, $4\pi r^2$ is a constant, and measures the surface of a particular sphere; but when r is variable, $4\pi r^2$ is also variable, and represents the surface of any sphere no matter how much it may increase or diminish.

It should not be understood, however, that we may in all cases treat quantities as constants or variables at pleasure without affecting the character of the magnitude represented by the expression or equation. For example, π is generally assumed to represent the ratio of the circumference of any circle to its diameter, which ratio is invariable. If a different value be assigned to π, the expression $4\pi r^2$ will not measure the surface of a sphere whose radius is r.

In some cases variation in a quantity changes the dimensions of the magnitude represented by the expression or equation; in others it changes the position only; and again it may change the character of the magnitude. Thus, if we suppose R to vary in the equation

$$(x - \alpha)^2 + (y - \beta)^2 = R^2,$$

we shall have a series of circles differing in size; but by changing α or β and not R the position only will be affected.

By changing b^2 within positive limits, the equation $a^2y^2 + b^2x^2 = a^2b^2$ represents different ellipses, but negative values for b^2 cause the equation to represent hyperbolas. In general, however, constants are supposed to have fixed values in the same expression, unless for a particular discussion it is otherwise stated.

FUNCTIONS.

2. A quantity is a *function* of another quantity when its value depends upon that of the second quantity. Thus, $4ax$ is a function of 4, a, and x. In general, any mathematical expression which contains a quantity is a function of that quantity. If, however, a quantity disappears from an expression by reduction or simplification the expression is not a function of that quantity. Thus, $x^2 + (c + x)(c - x)$, ax/bx, and $\tan x \cot x$, are not functions of x.

3. *A function of a single variable* is one whose value depends upon that of a single variable and varies with it. Thus,

$$4x^2/(1 - x^4), \quad \sqrt{r^2x^2 + 2px}, \quad \log(a + x), \quad \sec x,$$

in which x is the only variable, are functions of a single variable.

Any function of a single variable is also a variable, and varies simultaneously with the variable.

4. The relation between a function of a variable and its variable is one of mutual dependence. Any change in the value of one causes a dependent variation in that of the other. Either may, therefore, be regarded as a function of the other; and they are called *inverse functions*. Thus, if x passes from the value 2 to 3, the function $2x^2$ will vary from 8 to 18; and conversely, x will increase from 2 to 3, if $2x^2$ changes from 8 to 18. If x be again increased the same amount, that is from 3 to 4, the function will increase from 18 to 32. Similarly, with other functions we shall find that, in general, equal changes in the variables do not give equal changes in the corresponding functions.

It is therefore necessary, in referring to a change in a function *corresponding* to a change in the variable, to consider the states from which and to which the function and variable change, as well as the amount of change in each. With that understanding, *corresponding* changes in a function and its variable are mutually dependent.

In the equation of a curve, the ordinate of any point is a function of the abscissa, and the abscissa is the inverse function of the ordinate.

The function is considered as dependent, and the variable as independent; for which reason, the latter is called *the independent variable*, or simply *the variable.*

Representing a function of x, as x^3, by y, we have $y = x^3$; solving with respect to x, we have $x = \sqrt[3]{y}$; a form expressing directly x as a function of y.

The difference in form in the following important examples of direct and inverse functions should be observed.

Having, $y = x^n$; then $x = \sqrt[n]{y}$.

" $y = a + x$; " $x = y - a$.

" $y = ax$; " $x = y/a$.

" $y = a^x$; " $x = \log_a y$.

5. A *state* of a function *corresponding* to a value or expression for the variable is a *result* obtained by substituting the value or expression for the variable in the function. Thus,

$$-\infty, \quad -16a, \quad -2a, \quad 0, \quad 2a, \quad 16a. \quad \infty,$$

are the states of the function $2ax^3$ corresponding, respectively, to the values or expressions for x,

$$-\infty, \quad -2, \quad -1, \quad 0, \quad 1, \quad 2, \quad \infty,$$

and

$$0, \quad 1/2, \quad \sqrt{1/2}, \quad 1, \quad 0, \quad -1, \quad 0,$$

are the states of the function $\sin\phi$ corresponding, respectively, to the expressions or values of ϕ,

$$0, \quad \pi/6, \quad \pi/4, \quad \pi/2, \quad \pi, \quad 3\pi/2, \quad 2\pi.$$

A function of a variable has an unlimited number of states. It may have equal states corresponding to different values of the variable; and it may have two or more states corresponding to the same value of the variable. Thus,

$$5 \text{ and } 1, \quad 7\pm\sqrt{12}, \quad 13 \text{ and } 5, \quad 13\pm\sqrt{24}, \quad 25 \text{ and } 13,$$

are the states of the function $2x+1\pm\sqrt{4x}$, corresponding, respectively, to the values of x,

$$1, \quad 3, \quad 4, \quad 6, \quad 9.$$

Trigonometric functions have equal states for all angles differing by any entire multiple of 2π.

In connection with *any* state of a function corresponding to any value of the variable, it is frequently necessary to consider another state of the function, which results from increasing the value of the variable corresponding to the first state by some convenient arbitrary amount In order to distinguish between these two states of the function, the first is designated as a *primitive state*, and the other as *its new* or *second state.*

Any arbitrary amount by which the variable is increased from any assumed value is called *an increment of the variable.* It is generally represented by the letter h, or k, or by Δ written before the variable; as, Δx, read "increment of x" *

Let x' represent any particular value of x, and h, or $\Delta x'$, its increment; then will $2ax'^2$ and $2a(x' + h)^2$, or $2a(x' + \Delta x')^2$, represent, respectively, the primitive and new states of the function $2ax^2$, corresponding to x' and its increment h, or $\Delta x'$. The general expression $2a(x + h)^2$ represents the second state of *any* primitive state of the function $2ax^2$, and from it we obtain the second state corresponding to any *particular* primitive state by substituting the proper value of x.

6. *A function of two or more variables* is one which depends upon two or more variables and varies with each. Thus,

$$x \sin y, \qquad xy, \qquad x^y, \qquad y \log x, \qquad x^2 + \sqrt{xy - 3y},$$

are functions of x and y; and

$$x + y + z, \quad y^2 + \tan x/z, \quad z \sin^2(x^2y), \quad \sqrt[3]{x^2 + y^2} + \log z,$$

are functions of x, y and z. Each variable is independent of the others. Particular values or expressions may be

---

* Increment as here used is in an algebraic sense, and includes a decrement, which is a negative increment. In general, an assumed increment of a variable is regarded as positive.

assigned to one or more of the variables, and the result discussed as a function of the remaining variables. A function of two or more variables possesses all of its properties as a function of each variable. By substituting in the function $2x^2 + y$, any assumed value for y, as 5, the result $2x^2 + 5$ is a function of a single variable.

7. A quantity is a function of the *sum of two variables* when every operation indicated upon either variable includes the *sum* of the two. Thus,

$$3c\sqrt{x \pm y}, \qquad \sin(x \pm y), \qquad \log(x \pm y), \qquad a^{x \pm y},$$

and all algebraic expressions which may be written in the form

$$A(x \pm y)^n + B(x \pm y)^{n-1} + \ldots. + H,$$

in which A, B, etc., are constants, are functions of the sum of the two variables x and $\pm y$.

$$8ax(x + y)^n, \quad \sqrt{x - y} - 2y, \quad \sqrt{x} + y, \quad x^{x+y}, \quad x \sin(x - y),$$

are not functions of the sum of x and y.

$$\sin(x^2 \pm y^2), \quad A(x^2 \pm y^2)^n, \quad 3 \log(x^2 \pm y^2), \quad \sqrt[3]{2(x^2 \pm y^2)} + 7a,$$

are functions of the *sum* of the two variables x^2 and $\pm y^2$, but not of the sum of x and $\pm y$.

$$2(b\sqrt{x} + ay^2), \quad \cos^2(b\sqrt{x} + ay^2), \quad 2\sqrt{\log(b\sqrt{x} + ay^2 - 3c)},$$

are functions of the sum of the two variables $b\sqrt{x}$ and ay^2.

In any function of the sum of two variables, a single variable may be substituted for the sum, and the original function expressed as a function of the new variable.

Thus, z may be substituted for $(x+y)$ in the function $a(x+y)^n$, giving the function in the form az^n. In a similar manner we may write

$$\tan(x-y) = \tan z, \quad a^{x+y} = a^z, \quad 2a\sqrt{\log(x-y)} = 2a\sqrt{\log z};$$

but it must be remembered that z in the new form is a function of the two variables x and y.

8. A *state* of a function of two or more variables, corresponding to a set of values or expressions for the variables, is the *result* obtained by substituting those values or expressions for the corresponding variables. Thus,

$$-20, \quad -6, \quad 0, \quad 5, \quad 25,$$

are states of the function $4x + 3y + 2$ corresponding, respectively, to the values or expressions for x and y,

$$(-4, -2), \quad (-2, 0), \quad (-8, +10), \quad (0, 1), \quad (2, 5);$$

and

$$0, \quad \sqrt{1/3}, \quad 1, \quad \sqrt{3}, \quad \infty,$$

are states of the function $\tan(x+y)$ corresponding, respectively to the values or expressions for x and y,

$$(0, 0), \quad (\pi/18, \pi/9), \quad (\pi/12, \pi/6), \quad (2\pi/9, \pi/9), \quad (0, \pi/2).$$

Any function, in which all of the variables are independent, is a variable, and has an unlimited number of states.

9. A function of several variables may be equal to some constant value or expression; in which case one of the variables is dependent upon the others. Thus, the first member of the equation $2x + 3y = 7$ is a function of the two variables, x and y; but x and y are mutually dependent.

Any equation containing n variables expresses a dependence of each variable upon the others; and there are only $n - 1$ *independent* variables in such an equation. In other words, the number of *independent* variables in any equation is one less than the total number of variables.

In any group of equations, the number of *independent* variables is equal to the total number of variables less the number of independent equations.

10. An **Algebraic** function is one that can be expressed *definitely* by the ordinary operations of Algebra; that is, by addition, subtraction, multiplication, division, formation of powers with constant commensurable exponents, and extraction of roots with constant commensurable indices.

Algebraic functions have particular names based upon peculiarities of form.

A *rational* function of a variable is one in which the variable is not affected by a fractional exponent.

An *integral* function of a variable is one in which the variable does not enter the denominator of a fraction, or in other words, is not affected by a negative exponent.

$$x^m + Ax^{m-1} + Bx^{m-2} + \dots\dots Gx + H,$$

in which m is a positive integer, and A, B, etc., do not contain x, is a *rational* and *integral* function of x. The coefficients A, B, etc., may be irrational or fractional.

A rational integral function of a variable is also called an *entire* function of that variable.

A *linear* function of two or more variables is one in which each term is of the first degree with respect to the variables.

Thus, $2x + 3y + 7z$ is a linear function of x, y and z.

A function is homogeneous with respect to its variables

when all of its terms are of the same degree with respect to them.

A linear function is a homogeneous function of the first degree.

11. A **Transcendental** function is one that cannot be expressed *definitely* by the ordinary operations of Algebra.

In general, a transcendental function may be expressed algebraically by an infinite series.

Transcendental functions include *exponential*, *logarithmic*, *trigonometric*, *inverse trigonometric*, *hyperbolic* and *inverse hyperbolic* functions.

An **Exponential** function is one with a variable, or an incommensurable constant, exponent; as,

$$a^x, \qquad (a/x + 1)^x, \qquad e^x - 2cx, \qquad x^{\log a}.$$

A **Logarithmic** function is one that contains a logarithm of a variable; as,

$$\log x, \qquad \log (a + y), \qquad 2ax^2 - x/\log x.*$$

A **Trigonometric** function is one that involves the sine, or cosine, or tangent, etc., of a variable angle; as,

$$\cot x, \qquad \sec 2x^2, \qquad (x - \sin x)/x^3, \qquad \operatorname{versin}^2 x.$$

An **Inverse Trigonometric** function is one that contains an angle regarded as a function of a variable sine, or cosine, or tangent, etc.

$\operatorname{Sin}^{-1}y$, $\tan^{-1}y$, read "the angle whose sine is y"; "whose tangent is y"; are symbols used to denote such functions. Having given $y = \operatorname{versin} x$, then $x = \operatorname{versin}^{-1}y$; and if $u = \cos y$, then $y = \cos^{-1}u$, etc.

* Napierian logarithms will always be considered unless some other base, as a, is indicated by $\log_a$.

12. Hyperbolic Functions.—From Trigonometry, we have

$$\sin x = x - \frac{x^3}{\underline{|3}} + \frac{x^5}{\underline{|5}} - \frac{x^7}{\underline{|7}} + \text{etc.};$$

$$\cos x = 1 - \frac{x^2}{2} + \frac{x^4}{\underline{|4}} - \frac{x^6}{\underline{|6}} + \text{etc.}$$

Placing $\sqrt{-1} = i$, and substituting xi for x, we have

$$\sin xi = i\left(x + \frac{x^3}{\underline{|3}} + \frac{x^5}{\underline{|5}} + \frac{x^7}{\underline{|7}} + \text{etc.}\right);$$

$$\cos xi = 1 + \frac{x^2}{2} + \frac{x^4}{\underline{|4}} + \frac{x^6}{\underline{|6}} + \text{etc.}$$

From Algebra, we have

$$e^x = 1 + x + \frac{x^2}{2} + \frac{x^3}{\underline{|3}} + \frac{x^4}{\underline{|4}} + \text{etc.};$$

$$e^{-x} = 1 - x + \frac{x^2}{2} - \frac{x^3}{\underline{|3}} + \frac{x^4}{\underline{|4}} - \text{etc.}$$

Hence,

$$\frac{e^x + e^{-x}}{2} = 1 + \frac{x^2}{2} + \frac{x^4}{\underline{|4}} + \frac{x^6}{\underline{|6}} + \text{etc.};$$

$$\frac{e^x - e^{-x}}{2} = x + \frac{x^3}{\underline{|3}} + \frac{x^5}{\underline{|5}} + \text{etc.}$$

Therefore,

$$\sin xi = i(e^x - e^{-x})/2; \qquad \cos xi = (e^x + e^{-x})/2.$$

$\cos xi$ is real, and is called the *hyperbolic cosine* of x. It is generally written $\cosh x$. The real factor in the sine of xi is called the *hyperbolic sine* of x, and is written $\sinh x$. Thus,

$$\sinh x = (e^x - e^{-x})/2; \qquad \cosh x = (e^x + e^{-x})/2.$$

From which,

$$\cosh^2 x - \sinh^2 x = 1.$$

Comparing with $x^2 - y^2 = 1$, we see that $\cosh x$ and $\sinh x$ may be represented by the coördinates of points of an equilateral hyperbola, referred to rectangular coördinate axes, with the origin at its centre. Hence the name hyperbolic functions. The functions $\sinh x$ and $\cosh x$ are not periodic functions for real values of x, but increase with x indefinitely.

By analogy with trigonometric functions, other hyperbolic functions are defined, and written, as

$$\tanh x = \frac{\sinh x}{\cosh x} = \frac{e^x - e^{-x}}{e^x + e^{-x}};$$

$$\coth x = \frac{1}{\tanh x} = \frac{e^x + e^{-x}}{e^x - e^{-x}};$$

$$\operatorname{sech} x = \frac{1}{\cosh x} = \frac{2}{e^x + e^{-x}};$$

$$\operatorname{cosech} x = \frac{1}{\sinh x} = \frac{2}{e^x - e^{-x}}.$$

It follows that

$$\tanh^2 x + \operatorname{sech}^2 x = 1 = \coth^2 x - \operatorname{cosech}^2 x.$$

13. Inverse Hyperbolic Functions.—Writing $y = \sinh x$, we have $x = \sinh^{-1} y$.

From $y = \sinh x = (e^x - e^{-x})/2$, we find $e^x = y \pm \sqrt{1 + y^2}$; hence, $\sinh^{-1} y = x = \log(y \pm \sqrt{y^2 + 1})$. Similarly,

$y = \cosh x = (e^x + e^{-x})/2$ gives $\cosh^{-1} y = \log(y \pm \sqrt{y^2 - 1})$;

$y = \tanh x = \dfrac{e^x - e^{-x}}{e^x + e^{-x}}$ gives $\tanh^{-1} y = \frac{1}{2} \log \dfrac{1 + y}{1 - y}$;

$$y = \coth x = \frac{e^x + e^{-x}}{e^x - e^{-x}} \text{ gives } \coth^{-1} y = \tfrac{1}{2} \log \frac{y+1}{y-1};$$

$$y = \operatorname{sech} x = \frac{2}{e^x + e^{-x}} \text{ gives } \operatorname{sech}^{-1} y = \log \frac{1 \pm \sqrt{1 - y^2}}{y};$$

$$y = \operatorname{cosech} x = \frac{2}{e^x - e^{-x}} \text{ gives } \operatorname{cosech}^{-1} y = \log \frac{1 \pm \sqrt{1 + y^2}}{y}.$$

14. Explicit and Implicit Functions.—When a function is expressed directly in terms of its variable or variables, it is an explicit function; otherwise it is an implicit function.

Thus, in the equations

$$y = 2x^2 + 3z,\ y = \tan^2 x,\quad y = 3^x,\ y = \log 2ax^2,\ \ y = f(x, z),$$

y is an explicit function of the variables in the second members, and in the equations

$$a^2y^2 + b^2x^2 = a^2b^2,\quad y^{\frac{2}{3}} = \log x^2,\quad y^2 = r^2 - x^2,\ \ f(y, x) = 0,$$

y is an implicit function of x.

The relation between an implicit function and its variables is sometimes expressed by two equations. Thus, $y = 3u$, $u^2 = \sqrt{x}$, in which y is an implicit function of x.

$$y = f(u),\quad u = \phi(x);\quad \text{and } y = f(u),\quad x = \phi(u),$$

are forms expressing y as an implicit function of x.

15. Increasing and Decreasing Functions.—A function that increases when a variable increases, and decreases when that variable decreases, is an increasing function of that variable. Thus, $2x$, $7x^3$, 2^x, ax^3/b, $\tan x$, are increasing functions of x.

A function that decreases when a variable increases, and increases when that variable decreases, is a decreasing function of that variable. Thus, $1/x$, $(c - x)^3$, b/ax^3, $\cot x$, are decreasing functions of x.

Functions are sometimes increasing for certain values of the variable, and decreasing for others. Thus, $(c - x)^2$ is an increasing function for all values of x greater than c; but decreasing for all values of x less than c. $2ax^2$ is an increasing function when x is positive, and decreasing when x is negative. The positive value of $y = \pm \sqrt{r^2 - x^2}$ is an increasing function for values of x from $-r$ to 0, but decreasing for values of x from 0 to $+r$. The negative value of y is a decreasing function for negative values of x, and increasing for positive values of x. $\sin x$, $\cos x$, $\sec x$, $\text{vers}\, x$, are increasing functions for some values of the variable and decreasing for others.

16. Continuous and Discontinuous Functions.—A function is continuous between states corresponding to any two values of a variable when it has a real state for every intermediate value of the variable, and as the difference between any two intermediate values of the variable approaches zero, the difference between the corresponding states also approaches zero. Otherwise a function is discontinuous between the states considered.

The varying height of a growing plant is a continuous function of time.

If for any value of the variable a function is unlimited, imaginary or changes from one state to another without passing through all intermediate states, the continuity of the function is broken at the corresponding state.

$\pm \sqrt{2px}$ is continuous between states corresponding to $x = 0$ and $x = \infty$.

$\pm b/a \sqrt{a^2 - x^2}$ is continuous between states corresponding to $x = -a$ and $x = +a$.

$\pm b/a\sqrt{x^2 - a^2}$ is continuous between states corresponding to

$$x = -\infty \text{ and } x = -a, \qquad x = a \text{ and } x = \infty,$$

but is discontinuous between states corresponding to $x = -a$ and $x = a$.

Sin x, cos x, e^x, and all entire algebraic functions are always continuous.

A continuous function in passing from any assumed state to another must pass through all states intermediate to those assumed; but it may have intermediate states greater or less than the states assumed. Thus, the function $\sqrt{r^2 - x^2}$ is continuous between the states 0 and $r/2\sqrt{3}$, which correspond to $x = -r$ and $x = r/2$; but it is greater when $x = 0$ than either of the states considered.

A function always continuous changes its sign only by passing through zero; but a discontinuous function may change its sign without passing through zero.

Unless otherwise stated, functions will be regarded as continuous in the vicinity of states under consideration.

17. Functional Notation.—A function of any quantity, as x, is generally represented thus, $f(x)$, read "function of x." Other forms are also used; as, $f'(x)$, $F(x)$, $F_1(x)$, $\phi(x)$, $\phi'(x)$, $\psi(x)$, $\psi_1(x)$.

Thus, $ax/(1 + x)$ may be represented by $f(x)$. The f, or exterior symbol, is called the functional symbol, or symbol of operation. It represents the operations involved in any particular function. Thus, having $f(x) = ax/(1 + x)$, f indicates that x is to be multiplied by a, and that the product is to be divided by $1 + x$. Its significance remains unchanged throughout the same discussion or subject, and placed before the parenthesis enclosing any other quantity it indicates that the quantity enclosed is to be subjected to the same operations that x is in the expression $ax/(1+x)$. Thus,

$$f(y) = \frac{ay}{1+y}, \quad f(z) = \frac{az}{1+z}, \quad f(m) = \frac{am}{1+m},$$

$$f(2) = \frac{2a}{1+2}, \quad f(\sin\phi) = \frac{a \sin\phi}{1+\sin\phi}.$$

In order to represent different functions of the same quantity the functional symbol only is changed. Thus, if $F(x)$ is selected to represent $2\sqrt{bx}$, then some other form as $F_1(x)$, or $\phi(x)$, etc., should be taken to denote $4cx^3+2x$

Different functions of different quantities are represented by different symbols within and without the parentheses. Thus, $\sqrt{x^2-a^2}$ and $4y^2/(1-y^2)$ may be denoted by $f(x)$ and $\phi(y)$, respectively.

A function of x^2 is written $f(x^2)$ or $F(x^2)$, etc., and the square of a function of x is designated by $\overline{f(x)}^2$ or $\overline{\phi(x)}^2$, etc.

When the quantity is represented by a simple symbol, the notation fx, ϕx, Fx^2, ψx^3, etc., is frequently used.

$3c\sqrt{my^2}$ may be expressed as a function of my^2 by some form, as $f(my^2)$ or $f'(my^2)$, etc.

Having represented az^2 by $f(z)$, and $3c\sqrt{az^2}$ by $F(az^2)$, we may write $3c\sqrt{az^2} = F(az^2) = F[f(z)]$.

Having $a^x = \phi(x)$, and $b\sqrt{a^x} = \psi(a^x)$, we write

$$\frac{8b\sqrt{a^x} - 3cb\sqrt{a^x}}{2db\sqrt{a^x}+h} = f'\Big(\psi[\phi(x)]\Big), \text{in which } \psi[\phi(x)] = b\sqrt{a^x}.$$

An expression containing several different functions of a variable, as $2ax^2 - \log x + 3\sin x$, may be considered as a function of the several functions of the variable, and represented thus:

$$F[f(x),\quad \phi(x),\quad \psi(x)].$$

$\phi[F(y), \psi(x), f(z)]$ represents a function of three different functions of different variables.

Functions of two variables are denoted thus: $f(x, y)$, $f'(x, y)$, $F(y, z)$, $\phi(x, y)$, $\psi(x, z)$, $\psi_1(x, z)$, etc.; and functions of three variables by $F(x, y, z)$, $\psi(r, s, t)$, etc.

Functions of any number of variables are indicated similarly by writing all the variables within the parentheses.

In all cases a functional symbol indicates the same operations in any one subject.

Thus, if $f(x, y) = ax + by$, then $f(s, t) = as + bt$; $f(2, 3) = 2a + 3b$; $f(0, m) = bm$.

Having $\phi(x, z, y) = 2x - cz + y^2$, then $\phi(r, s, t) = 2r - cs + t^2$.

Functions are frequently represented by single letters; thus, $\pm\sqrt{R^2 - x^2}$ may be represented by y, giving $y = \pm\sqrt{R^2 - x^2}$; and $f(x, y)$ by z, giving $z = f(x, y)$.

$F(x + y)$, $f(x + h)$, $\phi(s^2 + r^2)$ are forms denoting functions of the sum of two variables. §7.

ILLUSTRATIONS.

Having $f(x) = x^m + Px^{m-1} + Qx^{m-2} + \ldots + U$, in which P, Q, etc., do not contain x, then

$$f(5) = 5^m + P5^{m-2} + Q5^{m-2} + \ldots + U.$$

$$f(3bc) = (3bc)^m + P(3bc)^{m-1} + \ldots + U.$$

$$f(a - x) = (a - x)^m + P(a - x)^{m-1} + \ldots + U$$

$$f(0) = 0^m + P0^{m-1} + \ldots + U$$

$$f(c^2) = (c^2)^m + P(c^2)^{m-1} + \ldots + U.$$

Having	then
$\phi(a) = 4a^2 + ca,$	$\phi(x+y) = 4(x+y)^2 + c(x+y).$
$\psi(ax^3) = 4(ax^3)^2 + c(ax^3),$	$\psi(\sin\theta) = 4\sin^2\theta + c\sin\theta.$
$F(x) = a^x,$	$F(x+y) = a^{x+y} = a^x \times a^y = F(x)\times F(y).$
$F(xy) = a^{xy},$	$\overline{F(x)}^y = \overline{F(y)}^x = (a^x)^y = (a^y)^x = F(xy).$
$fx = \log x,$	$f(\text{limit } x) = \log(\text{limit } x).$
$f_n x = a^x,$	$f_n(\text{limit } x) = a^{\text{limit } x}.$

If $\phi(z) = \sqrt[n]{z}$, $\psi(x) = 5ax$, and $F(w) = \dfrac{c\sqrt{w} - (w)^{\frac{1}{2}}}{3w - 2p}$,

$$\text{then} \quad F(\psi[\phi(z)]) = \frac{c\sqrt{5a\sqrt[n]{z}} - (5a\sqrt[n]{z})^{\frac{1}{2}}}{15a\sqrt[n]{z} - 2p} = f(z).$$

If $\phi(x, y) = 2x + \sin y$, and $\psi(z) = 3\sqrt{z}$, then $\psi[\phi(x, y)] = 3\sqrt{2x + \sin y}$.

If $f(x, y, z) = 7ax^2yz$, and $F(y) = \sqrt[3]{y^2}$, and $\phi(x) = a^x$, and $\psi(z) = 2z$,

$$\text{then } \psi\Big[\phi\Big(F\big[f(x, y, z)\big]\Big)\Big] = 2a^{\sqrt[3]{(7ax^2yz)^2}}.$$

18. Lines are classed as algebraic or transcendental according as their equations involve algebraic functions only or contain transcendental functions.

Any portion of any line may be considered as generated by the continuous motion of a point. The law of its motion determines the nature and class of the line generated.

Let s represent the length of a varying portion of any line in the coördinate plane XY, of which the equation in x and y is given. s depends upon the coördinates of its variable extremities, and varies with each; but the equation of the line establishes a dependence between these coördinates. Hence, s is a function of one *independent* variable only.

If the line is in space, its two equations establish a dependence between the three coördinates of its extremities, so that one only is independent.

The same result will follow if a system of polar coördinates is used.

19. Convexity and Concavity.—The side of an arc of any curve upon which adjacent tangents, in general, inter-

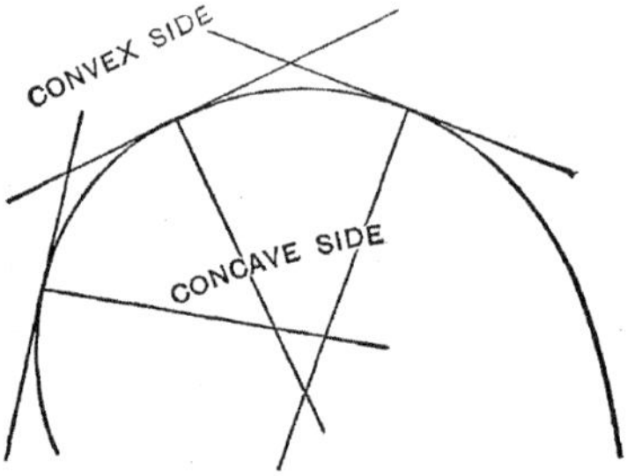

sect is called the convex, and the other, or that upon which adjacent normals intersect, the concave side. A curve, at any point, is said to be convex towards the convex side and concave in the opposite direction.

20. Graphic Representative of a Function of a Single Variable.—The relation between any function and its variable may be expressed by the equation formed by placing the function equal to a symbol. Thus, placing fx equal to y, we have $y = fx$, which expresses the relations between y and x, and therefore between the function fx and its variable x. $y = fx$ is also the equation of a locus, the coördinates of whose points bear the same relations to each other as those existing between the corresponding states of the function and variable. Therefore, by constructing, as in Analytic Geometry, any point of this locus, its *ordinate* will represent graphically the state of the function corresponding to the state of the variable similarly represented by its *abscissa*. The *locus* thus determined is called the *graph* of the function. It is important to notice that it is the *ordinate* of the graph, not the graph itself, that represents the function.

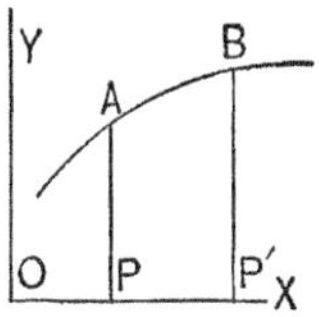

To illustrate, let the line AB be the graph of fx. Then the ordinate PA is the graphic representative of fx, corresponding to a value of x represented by OP. Similarly, $P'B$ represents fx when $x = OP'$.

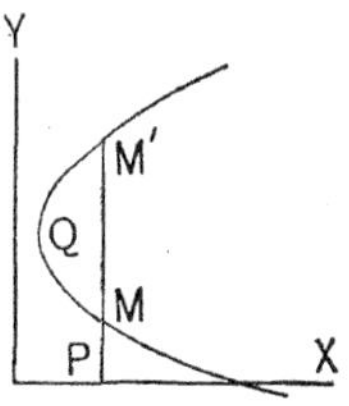

The ordinates PM and PM' of the graph MQM' represent two different states of the function corresponding to the same value of the variable, § 5.

The ordinates PM, RN, and SO of the graph MNO represent equal states of the function corresponding to different values of the variable, § 5.

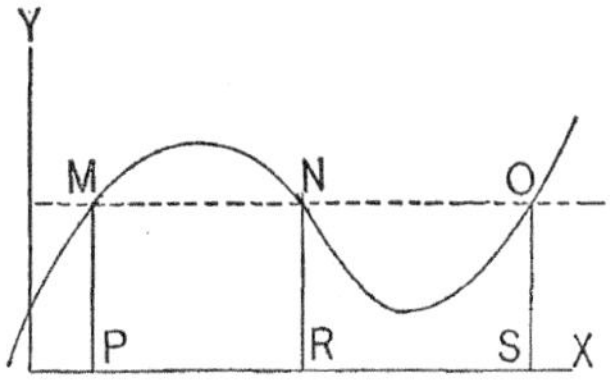

The graph of a function which is of the first degree with respect to its variable is a right line, otherwise not.

The graph of a continuous function is a continuous line.

21. Surfaces.—Any portion of any surface may be considered as generated by the continuous motion of a line. The form of the line and the law of its motion determine the nature and class of the surface generated.

22. Let u represent the area of a varying portion of the surface generated by the continuous motion of the ordinate of any given line in the plane XY.

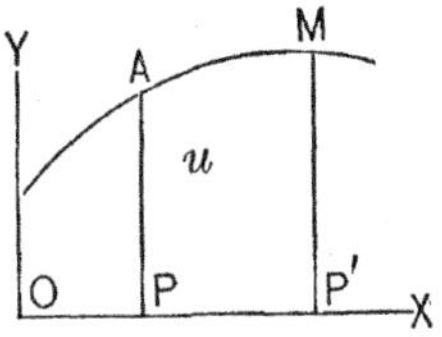

u depends upon the coördinates of the variable extremities of that portion of the given line which limits it, and varies with each; but the equation of the given line establishes a dependence between these coördinates. Hence, u is a function of but one *independent* variable.

23. Let $r = f(v)$ be the polar equation of any plane curve, as DM, referred to the pole P, and the right line PS. Let u represent the area of a varying portion of the surface, generated by the radius vector revolving about the pole. u will change with v and r; but v and r are mutually dependent. Hence, u is a function of but one *independent* variable.

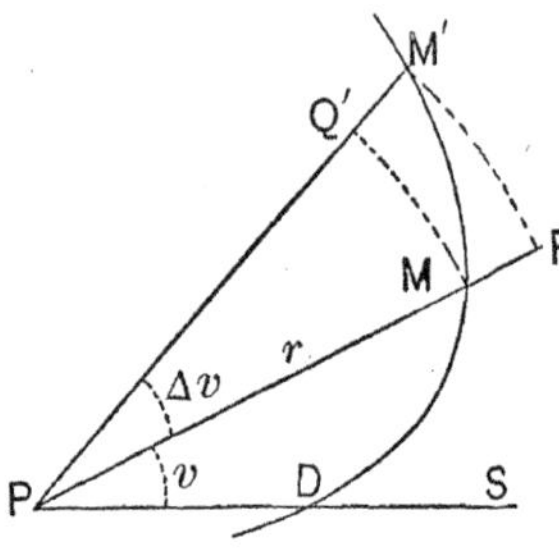

24. Let any line in the plane XY, as AM, revolve about the axis of X. It will generate a surface of revolution.

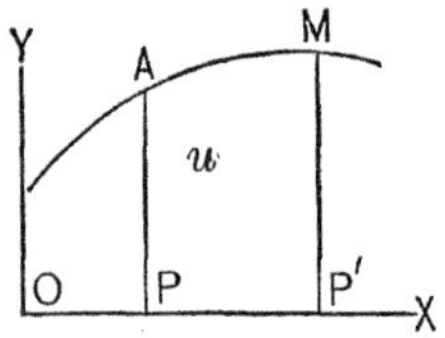

The same surface may be generated by the circumference of a circle, whose centre moves along the axis X, with its plane perpendicular to it; and whose radius changes with the abscissa of the centre of the circle, so as to always equal the corresponding ordinate of the curve AM. The radius of the generating circumference is, therefore, a function of the abscissa of its centre. Hence, the generating circumference, and any varying zone of the surface generated as described, is a function of but one *independent* variable.

25. The area of any surface with two independent variable dimensions is a function of two independent variables. For example, the area of any rectangle with variable sides, parallel respectively to the coördinate axes X and Y, is a function of the two independent variables x and y.

26. Having any surface, as ATL, let $ABCD = u$ be a portion included between the coördinate planes XZ, YZ, and the planes DQR and BPS, parallel to them respectively. Let $OP = x$ and $OQ = y$ be independent varia-

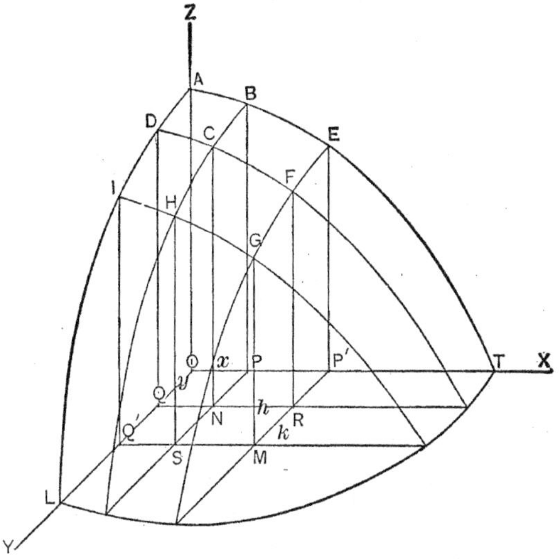

bles. u will depend upon x, y, and z; but the equation of the surface makes z dependent upon x and y. Hence, u is a function of but two *independent* variables. Similarly, it may be shown that any varying portion of the surface included between any four planes, parallel two and two, to the coördinate planes XZ and YZ, is a function of but two *independent* variables.

27. Graphic Representative of a Function of Two Variables.—Placing any function of two variables, as $f(x\ y)$, equal to z, we have $z = f(x, y)$ which expresses the relations between the function and its variables.

The locus whose equation is $z = f(x, y)$, is called the graphic surface of $f(x, y)$, for the reason that the *ordinate* of any of its points will represent graphically the state of the function corresponding to the states of the variables similarly represented by their respective abscissas.

It is important to notice that it is the *ordinate* of the graphic surface that represents the function, and not a portion of the surface as in the case described in § 26.

The graphic surface of a function which is of the first degree with respect to each of two variables is a plane, otherwise not.

28. Volumes.—Any portion of any volume may be considered as generated by the continuous motion of a surface. The form of the surface and the law of its motion determine the nature and class of the volume.

29. Let any plane surface included between any line in the plane XY, as AM, and the axis of X be revolved about X. It will generate a volume of revolution. The same volume may be generated by the circle, whose centre moves along the axis X, with its plane perpendicular to it; and whose radius changes with the abscissa of the centre of the circle, so as to always equal the corresponding ordinate of the curve AM. The radius of the generating circle is, therefore, a function of the abscissa of its centre. Hence, the generating circle, and any varying segment of the volume generated as described, is a function of but one *independent* variable.

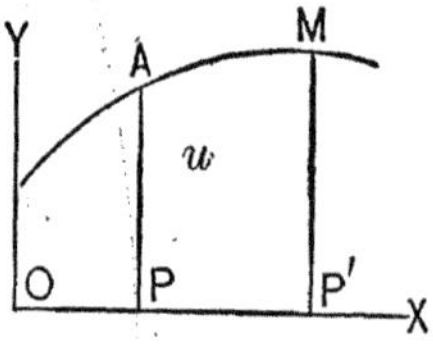

It is important to notice in this case, that the generating surface is limited by the ordinates PA and $P'M$, corresponding to the extremities of the limiting curve, which ordinates are perpendicular to the axis of revolution.

30. Having any volume, as ATL, bounded by a surface whose equation is given, and the coördinate planes, let

$ABCD\,ON = V$ be a portion included between the coördinate planes XZ, YZ, and let the planes DQR and BPS be parallel to them respectively.

Let $OP = x$ and $OQ = y$ be independent variables. V will depend upon x, y and z; but the equation of the surface makes z dependent upon x and y. Hence, V is a function of but two *independent* variables.

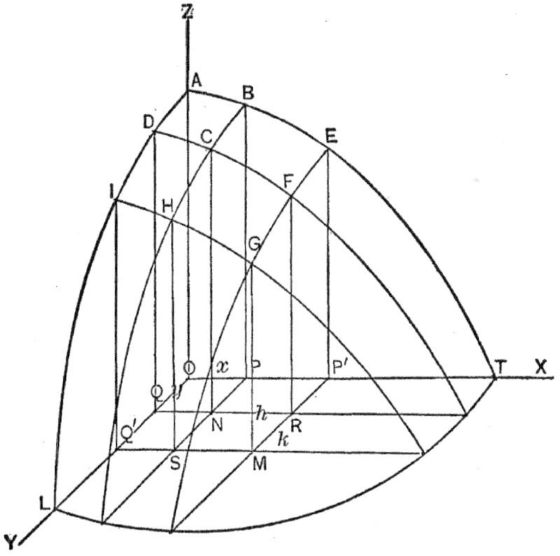

In a similar manner it may be shown that any varying portion of the volume included between any four planes, parallel two and two, to the coördinate planes XZ and YZ, is a function of but two independent variables.

31. Any volume with three independent variable dimensions is a function of three independent variables. For example, the volume of any parallelopipedon with variable edges parallel, respectively, to the coördinate axes X, Y and Z, is a function of x, y and z; all of which are independent.

CHAPTER II.

PRINCIPLES OF LIMITS.

32. **The Limit** of a variable* is a fixed finite quantity or expression which the variable, in accordance with a law of change, continually approaches, and from which it may be made to differ by a quantity less numerically than any assumed quantity however small.

Thus, any constant, as C, is the limit of any variable, as $f(x)$, when, under a law, $f(x)$ approaches C to within less than any assumed value however small it may be.

Various symbols are used to indicate a limit under a law. Thus, assuming that $f(x)$ approaches a limit C as x approaches a, we write

$$\operatorname*{limit}_{x \doteq a} f(x) = \operatorname*{Lt.}_{x=a} f(x) = \lim_{x=a} f(x) = \operatorname*{limit}_{x \ggg\!\!\rightarrow a} f(x) = C.$$

Each form is read, "the limit of $f(x)$ as x approaches a."

Any variable which under a law approaches zero as a limit is called an *infinitesimal*. Thus,

$$\operatorname*{limit}_{x \ggg\!\!\rightarrow 0} [1 - \cos x] = 0.$$

Any variable which under a law can exceed all assumed values, however great, is called an *infinite*. It is not a definite quantity.

---

* In this chapter the term *variable* is used in its general sense (§ 1) and includes all functions of variables.

An infinite cannot be a limit. Thus,

$$\underset{x \ggg\to 0}{\text{limit}} \frac{1}{x} = \infty$$

is a form indicating that as x approaches zero, $1/x$ is unlimited.

A tangent to any curve is a limiting position of a secant through the point of tangency, under the law that one or more of its points of intersection with the curve approach coincidence with the point of tangency.

In some cases, due to the form of the function or to the law of change, the variable can never become equal to its limit. Thus,

$$\underset{x \ggg\to \infty}{\text{limit}} \frac{x+1}{x} = \text{lim.}\left[1 + \frac{1}{x}\right] = 1.$$

But $\dfrac{x+1}{x} \lessgtr 1$, for all values of x.

The circumference of a circle is the limit of the perimeter of an inscribed regular polygon as the number of its sides is continually increased. The radius is the limit of the apothem, and the circle that of the polygon, under the same law.

An incommensurable number is the limit of its successive commensurable approximating values. Thus, the terms of the series 1.7, 1.73, 1.732, etc., taken in order, are approaching $\sqrt{3}$ as a limit.

In all cases, whether a variable becomes equal or not to its limit, the important property is that their difference is an infinitesimal.

An infinitesimal is not necessarily a small quantity in any sense. Its essence lies in its power of decreasing numerically; in other words, in having zero as a limit, and not in any small value that it may have. It is frequently defined as "*an infinitely small quantity*"; that is not, however, its significance as here used.

In representing infinitesimals by geometric figures they should be drawn of convenient size; and it is useless to strain the imagination in vain efforts to conceive of the appearance of the figure when the infinitesimals decrease beyond our perceptive faculties. Usually one or two auxiliary figures representing the magnitudes at one or two of their states under the law give all the assistance that can be derived from figures.

In all cases, when referring to the limit of a variable, it is necessary to give the law; for the limit depends not only upon the variable, but also upon the law by which it changes. Under a law, a determinate variable has but one limit; but it may have different limits under different laws.

An important consequence of the definition of a limit is that if two variables, in approaching limits under a law, have their corresponding values always equal, their limits will be equal. Thus, for all values of x, we have

$$(a^2 - x^2)/(a - x) = a + x,$$

hence

$$\underset{x \ggg\!\!\to a}{\text{limit}} (a^2 - x^2)/(a - x) = \lim [a + x] = 2a.$$

33. *A variable which, in approaching a limit, ultimately has and retains a constant sign cannot have a limit with a contrary sign.*

For suppose $f(x)$ becomes and remains positive, and that limit $f(x) = -C$. From the definition of a limit, $f(x)$ may be made to differ from $-C$ by a value numerically less than C. It would therefore become negative, which is contrary to the hypothesis. In a similar manner, it may be shown that a variable always negative cannot have a positive limit.

34. *If the difference between the corresponding values of any two variables, approaching limits, is an infinitesimal, the va iables have the same limit.**

Let U and V represent any two variables giving

$$U - V = \delta, \quad \text{or} \quad U = V + \delta,$$

in which δ is an infinitesimal.

Let C be the limit of U, then $U = C - \epsilon$, in which ϵ is an infinitesimal.

Substituting we have

$$C - \epsilon = V + \delta, \quad \text{or} \quad C - V = \delta + \epsilon,$$

the second member of which is an infinitesimal. Hence, C is the limit of V.

35. *The limit of the sum of any finite number of variables is the sum of their limits.*

Let U, $-V$, W, etc., represent any variables, and A, $-B$, C, etc., their respective limits; then

$$U = A - \epsilon, \quad -V = -B + \delta, \quad W = C - \omega, \text{ etc.,}$$

in which ϵ, δ, ω, etc., are infinitesimals.

Adding the corresponding members we have

$$U - V + W + \text{etc.} = A - B + C + \text{etc.} - \epsilon + \delta - \omega + \text{etc.}$$

* In order to avoid the frequent repetition of the expression "under the law," it will be assumed, unless otherwise stated, that the changes in all the variables considered together, or in the same discussion, are due to one and the same law; that all variables and their functions are continuous between all states considered, and that they have limits under the law.

Hence,

$$\text{limit}\,[U - V + W + \text{etc.}] = A - B + C + \text{etc.}$$
$$= \lim U - \lim V + \lim W + \text{etc.}$$

36. In general, *the limit of the product of any two variables is the product of their limits.*

Let U and V represent any two variables having A and B, respectively, as limits.

Then $U = A - \epsilon$ and $V = B - \delta$, in which ϵ and δ are infinitesimals. Multiplying member by member, we have

$$UV = AB - B\epsilon - A\delta + \epsilon\delta,$$

and

$$\text{limit}\,[UV] = AB = \text{limit}\ U\ \text{limit}\ V.$$

It follows that, in general, *the limit of any power or root of any variable is the corresponding power or root of its limit.*

Thus, limit $U^a = (\text{limit}\ U)^a$, and limit $U^{\frac{1}{a}} = (\text{limit}\,U)^{\frac{1}{a}}$,

Having $a^x = N$, x and N approach corresponding limits together; hence $a^{\lim x} = \lim N = \lim a^x$, and $\lim x = \log \lim N$. Also, since $x = \log N$, we have $\lim x = \lim \log N$. Therefore $\log \lim N = \lim \log N$.

37. In general, *the limit of the quotient of any variables is the quotient of their limits.*

With the same notation as in § 36, we have

$$\text{limit}\ \frac{U}{V} = \lim\,[UV^{-1}] = \lim U[\lim\ V]^{-1} = \frac{\lim U}{\lim V} = \frac{A}{B}.$$

When $B = 0$, and $A \neq 0$, U/V is unlimited.

When $B = 0 = A$, the principle fails to determine the limit which by definition is determinate.

38. It follows from §§ 35, 36, 37, that, in general, ***the***

limit of any function of any variables is the same function of their respective limits.

Thus, in general,

$$\text{limit } f(U\ V, \ldots .) = f(\lim U,\ \lim V, \ldots .),$$

and to obtain the limit of any function of variables we, in general, substitute for each variable its limit.

39. Exceptions to the above general rule arise, and are indicated by the occurrence of some indeterminate form, as $0/0$, ∞/∞, 0∞, $\infty - \infty$, 0^0, ∞^0, 1^∞.

To illustrate, having $f(x) = (x^2 - 1)/(x - 1)$, the general rule gives $\underset{x \ggg\!\rightarrow 1}{\text{limit}} f(x) = 0/0$, whereas we find

$$f(2) = 3, \quad f(1.5) = 2.5, \quad f(1.1) = 2.1, \quad f(1.01) = 2.01,$$
$$f(1.001) = 2.001, \quad \text{etc.},$$

and the nearer we take x to 1, the nearer will $f(x)$ approach to 2. By taking x sufficiently near to 1, $f(x)$ may be made to differ from 2 by a number less numerically than any assumed number however small. Hence, 2 is (§ 32) the limit of $f(x)$ as $x \ggg\!\rightarrow 1$. It should also be observed that 2, considered with the states of $f(x)$ which immediately precede and follow it, conforms to the law of continuity.

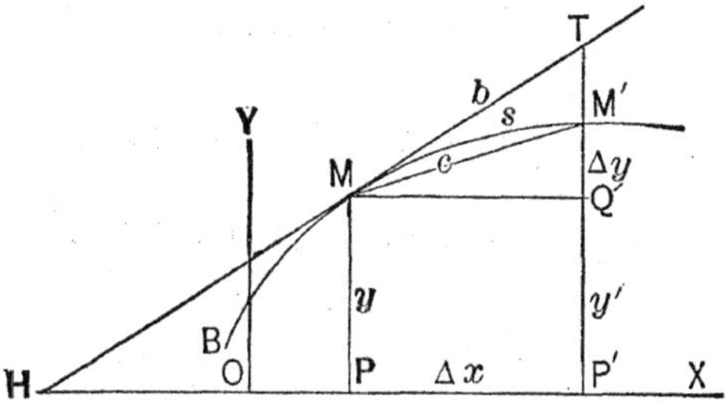

To illustrate a failing case geometrically, let the curve BMM' be the graph of a function. Take any state, as

PM corresponding to $x = OP$, and increase x by PP' represented by Δx. Draw the ordinate $P'M'$ and the secant MM'. Through M draw MQ' parallel to X. $Q'M'$, denoted by Δy, will represent the increment of the function corresponding to Δx.

$Q'M'/PP' = \Delta y/\Delta x = \tan Q'MM'$ will be the ratio of the increment of the function to the corresponding increment of the variable.

At M draw MT tangent to the curve. Then, under the law that Δx approaches zero, the secant MM' will approach coincidence with the tangent MT, and the angle $Q'MM'$ will approach the angle $Q'MT$, or its equal XHT, as a limit.

Hence

$$\underset{\Delta x \ggg\!\rightarrow 0}{\text{limit}} (\Delta y/\Delta x) = \text{lim. } \tan Q'MM' = \tan XHT,$$

whereas the general rule gives 0/0 as a result.

We observe from the above illustration that *the limit of the ratio of any increment of any function of a single variable to the corresponding increment of the variable, under the law that the increment of the variable approaches zero, is equal to the tangent of the angle made with the axis of abscissas by a tangent, to the graph of the function, at the point corresponding to the state considered.*

When M' coincides with M the secant may have any one of an infinite number of positions other than that of the tangent line MT, for the only condition then imposed is that it shall pass through M.

Therefore, while $\underset{\Delta x \ggg\!\rightarrow 0}{\text{limit}} (\Delta y/\Delta x)$ is definite, and equal to the tangent of the angle that the tangent line at M makes with X, limit Δy/limit $\Delta x = 0/0$ indicates that the tangent of the angle which the secant makes with X becomes indeterminate when M' coincides with M.

Limit $(\Delta y/\Delta x)$ is, therefore, one of the many values that limit Δy/limit Δx may have under the law.

The exceptional cases, in general, require transformation in order that factors common to the numerator and denominator may be cancelled, or from which the limit may otherwise be determined. They are of the highest importance, for the Differential Calculus, as it will be seen, is based upon the limit of the ratio of the increment of the function to the corresponding increment of the variable under the law that the increment of the variable vanishes. The remainder of this chapter will, therefore, be devoted to certain important exceptional cases and methods.

40. $$\underset{x \ggg\!\rightarrow y}{\textbf{Limit}} \frac{\mathbf{x}^m - \mathbf{y}^m}{\mathbf{x} - \mathbf{y}} = \mathbf{m}\mathbf{y}^{m-1}.$$

This formula is deduced in Algebra for all commensurable values of m. Since (§ 32) any incommensurable number is the limit of its successive commensurable approximating values, the formula holds true when m is incommensurable.

41. $$\underset{x \ggg\!\rightarrow \infty}{\textbf{Limit}} \frac{\mathbf{a}^x}{\underline{|\mathbf{x}}} = \mathbf{0}.$$

$$\underset{x \ggg\!\rightarrow \infty}{\text{Limit}} \left[\frac{a^x}{\underline{|x}} \Big/ \frac{a^{x-1}}{\underline{|x-1}} \right] = \lim \frac{a}{x} = 0.$$

That is, as $x \ggg\!\rightarrow \infty$, 0 is the limit of the ratio of $a^x/\underline{|x}$ to its preceding value; consequently 0 is the limit of $a^x/\underline{|x}$.

42. $\underset{y \ggg\to 0}{\textbf{Limit}}(1 + y)^{1/y} = e.$

$$(1+y)^{1/y} = 1 + \frac{y}{y} + \frac{1}{y}\left(\frac{1}{y} - 1\right)\frac{y^2}{2} + \dots$$
$$+ \frac{1}{y}\left(\frac{1}{y} - 1\right)\dots\left(\frac{1}{y} - n + 1\right)\frac{y^n}{\lfloor n} + \text{etc.}$$
$$= 1 + 1 + \frac{1-y}{2} + \dots$$
$$+ \frac{(1-y)(1-2y)\dots[1-(n-1)y]}{\lfloor n} + \text{etc.}$$

As $y \ggg\to 0$ each term approaches, as a limit, the corresponding term of the series

$$1 + 1 + \frac{1}{2} + \frac{1}{\lfloor 3} + \dots + \frac{1}{\lfloor n} + \text{etc.},$$

the sum of which is shown in Algebra to be $e = 2.71828\dots$

43. $\underset{h \ggg\to 0}{\textbf{Limit}} \dfrac{a^h - 1}{h} = \log a.$

Place $a^h = 1 + y$, whence $h = \log_a(1+y)$, and $y \ggg\to h \ggg\to 0$; giving

$$\underset{h \ggg\to 0}{\text{limit}} \frac{a^h - 1}{h} = \underset{y \ggg\to 0}{\text{limit}} \frac{y}{\log_a(1+y)}$$
$$= \lim \frac{1}{\log_a(1+y)^{1/y}} = \frac{1}{\log_a e} = \log a.$$

44. *If unity is the limit of the ratio of any two variables, the limit of any function of one will be equal to the limit of the same function of the other.*

Let U and V represent any two variables, giving limit $U/V = 1$. Then

$$\text{limit}\, f(U) = \lim f\left[\frac{UV}{V}\right] = f\left(\lim\left[\frac{U}{V}\right]\lim V\right)$$
$$= f(\lim V) = \lim f(V).$$

Thus,

$$\lim[C+U]=\lim[C+V];\quad \lim[CU]=\lim[CV];$$

$$\lim C^U=\lim C^V;\quad \lim[U/W]=\lim[V/W].$$

Therefore, *in searching for the limit of any function under a law, we may replace any variable entering it by another variable, provided that, under the same law, unity is the limit of the ratio of the two variables interchanged.* The great advantage in so doing arises when it enables us to determine the required limit more readily. Thus, in the last example above we may be able to determine the limit of V/W more readily than that of U/W.

In making the above substitution it is important to notice that the limits only are equal, and that corresponding values of the quantities interchanged, in general, are not equal to each other.

The privilege of replacing one variable by another under the conditions described, so facilitates the determination of limits in certain exceptional cases, that it is important to determine under what circumstances the limit of the ratio of two variables is equal to unity.

45. In general when $\lim U=\lim V$,

$$\lim[U/V]=\lim U/\lim V=1. \quad \S\,37.$$

That is, in general, unity is the limit of the ratio of any two variables when, under the same law, they have the same limit, or, what is equivalent, when the difference between their corresponding values is an infinitesimal.

If, however, $\lim U=\lim V=0$, it does not follow that $\lim[U/V]=1$ (§ 39). Such cases require special investigation, and the following are selected on account of their subsequent importance.

46. Take any plane surface, as $PMM'P'$, included between any arc, as MM', the ordinates of its extremities, and the axis of X.

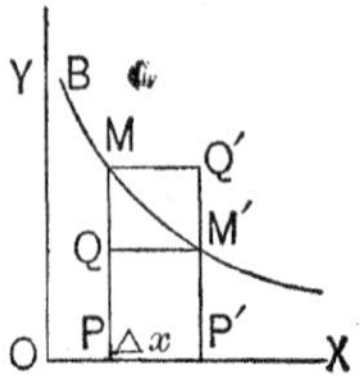

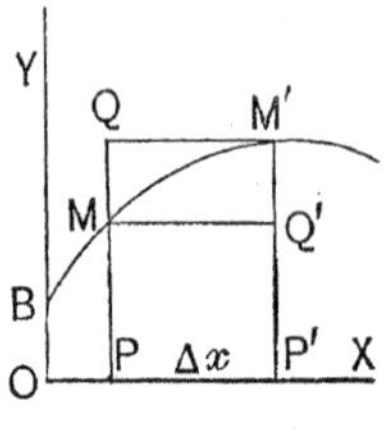

Through M and M', respectively, draw MQ' and $M'Q$ parallel to X, and complete the rectangle $MQM'Q'$. Let $y = PM$, and $y' = P'M'$.

Then as $PP' = \Delta x$, $\ggg\!\!\rightarrow 0$, we ultimately have

$$PQM'P' \gtrless PMM'P' \gtrless PMQ'P',$$

and $y' \ggg\!\!\rightarrow y$.

Therefore $$\underset{\Delta x \ggg\rightarrow 0}{\text{limit}} \frac{PQM'P'}{PMQ'P'} = \lim \frac{y' \Delta x}{y \Delta x} = 1,$$

and $$\underset{\Delta x \ggg\rightarrow 0}{\text{limit}} [PMM'P'/PMQ'P'] = 1.$$

Hence (§ 44)

$$\underset{\Delta x \ggg\rightarrow 0}{\text{limit}} \frac{PMM'P'}{\Delta x} = \lim \frac{PMQ'P'}{\Delta x} = \lim \frac{y \Delta x}{\Delta x} = y.$$

If the coördinate axes make an angle θ with each other, then

$$\underset{\Delta x \ggg\rightarrow 0}{\text{limit}} \frac{PMM'P'}{\Delta x} = \lim \frac{y \sin \theta \Delta x}{\Delta x} = y \sin \theta.$$

47. Let MPM' be the surface generated by the radius vector $PM = r$, revolving about P, as a pole, from any assumed position, as PM, to any other, as PM'. Let Δv represent the corresponding angle MPM'. With P as a centre, and the radii PM and PM', describe the arcs MQ' and $M'R$ respectively.

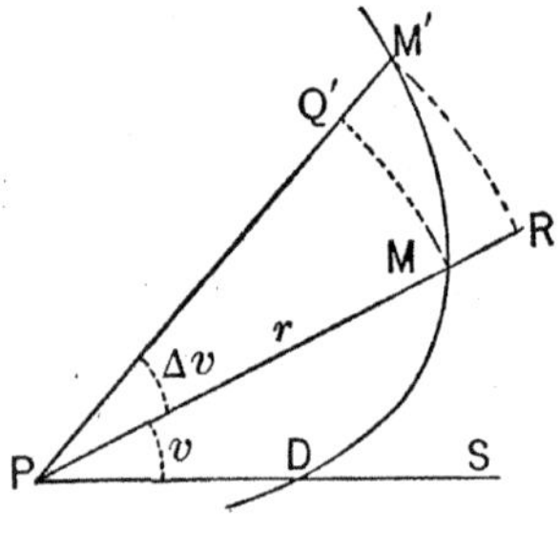

Then, as $\Delta v \ggg\!\to 0$, we ultimately have, in any case,

$$\text{area } RPM' \gtrless \text{area } MPM' \gtrless \text{area } MPQ',$$

and $$\lim_{\Delta v \ggg\to 0} [\text{area } RPM'/\text{area } MPQ'] = 1.$$

Hence, $$\lim_{\Delta v \ggg\to 0} [\text{area } MPM'/\text{area } MPQ'] = 1.$$

Therefore (§ 44)

$$\lim_{\Delta v \ggg\to 0} \frac{\text{area } MPM'}{\Delta v} = \lim \frac{\text{area } MPQ'}{\Delta v} = \lim \frac{r^2 \Delta v/2}{\Delta v} = \frac{r^2}{2}.$$

48. Let $PMM'P'$ be any plane figure as described in §46, and $MQM'Q'$ the corresponding rectangle.

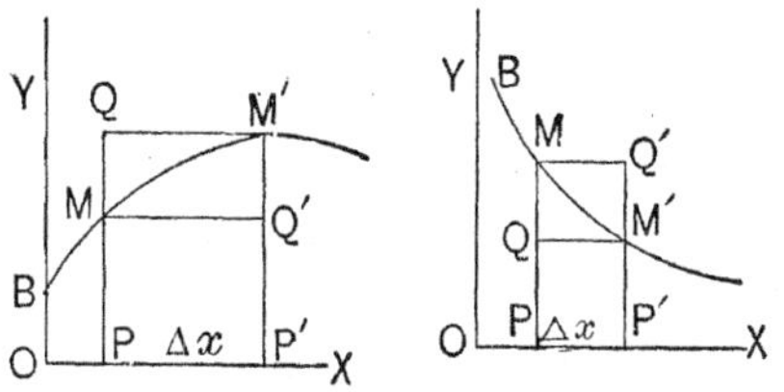

Revolve the entire figure about X.

Then as $\Delta x \ggg\!\to 0$, we ultimately have

$$\begin{matrix}\text{Vol. gen. by} \\ PQM'P'\end{matrix} \gtrless \begin{matrix}\text{Vol. gen. by} \\ PMM'P'\end{matrix} \gtrless \begin{matrix}\text{Vol. gen. by} \\ PMQ'P'.\end{matrix}$$

But

$$\lim_{\Delta x \ggg\to 0} \left[\begin{matrix}\text{Vol. gen by} \\ PQM'P'\end{matrix} \Big/ \begin{matrix}\text{Vol. gen. by} \\ PMQ'P'\end{matrix}\right] = \lim \frac{\pi y'^2 \Delta x}{\pi y^2 \Delta x} = 1.$$

Hence, $$\lim_{\Delta x \ggg\to 0} \left[\begin{matrix}\text{Vol. gen. by} \\ PMM'P'\end{matrix} \Big/ \begin{matrix}\text{Vol. gen. by} \\ PMQ'P'\end{matrix}\right] = 1.$$

Therefore (§ 44)

$$\lim_{\Delta x \ggg\to 0} \left(\frac{\begin{matrix}\text{Vol. gen. by} \\ PMM'P'\end{matrix}}{\Delta x}\right) = \lim \left(\frac{\begin{matrix}\text{Vol. gen} \\ PMQ'P'\end{matrix}}{\Delta x}\right)$$

$$= \lim [\pi y^2 \Delta x/\Delta x] = \pi y^2.$$

49. Let $MNM'N'$ be a portion of any surface included between the coördinate planes ZX, ZY and the two planes $N'SE$ and NPD parallel to them respectively.

Denote the corresponding volume $MNM'N'$, $OPFS$ by V. Construct the parallelopipedons $OPFS$-OM and

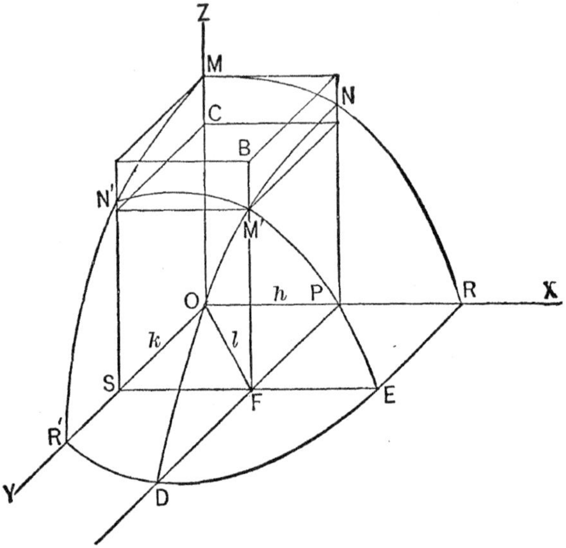

$OPFS$-FM', and represent their volumes by P and P' respectively. Let $OP = h$, $OS = k$, $OF = l$, $OM = z$, and $FM' = z'$.

Then as $h \ggg\!\!\rightarrow k \ggg\!\!\rightarrow 0$, or what is equivalent, as $l \ggg\!\!\rightarrow 0$, whence $z' \ggg\!\!\rightarrow z$, V will, in any case, ultimately be, and remain, between P and P'.

$$\text{But} \qquad \underset{l \ggg\!\!\rightarrow 0}{\text{limit}} [P/P'] = \underset{z' \ggg\!\!\rightarrow z}{\text{limit}} [zhk/z'hk] = 1.$$

$$\text{Hence,} \quad \underset{l \ggg\!\!\rightarrow 0}{\text{limit}} [V/P] = 1, \quad \text{and} \quad (\S\ 44)$$

$$\underset{l \ggg\!\!\rightarrow 0}{\text{limit}} [V/\text{rect. } OPFS] = \lim [P/hk] = \lim [zhk/hk] = z.$$

50. Let $MNM'N'$, denoted by S, be a portion of any surface included between the coördinate planes ZX, ZY and the two planes $N'SE$ and NPD parallel to them respectively.

Let $OP = h$, and $OS = k$.

At M draw the tangents MB and MB' to the curves MN and MN' respectively, complete the parallelogram $MBQB'$, and denote it by T. T is the portion of the tangent plane to the surface at M included between the

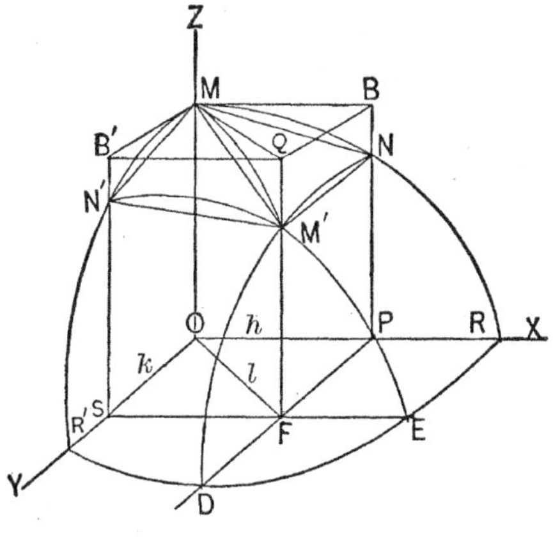

planes which limit S. Let β equal the angle which T makes with XY, giving $T \cos \beta = OPFS$. Inscribed in S, conceive an auxiliary surface, composed of n plane triangles the sum of which, as $n \to \infty$, will have S as a limit and such that the sum of their projections upon XY will equal $OPFS$. The two triangles $MN'M'$ and MNM' in the figure illustrate a set fulfilling the conditions. Let t, t', etc., represent the areas of the triangles and θ, θ', etc., the angles which their planes respectively make with XY.

Then $\qquad OPFS = \Sigma t \cos\theta = T\cos\beta,$

and $\qquad \Sigma t \cos\theta / T\cos\beta = 1. \quad . \quad . \quad . \quad . \quad (1)$

As $n \ggg\!\rightarrow \infty$, S remaining constant, each triangle is an infinitesimal, and we have

$$\lim_{n \ggg\rightarrow \infty} [S/\Sigma t] = 1.$$

The same effect and result follows if S is made infinitesimal and n remains constant. Hence, under the law that h and k vanish, or, what is equivalent, that OF, represented by l, is infinitesimal, we have

$$\lim_{l \ggg\rightarrow 0} [S/\Sigma t] = 1. \quad . \quad . \quad . \quad . \quad . \quad (2)$$

Under the same law, β is the common limit of θ, θ', etc. Hence (1)

$$\lim_{l \ggg\rightarrow 0} [\Sigma t \cos\theta / T\cos\beta] = \lim\,[\Sigma t/T] = 1. \quad . \quad (3)$$

Therefore (2), (3), $\lim_{l \ggg\rightarrow 0} [S/T] = 1$, and (§ 44)

$$\lim_{h \ggg\rightarrow k \ggg\rightarrow 0} [S/hk] = \lim\,[T/hk] = \lim\,[(hk/\cos\beta)/hk]$$
$$= 1/\cos\beta.$$

51. *Unity is the limit of the ratio of an angle to its sine, of an angle to its tangent, and of the tangent of an angle to its sine, as the angle approaches zero.*

Let $OCM = \phi$ given in radians be any angle less than $\pi/2$; then $\tan\phi > \phi > \sin\phi$, and as $\phi \ggg\rightarrow 0$, we have always

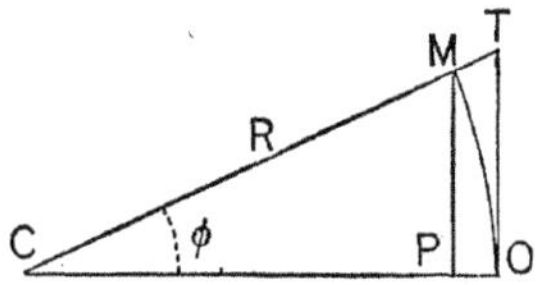

$$\frac{\tan\phi}{\sin\phi} > \frac{\phi}{\sin\phi} > 1.$$

But $\qquad \lim_{\phi \ggg\rightarrow 0} \frac{\tan\phi}{\sin\phi} = \lim \frac{1}{\cos\phi} = 1.$

Hence, since $\phi/\sin\phi$ is always between $\tan\phi/\sin\phi$ and unity, we have

$$\underset{\phi \ggg\!\rightarrow 0}{\text{limit}} [\phi/\sin\phi] = 1.$$

Similarly, since $1 > \phi/\tan\phi > \sin\phi/\tan\phi$, we have

$$\underset{\phi \ggg\!\rightarrow 0}{\text{limit}} [\phi/\tan\phi] = 1.$$

52. *Unity is the limit of the ratio of any arc of any curve to its chord, as the arc approaches zero.*

Let s denote the length of any arc of any curve, and conceive it to be divided into n equal parts, the consecutive points of division being joined by chords forming an inscribed broken line whose length is designated by p.

Then $$\underset{n \ggg\!\rightarrow \infty}{\text{limit}} [s/p] = 1.$$

Under the above law the equal arcs of s vary inversely with n; hence the same effect and result may be caused by retaining any fixed value for n, and making s approach zero. Hence,

$$\underset{s \ggg\!\rightarrow 0}{\text{limit}} [s/p] = 1,$$

and if $n = 1$, $$\underset{\text{arc} \ggg\!\rightarrow 0}{\text{limit}} [\text{arc}/\text{chord}] = 1.$$

53. *Unity is the limit of the ratio of the surface generated by any arc of any curve under a law to that generated by its chord as the arc approaches zero.*

Let s, p and n denote, respectively, the same quantities as in the last article, and conceive s and p to move together, under a law, so as to generate two surfaces represented by S and P respectively.

As $n \ggg\!\rightarrow \infty$, any state of s without regard to form is the limit of the corresponding state of p, and S is the limit of P.

That is, $$\lim_{n \to \infty} [S/P] = 1.$$

As $n \to \infty$, the equal arcs of s approach zero; hence the same effect and result may be caused by retaining any fixed value for n and making $s \to 0$. Therefore

$$\lim_{s \to 0} [S/P] = 1.$$

If $n = 1$, $$\lim_{s \to 0} \left[\frac{\text{Sur. gen. by an arc}}{\text{Sur. gen. by its chord}}\right] = 1.$$

The results determined in §§ 51, 52 and 53, with the principle § 44, are of great value in finding the limits in the following exceptional cases.

54. Let $MM' = s$ be any arc of any plane curve, PM and $P'M'$ the ordinates of its extremities.

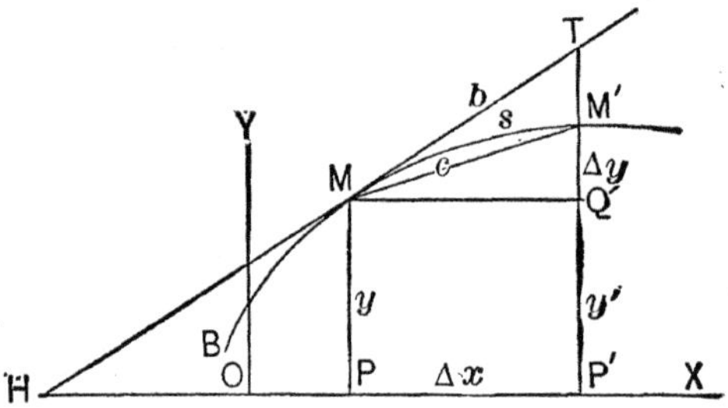

Through M draw the chord $MM' = c$, the tangent $MT = b$, and $MQ' = PP' = \Delta x$, parallel to X.

From the triangle $MM'T$, we have $\dfrac{b}{c} = \dfrac{\sin MM'T}{\sin MTM'}$.

As Δx approaches zero, the arc s and the angle $M'MT$ vanish, but the angle T remains constant. Hence, the angle $MM'T$ approaches $[180° - T]$, and

$$\lim_{s \to 0} \left[\frac{b}{c}\right] = \lim_{s \to 0} \left[\frac{\sin MM'T}{\sin T}\right] = \frac{\sin (180° - T)}{\sin T} = 1.$$

Hence, since (§ 52)

$$\lim_{s \ggg\to 0} [s/c] = 1, \quad \text{we have (§ 44)} \quad \lim_{s \ggg\to 0} [b/s] = 1.$$

Therefore (§ 44)

$$\lim_{\Delta x \ggg\to 0} \frac{s}{\Delta x} = \lim \frac{b}{\Delta x} = \lim \left[\frac{\Delta x / \cos Q'MT}{\Delta x} \right] = \frac{1}{\cos Q'MT}.$$

55. Let $r = f(v)$ be the polar equation of any plane curve, as AMM', referred to the right line PD, and pole P.

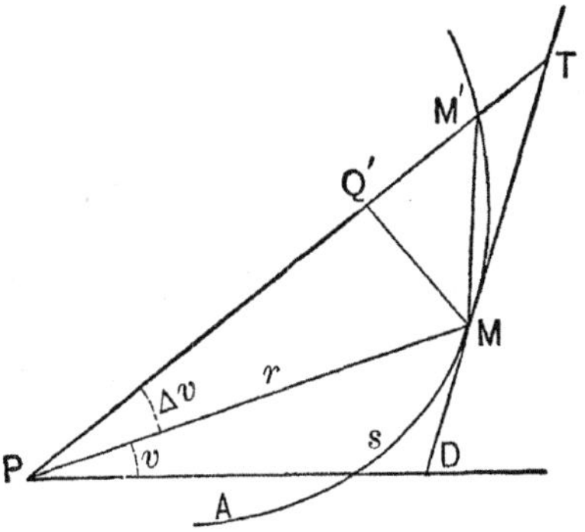

Let $AM = s$ be any portion of the curve, and $PM = r$ the radius vector corresponding to M.

Regarding s as a function of v (§ 18) let v be increased by $MPM' = \Delta v$. The arc MM' will be the corresponding increment of s. Draw MQ' perpendicular to PM', and denote PM' by r'. Then (§§ 44, 52) we have

$$\lim_{\Delta v \ggg\to 0} \frac{\text{arc } MM'}{\Delta v} = \lim \frac{\text{ch. } MM'}{\Delta v} = \lim \sqrt{\frac{\overline{MQ'}^2 + \overline{Q'M'}^2}{(\Delta v)^2}}$$

$$= \lim \sqrt{\frac{(r \sin \Delta v)^2 + (r' - r \cos \Delta v)^2}{(\Delta v)^2}}$$

$$= \lim \sqrt{r^2 \left(\frac{\sin \Delta v}{\Delta v}\right)^2 + \left(\frac{r' - r}{\Delta v}\right)^2}.$$

Hence (§ 51)

$$\underset{\Delta v \ggg 0}{\text{limit}} \frac{\text{arc } MM'}{\Delta v} = \lim \sqrt{r^2 + \left(\frac{r' - r}{\Delta v}\right)^2}. \qquad (1)$$

We also have

$$\tan PMD = \underset{\Delta v \ggg 0}{\text{limit}} \tan Q'M'M = \lim \frac{Q'M}{Q'M'}$$

$$= \lim \frac{r \sin \Delta v}{r' - r \cos \Delta v} = \lim \frac{r \Delta v}{r' - r}.$$

Hence, $$\frac{r}{\tan PMD} = \underset{\Delta v \ggg 0}{\text{limit}} \frac{r' - r}{\Delta v},$$

which, substituted in (1), gives

$$\underset{\Delta v \ggg 0}{\text{limit}} \frac{\text{arc } MM'}{\Delta v} = \sqrt{r^2 + \frac{r^2}{\tan^2 PMD}}.$$

If the radius vector PM coincides with the normal to the curve at M, we have

$$\tan PMD = \infty, \quad \text{and} \quad \underset{\Delta v \ggg 0}{\text{limit}} [\text{arc } MM'/\Delta v] = r.$$

56. Let any plane figure, as $PMM'P'$, included between any plane arc, as MM', the ordinates of its extremities and the axis of X, be revolved about X. Then (§§ 44, 53)

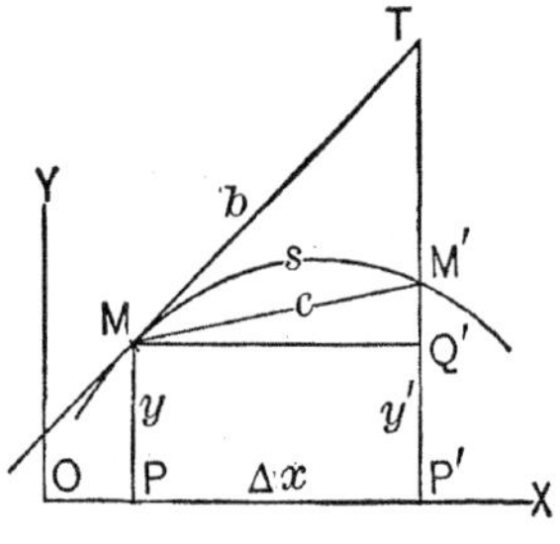

$$\underset{\Delta x \ggg 0}{\text{limit}} \frac{\text{Sur. gen. by arc } MM'}{\Delta x}$$

$$= \lim \frac{\text{Sur. gen. by ch. } MM'}{\Delta x} = \lim \frac{\pi(y + y')c}{\Delta x}$$

$$= \lim \frac{\pi(y + y') \Delta x / \cos Q'MM'}{\Delta x} = \frac{2\pi y}{\cos Q'MT}.$$

CHAPTER III.

RATE OF CHANGE OF A FUNCTION.

57. A function changes uniformly with respect to a variable when from each state all increments of the variable are directly proportional to the corresponding increments of the function.

It follows that from all states equal increments of the function correspond to equal increments of the variable; also that the ratio of any increment of the function to the corresponding increment of the variable is constant.

Thus, having $2ax$, increase x by any amount denoted by h, then $2a(x+h) - 2ax = 2ah$ will be the corresponding increment of $2ax$. It varies directly with h, it is the same for all states of the function, and $2ah/h = 2a$ is a constant. Hence, $2ax$ changes uniformly with respect to x.

Let fx be any uniformly varying function, and h any increment of x. $f(x+h) - fx$ will be the corresponding increment of the function, and

$$[f(x+h) - fx]/h = \text{constant} = A.$$

Hence $$f(x+h) = Ah + f(x),$$

in which $x = 0$, gives $$f(h) = Ah + f(0).$$

Therefore, $$fx = Ax + f(0).$$

Hence, all functions which change uniformly with respect to a variable are *algebraic*, and of the *first degree* with respect to that variable, and all *algebraic* functions of the *first degree* with respect to a variable change uniformly with that variable.

The graphs of such functions are right lines, and any function whose graph is a right line changes uniformly with respect to the variable.

To illustrate, the right line AD is the graph of a function. Consider any state, as that represented by the ordinate PA.

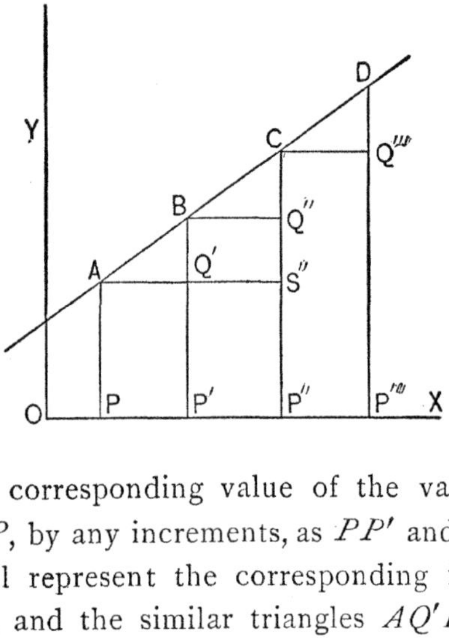

Increase the corresponding value of the variable, represented by OP, by any increments, as PP' and PP''. $Q'B$ and $S''C$ will represent the corresponding increments of the function, and the similar triangles $AQ'B$ and $AS''C$ give

$$AQ' : AS'' :: Q'B : S''C.$$

That is, the corresponding increments of the variable and function are proportional.

By giving to $x = OP$ any equal increments, as PP', $P'P''$, $P''P'''$, in succession, the corresponding increments of the function, $Q'B$, $Q''C$, and $Q'''D$, are equal to each other.

It is also evident that the ratio of any increment of the function to the corresponding increment of the variable, as $Q'B/PP'$, or $S''C/PP''$, or $Q'''D/P''P'''$, is constant.

53. Having the function $2x$,—

$x = 1$		gives	$2x = 2.$	
	<1			<2
$x = 2$		"	$2x = 4.$	
	<1			<2
$x = 3$		"	$2x = 6.$	
etc.			etc.	

Hence, the function $2x$ increases two units while the variable increases one; in other words, twice as fast.

Having the function $5x$,—

$x = 1$		gives	$5x = 5.$	
	<1			<5
$x = 2$		"	$5x = 10.$	
	<1			<5
$x = 3$		"	$5x = 15.$	
etc.			etc.	

Therefore, the function $5x$ changes fives times as fast as the variable.

In general, different functions change, with respect to their variables, with different degrees of rapidity.

The measure of the relative degree of rapidity of change of a function and its variable at any state is called the rate of change of the function, with respect to the variable, corresponding to the state.

A rate of change of a function with respect to a variable, corresponding to a state, is an answer to the question: At the state considered, how many times as fast as the variable is the function changing?

59. Since, from all states, any uniformly varying function receives equal increments for equal increments of the variable, *its rate, from state to state, is constant, and equal to the ratio of any increment of the function to the corresponding increment of the variable.*

$$\text{Thus, rate of } 2x = \frac{2(x+h) - 2x}{h} = 2.$$

$$\text{Rate of } 3x + 2 = \frac{[3(x+h)+2] - [3x+2]}{h} = 3.$$

$$\text{Rate of } 5x - 3 = \frac{[5(x+h)-3] - [5x-3]}{h} = 5.$$

It follows that the rate of any uniformly varying function is equal to the tangent of the angle which its graph makes with the axis of the variable, also that the product of the rate by any increment of the variable is equal to the corresponding increment of the function.

60. It follows from § 57 that algebraic functions not of the first degree with respect to the variable, and all transcendental functions, are, with respect to the variable, non-uniformly varying functions. Thus, having ax^2, increase x by any amount as h. $2axh + ah^2$, which varies with x, is the corresponding increment of ax^2. Also the ratio $(2axh + ah^2)/h = 2ax + ah = \phi(x, h)$. Hence, ax^2 does not change uniformly with respect to x.

Having $f(x^n)$, in which n is not equal to 1, then $[f(x+h)^n - f(x^n)]/h = \phi(x, h)$, and $f(x^n)$ is therefore a non-uniformly varying function with respect to x.

The graphs of such functions are curves, and any function whose graph is a curve does not change uniformly with respect to the variable.

To illustrate, the curve MNT is the graph of a function.

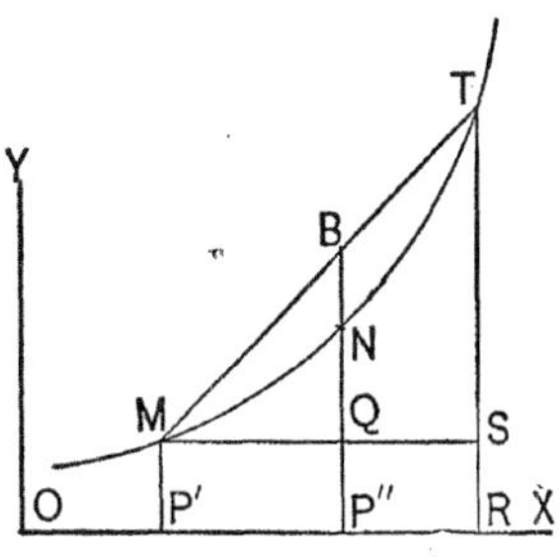

Increase, in succession, any value of x, as OP', by any amounts, as $P'P''$ and $P'R$. $P'P''/P'R = QB/ST$ is the ratio of the increments of the variable, and it differs from QN/ST, which is the ratio of the corresponding increments of the function. Hence (§ 57), the ordinate of MNT, and the function represented by it, do not change uniformly with x.

61. In the function $2x^2$,—

$x = 1$		gives	$2x^2 = 2.$	
	< 1			< 6
$x = 2$		"	$2x^2 = 8.$	
	< 1			< 10
$x = 3$		"	$2x^2 = 18.$	
	< 1			< 14
$x = 4$		"	$2x^2 = 32.$	

Which shows that at different states the function $2x^2$ has different rates with respect to x.

A non-uniformly varying function, in general, receives unequal increments for equal increments of the variable. It follows that its rate varies from state to state.

Any particular rate is, therefore, designated as the rate corresponding to a particular state.

If a function has two or more states corresponding to any value of the variable, each state will have a rate.

If a function has equal states for different values of the variable, it may have a different rate at each ; in which case it is necessary to indicate the value of the variable corresponding to the state considered.

62. Rate of Change of any Function. Let fx be any function. Denote by R its rate, with respect to x, corresponding to any state, as PA. Increase $x = OP$ by $h = PP'$, and let R' represent the rate of the function at the new state $f(x+h) = P'B$. $QB = f(x+h) - fx$ will represent the corresponding increment of the function.

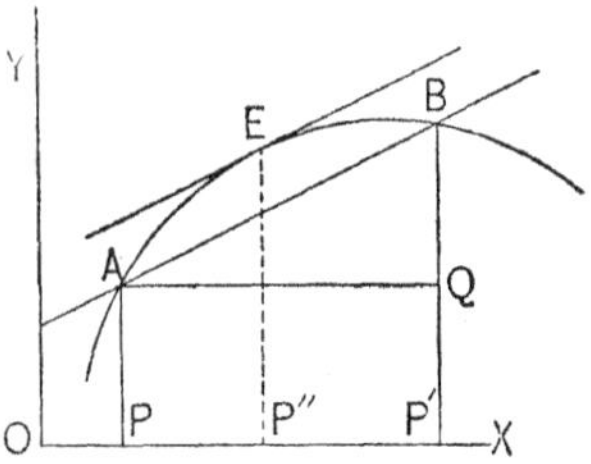

If fx changes uniformly,

$$\frac{f(x+h)-fx}{h} = \frac{QB}{PP'} = \tan QAB$$

will (§ 59) be its constant rate, $R = R'$, and its graph will be the right line AB (§ 57).

If fx is a non-uniformly varying function, its graph is a curve, as AEB, and in general $R \neq R'$.

In this case let $PP' = h$ be taken so small that while x varies from OP to OP', the *rate* of fx will either decrease or increase in order from R to R'.

In either case, $[f(x+h) - fx]/h$, which is the constant rate of the uniformly varying function whose graph is the right line AB, will be between R and R'; for otherwise the uniformly varying function would change between the states considered by an amount greater or less than QB.

In other words, $[f(x+h) - fx]/h$ is the rate of fx at some state, as $P''E$, between PA and $P'B$. Let $PP'' = \theta h$, in which θ is the proper fraction PP''/PP'.

Then $$P''E = f(x+\theta h),$$

and $[f(x+h)-fx]/h = \text{rate of } f(x+\theta h)$. . (1)

This relation will always exist as $h \Rrightarrow 0$,

$$\text{Hence,} \quad \underset{h \Rrightarrow 0}{\text{limit}} \left[\frac{f(x+h)-fx}{h}\right] = \text{rate of } fx. \quad (2)$$

That is, *the rate of change of any function with respect to a variable, corresponding to any state, is equal to the limit of the ratio of any increment of the function, from the state considered, to the corresponding increment of the variable, under the law that the increment of the variable approaches zero.*

The above principle enables us to find the rate of any function with respect to a variable, corresponding to any state, by the following general rule :

Give to the variable any variable increment, and from the corresponding state of the function subtract the primitive. Divide the remainder by the increment of the variable, and determine the limit of this ratio, under the law that the increment of the variable approaches zero. In the result substitute the value of the variable corresponding to the state.

It should be observed that *a rate*, determined by the above method, is equal to a *limit* of *a ratio* of two infinitesimals, which limit is determinate ; and that it is not equal to the *ratio* of their limits. § 39.

To illustrate, let the curve $AB''B'$ in the figure on page 51 be the graph of fx, and let PA be the state at which the rate is required.

When $h = PP'$,

$$\frac{f(x+h)-fx}{h} = \frac{Q'B'}{PP'} = \tan Q'AB',$$

which is the rate of the function whose graph is the secant AB'.

When $h = PP''$.

$$\frac{f(x+h) - fx}{h} = \frac{Q''B''}{PP''} = \tan Q''AB'',$$

which is the rate of the function whose graph is the secant AB''.

As $h \Rrightarrow 0$, the above ratio is always the rate of a function whose graph is a secant approaching the tangent AT as a

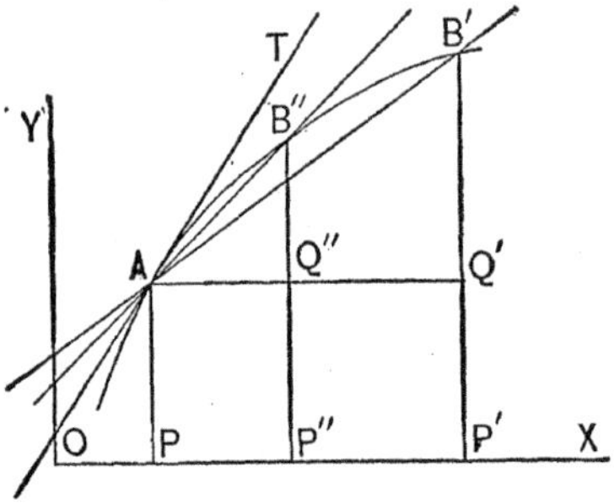

limit. Hence, *the limit of the ratio* is the rate of the function whose graph is the tangent AT, and is equal to the tangent of the angle $Q'AT$.

That is, *the limit of the ratio of any increment of any function of a single variable to the corresponding increment of the variable, under the law that the increment of the variable approaches zero, is equal to the tangent of the angle made with the axis of abscissas by a tangent at the corresponding point of the graph of the function.*

The same result was obtained in § 39. The angle which the tangent AT makes with X is called *the inclination*, and the numerical value of its tangent is called *the slope* of the graph at M.

The rate of any function at any state is therefore the same as that of a uniformly varying function whose graph is the tangent to the graph of the given function at the point corresponding to the state considered.

This agrees with previous conceptions and definitions, for the direction of the motion of the point generating the curve at any position is along the tangent at the point, and the ordinate of the curve representing the state considered is changing at the same rate as the ordinate of the corresponding tangent.

It follows that the product of the rate of a non-uniformly varying function by an increment of the variable is not, in general, equal to the corresponding increment of the function.

63. Let $y = PA$ represent any state of any increasing function of x; and y' the new state corresponding to an incre-

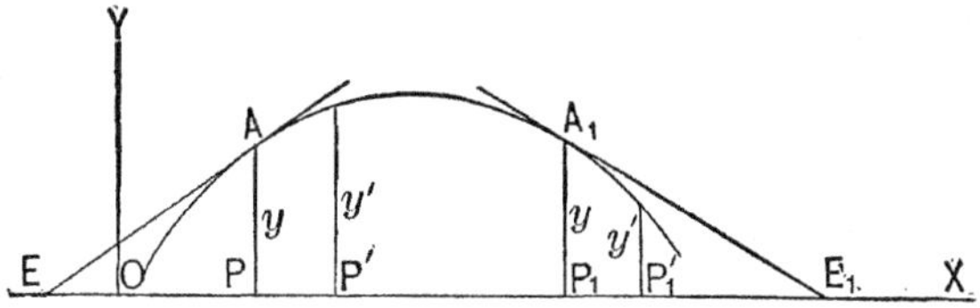

ment $PP' = h$, of the variable. $(y' - y)/h$ will be the ratio of the increment of the function to that of the variable. If h is assumed sufficiently small, this ratio will be positive and remain so as $h \ggg\!\!\rightarrow 0$. § 15.

$$\text{Hence (§ 33), } \lim_{h \ggg\!\!\rightarrow 0} [(y' - y)/h] = \tan XEA \text{ is positive.}$$

Let $y = P_1A_1$ represent any state of a decreasing function; and y' its new state due to an increment of the variable equal to $P_1P_1' = h$. Then $(y' - y)/h$ will be negative, if h is small enough, and will remain so as $h \ggg\!\!\rightarrow 0$.

$$\text{Hence, } \lim_{h \ggg\!\!\rightarrow 0} [(y' - y)/h] = \tan XE_1A \text{ is negative.}$$

Therefore, *the rate corresponding to any state of an increasing function is positive, and of a decreasing function is negative.*

It follows *that a function is an increasing one when its rate is positive, and a decreasing one when its rate is negative.*

When, for any state of a function, as the one represented by PA or $P'A'$, the function is neither increasing nor de-

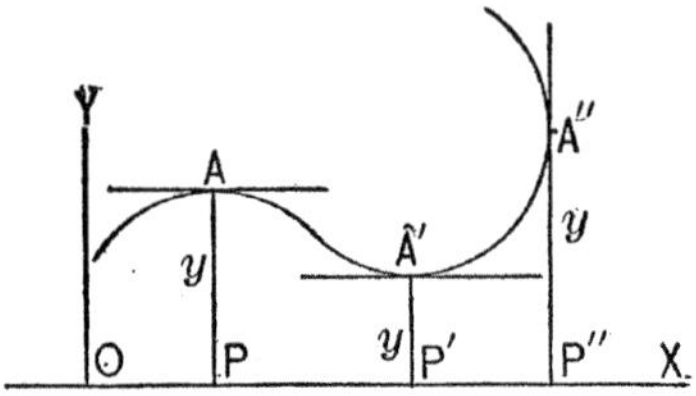

creasing, its rate is zero, and the tangent at the corresponding point of its graph is parallel to the axis of x.

If for any state of a function, as the one represented by $P''A''$, the rate is unlimited, the corresponding tangent to its graph is perpendicular to the axis of x.

EXERCISES.

Find the rate of change of each of the following functions:

1. $2ax$. Ans. $\underset{h \ggg\!\rightarrow 0}{\text{limit}} \left[\frac{2a(x+h) - 2ax}{h} \right] = 2a.$

2. x^2. Ans $\underset{h \ggg\!\rightarrow 0}{\text{limit}} \left[\frac{(x+h)^2 - x^2}{h} \right] = 2x.$

3. $ax^2 + bx$. Ans. $\underset{h \ggg\!\rightarrow 0}{\text{limit}} \left[\frac{a(x+h)^2 + b(x+h) - (ax^2+bx)}{h} \right] = 2ax + b$

4. $\frac{a}{x}$. Ans. $\underset{h \ggg\!\rightarrow 0}{\text{limit}} \left(\frac{\frac{a}{x+h} - \frac{a}{x}}{h} \right) = -\frac{a}{x^2}.$

5. $2ax^2$. Ans. $4ax$. 6. x^3. Ans. $3x^2$.

7. $4x^4$. Ans. $16x^3$. 8. $\frac{1}{1+x}$ Ans. $-\frac{1}{(1+x)^2}$.

9. $\frac{2x}{3+x}$. Ans. $\frac{6}{(3+x)^2}$.

10. How is the ordinate of a parabola, corresponding to $x = 3$, changing with respect to the abscissa?

Rate of

$$y = \pm \underset{h \ggg 0}{\text{limit}} \left[\frac{\sqrt{2p(x+h)} - \sqrt{2px}}{h}\right] = \pm\sqrt{2p}\, L\left[\frac{(x+h)^{\frac{1}{2}} - x^{\frac{1}{2}}}{h}\right]$$

$$= \pm\sqrt{2p}\, L\left[\frac{x^{\frac{1}{2}} + \frac{1}{2}x^{-\frac{1}{2}}h + \text{etc.} - x^{\frac{1}{2}}}{h}\right] = \pm\sqrt{2p}\left(\frac{1}{2\sqrt{x}}\right) = \pm\sqrt{\frac{p}{2x}}.$$

$$\text{Ans.} \left(\pm\sqrt{\frac{p}{2x}}\right)_{x=3} = \pm\sqrt{\frac{p}{6}}.$$

11. Same corresponding to focus? Ans. 1.

12. Find the abscissa of the point, of the parabola $y^2 = 4x$, where the ordinate is changing twice as fast as the abscissa. Ans. $x = 1/4$.

13. At the vertex of a parabola, how is the ordinate changing as compared with the abscissa?

14. Find the rate of change of the abscissa of a parabola with respect to the ordinate? Ans. $y/p = \pm\sqrt{2x/p}$.

15. Find the coördinates of the point of the parabola $y^2 = 8x$, where the abscissa is changing twice as fast as the ordinate. Ans. (8, 8).

16. Find the rate of change of the ordinate of the right line $2y - 3x = 12$, with respect to the abscissa. Ans. 3/2

17. A point moves from the origin so that y always increases 5/4 times as fast as x; find the equation of the line generated.

5/4 = tan of angle line makes with X. $\therefore$ Ans. $4y = 5x$.

18. Find the slope of the graph of $\pm\sqrt{12x}$ when $x = 1/2$.
Ans. ± 2.4495.

19. Find the abscissa of the point of the graph of $\sqrt{2px}$ when the slope is 1.
Ans. $x = p/2$.

20. Find the angles which the lines $y^2 = 8x$, and $3y - 2x = 8$, make with each other at their intersections.
Ans. $11^\circ\ 18'\ 35''$, and $7^\circ\ 7'\ 30''$.

21. Find the angles which the lines $y^2 = 4x$, and $2y = x + 2$, make with each other at their intersections.
Ans. $10^\circ\ 14'$, and $33^\circ\ 4'$.

64. A function of two or more variables is a uniformly varying function with respect to all of its variables when it changes uniformly with respect to each. It follows (§ 57) that all uniformly varying functions are algebraic, and of the first degree with respect to each variable, and all algebraic functions of the first degree with respect to each variable are uniformly varying functions.

Let $u = Ax + By + Cz +$ etc., in which A, B, C, etc., are constants, be any uniformly varying function. Increase the variables x, y, z, etc., respectively, by any increments, as h, k, l, etc., giving a new state,

$$u' = A(x+h) + B(y+k) + C(z+l) + \text{etc.}$$

$u' - u = Ah + Bk + Cl +$ etc., is the corresponding increment of the function. It is independent of the state of the function, dependent upon the increments of the variables, and is equal to the sum of the increments due to the increase of each variable separately.

65. A uniformly varying function of two variables is some particular case of the general expression $Ax + By + C$, in which A, B and C are constants. Its graphic surface is a plane (§ 27), and any function of two variables whose graphic surface is a plane is a uniformiy varying function.

To illustrate, take any ordinate, as NM, of any plane, as MLH. Through NM pass the planes MNR and MNP parallel, respectively, to ZX and ZY, intersecting the given plane in the lines MH and ML. Assume NR as the increment of x, and NP as the increment of y. Complete the

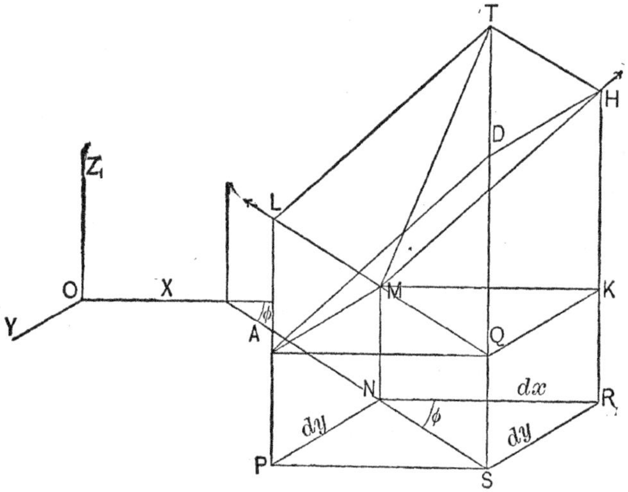

parallelogram NS, the parallelopipedon NQ, and in the given plane the parallelogram $MLTH$. Produce RK to H, SQ to T, and PA to L. KH is the increment of MN due to the increment, NR, of x alone, and AL is its increment due solely to NP, the increment of y. QT is the entire increment of MN due to the increase of both variables together.

Draw HD parallel to KQ, and draw AD; it will be parallel to LT, because it is parallel to MH, which is parallel to LT. Hence, $DT = AL$.

Therefore, $QT = QD + DT = KH + AL.$

That is, the total increment of any uniformly varying function of two variables from any state is equal to the sum of the increments from that state due to the increment of each variable separately.

It is important to notice that, while a uniformly varying function of two variables has a constant rate with respect to each variable alone, it has no fixed *total rate* with respect to both variables changing simultaneously. In the case illustrated,

$$\frac{QT}{QM} = \frac{\text{increment of } MN}{\sqrt{\overline{NR}^2 + \overline{RS}^2}} = \tan QMT.$$

is the corresponding total rate of MN, but in general any change in the relative value of NR and RS will cause a change in the total rate. Thus as the ratio NR/RS changes through all possible values, ϕ, the angle which the vertical plane MNS makes with the plane ZX changes, and the right line MT, cut from the given plane by the plane MNS revolving about MN as an axis, will in succession coincide with all right lines in the given plane which pass through M. Hence, depending upon the ratio of increments of the variables, the total rate of any uniformly varying function of two variables with respect to both variables changing simultaneously, may have any value from zero to the tangent of the angle made by the graphic plane of the function with XY, the numerical value of which is called the slope of the plane.

All functions of two variables not of the first degree with respect to each variable do not vary uniformly with respect to both variables, and their graphic surfaces are curved.

66. The Calculus is that branch of mathematics in which measurements, relations, and properties of functions and their states are determined from their rates of change. It is generally separated into two parts, called, respectively, Differential and Integral Calculus.

Differential Calculus embraces the deductions and applications of the rates of functions.

DIFFERENTIAL CALCULUS.

PART I.

DIFFERENTIALS AND DIFFERENTIATION.

CHAPTER IV.

THE DIFFERENTIAL AND DIFFERENTIAL COEFFICIENT.

FUNCTIONS OF A SINGLE VARIABLE.

67. An arbitrary amount of change assumed for the independent variable is called *the differential of the variable.*

It is represented by writing the letter d before the symbol for the variable; thus dx, read "differential of x," denotes the differential of x.

It is always assumed as positive, and remains constant throughout the same discussion unless otherwise stated.

68. The differential of a function of a single variable is *the change that the function would undergo from any state, were it to retain its rate at that state, while the variable changed by its differential.*

The differential of a function is denoted by writing the letter d before the function or its symbol.

Thus, $d\,2ax^3$, read "differential of $2ax^3$," indicates the differential of the function $2ax^3$.

Having $y = \log \sqrt{ax^2}$, we write $dy = d \log \sqrt{ax^2}$.

$\frac{dy}{dx}dx$ denotes the differential of y regarded as a function of x; and $\frac{dx}{dy}dy$ is a symbol for the differential of the inverse function; that is, of x regarded as a function of y.

The differential of a function which varies uniformly with its variable is equal to the change in the function corresponding to that assumed for the variable, because its rate is constant.

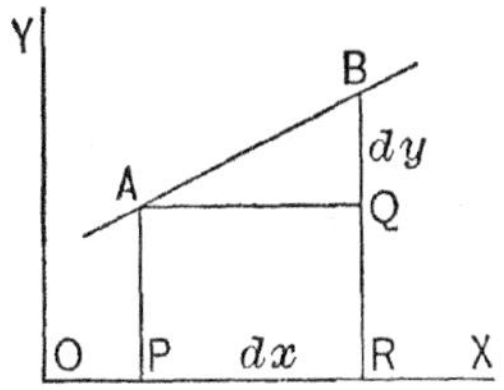

Thus, let PA be any state of the uniformly varying function whose graph is the right line AB. Assume $PR = dx$. Then $QB = dy$, the corresponding change in the function, is the differential of the function.

The differential of a function which does not vary uniformly with its variable is not, in general, equal to the corresponding change in the function, because its rate varies; but it is equal to the corresponding change of a function having a constant rate equal to that of the given function at the state considered; or, in other words, it is the change that the function would undergo were it to continue to change from any state, *as it is changing at that state, uniformly* with a change in the variable equal to its differential.

Thus, let PA be any state of a given function whose graph is the curve AM. Assume $PR = dx$.

QM is the corresponding change in the function; but QB, the corresponding change in the function represented by the ordinate of the right line AB drawn tangent to AM at A, is the differential of the given function corresponding to the state PA. The function whose graph is AB has a constant rate equal to that of the given function at PA, and QB is the change that the given function would undergo were it to continue to change from the state PA, *as it is changing at that state*, *uniformly* with a change in x equal to dx.

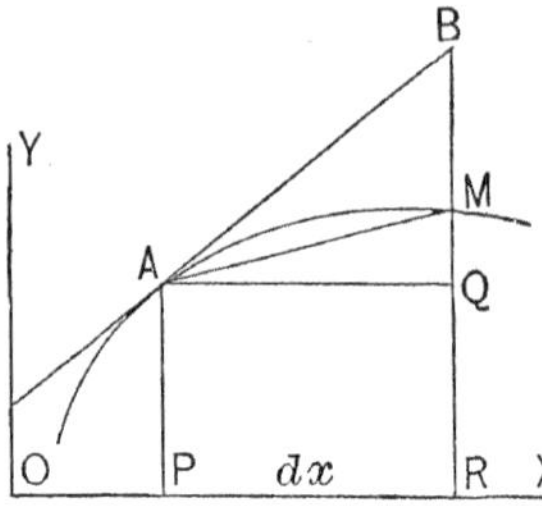

The differential of a function which does not vary uniformly with its variable may be less than the corresponding change in the function. Thus, $QB, < QM$, is the differential of the function represented by the ordinate of the curve AM, corresponding to PA.

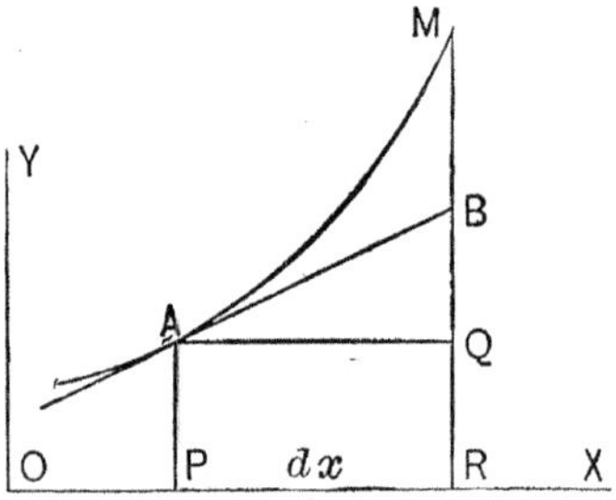

A train of cars in motion affords a familiar example of a differential of a function.

Suppose that a train of cars starts from the station A, and moves in the direction AE with a continuously increasing speed. Let x denote the variable distance of the train from A at any instant; it will be a function of the time, represented by t, during which the train has moved, giving $x = f(t)$.

Suppose the train to have arrived at B, for which point $x = AB$. Let BD represent the distance that the train will actually run in the next unit of time, say one second, with its rate constantly increasing.

Let BC represent the distance that the train would run if it were to move from B with its rate at that point unchanged, in a second,

Then will the distance BC represent the differential of x regarded as a function of t, corresponding to the state $x = AB$; and one second will be the differential of the variable.

69. From the definition of a differential of a function, and from § 59, it follows that a differential of a function is the product of two factors, one of which is *the rate of change of the function* at the state considered, and the other is the assumed *differential of the variable.* Hence, the differential of any given function may be determined by finding its rate, by the general rule, § 62, and multiplying it by the differential of the variable. Thus, having the function $2x^2$, we find

$$\underset{h \ggg\!\rightarrow 0}{\text{limit}} \left[\frac{2(x+h)^2 - 2x^2}{h} \right] = 4x = \text{rate corresp. to any state.}$$

$4xdx$ is, therefore, a general expression for the differential of $2x^2$, and is written $d\,2x^2 = 4xdx$.

Its value corresponding to any particular state is obtained by substituting the value of the variable corresponding to the state; thus, for $x = 2$, we have $(d\,2x^2)_{x=2} = 8dx$.

70. Since, in the expression for the differential of a function, *the rate of change of the function* is the coefficient of the differential of the variable, it is, in general, called the "*differential coefficient,*" and may be determined by the general rule, § 62.

The *differential of a function* is therefore equal to the product of the *differential coefficient* by the *differential of the variable.*

It follows that the differential coefficient is the quotient of the differential of the function by the differential of the variable. Thus, having $d\,2x^2 = 4xdx$, $4x$ is the differential coefficient. In general, having $y = f(x)$, and representing its differential by dy or $\frac{dy}{dx}dx$, its differential coefficient is dy/dx.

The expressions $(dy/dx)_{(x')}$ and dy'/dx' are used to denote the particular value of dy/dx corresponding to $x = x'$ and $y = y'$.

Thus, having $y = 2x^2$, then $dy/dx = 4x$, and $(dy/dx)_{(2)} = 8$.

Having $y = f(x)$, in which y is any function of any variable x, let y' denote the new state of the function corresponding to any increment of the variable, as h or Δx, and let $\Delta y = y' - y$ represent the corresponding increment of y. Then (§ 62)

$$\underset{h \ggg\!\!\rightarrow 0}{\text{limit}} \frac{f(x+h) - f(x)}{h} = \lim \frac{y' - y}{h} = \underset{\Delta x \ggg\!\!\rightarrow 0}{\text{limit}} \frac{\Delta y}{\Delta x} = \frac{dy}{dx}.$$

Since the increment of the variable, represented by h or Δx, varies, it may happen that $h = \Delta x = dx$. It is exceedingly important to observe, however, that the corresponding value of $y' - y$ or Δy *is not*, in general, equal to dy; for that would give

$$\left(\frac{y' - y}{h}\right)_{h = dx} = \left(\frac{\Delta y}{\Delta x}\right)_{\Delta x = dx} = \frac{dy}{dx};$$

which, in general, is impossible, since dy/dx is not a value of the ratio $(y' - y)/h$, but is its limit under the law that h vanishes.

If, however, the function changes uniformly with respect to the variable, $(y' - y)/h$ will be constant for all values of h (§ 59), and $y' - y$ will be equal to dy when h is equal to dx.

71. The following are important facts in regard to a differential coefficient:

It is zero for a constant quantity. In other words, a constant has no differential coefficient.

It is constant for any function which varies uniformly.

It varies from state to state for any function which does not vary uniformly.

In general, therefore, it is a function of the variable, and has a differential.

It is positive for an increasing function, and negative for a decreasing one. § 63.

It may have values from $-\infty$ to $+\infty$.

Having represented a function by the ordinate of a curve, the differential coefficient is equal to the tangent of the angle made with the axis of abscissas, by a tangent to the curve at the point corresponding to the state considered. § 62.

Thus, assuming $PR = dx$, the differential coefficient of the function whose graph is the curve AM, at the state PA, is

$$dy/dx = \tan XEA = \tan QAB.$$

It should be noticed that dy/dx is independent of the

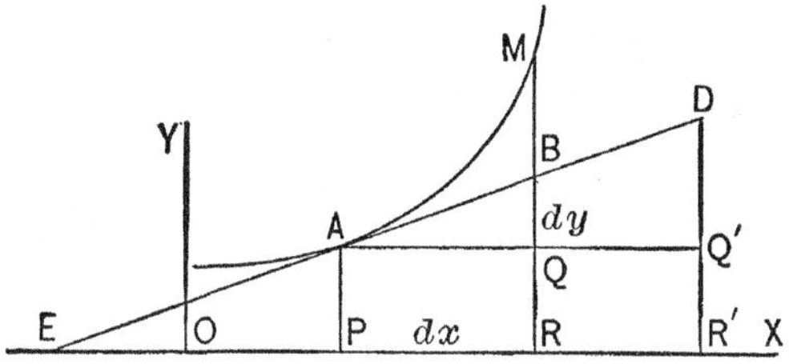

value assumed for the differential of the variable; for if $PR' = dx$, then $Q'D = dy$, and we have, as before, $dy/dx = \tan XEA$.

In this illustration the function is an increasing one, its

differential coefficient is positive, and the angle XEA is acute. § 63.

In case the function represented by the ordinate of AM is a decreasing one, its differential coefficient corresponding to PA is negative, and the angle XEA is then obtuse. § 63.

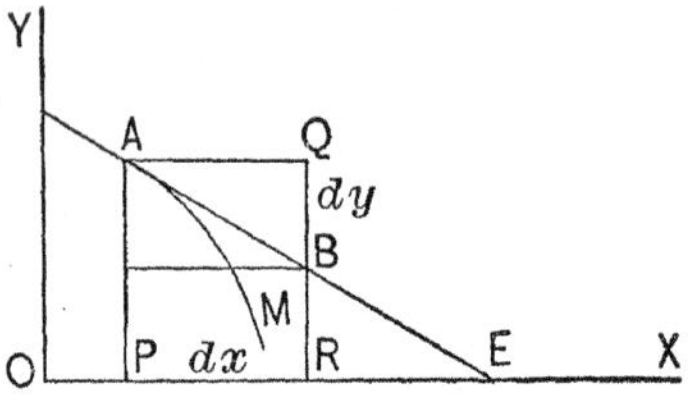

If for any value of the variable the differential coefficient is zero, the function is neither increasing nor decreasing, and the tangents at the corresponding points of the graph of the function are parallel to the axis of x.

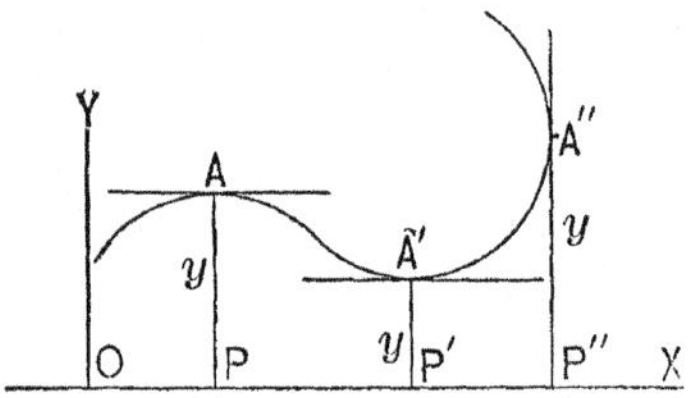

If the differential coefficient is infinite, the rate of the function is infinite; and the tangents at the corresponding points of the graph of the function are perpendicular to the axis of x. § 63.

If for a finite value of the variable the state of a function is unlimited, its corresponding differential coefficient will also be unlimited. Thus, having $y = f(x)$ and $f(a) = \infty$,

$$(dy/dx)_a = \underset{h \ggg\!\!\to 0}{\text{limit}} \left[\left(f(a+h) - f(a)\right)/h\right].$$

Hence,

$$(dy/dx)_a + e = [f(a+h) - f(a)]/h,$$

in which e vanishes with h. $f(a+h)$ is not, in general, unlimited, therefore $(dy/dx)_a = \infty - e = \infty$.

The principle is not necessarily true for an infinite state corresponding to an infinite value of the variable, for in that case $f(a+h)$ will also be unlimited.

72. The following facts concerning a differential of a function should now be apparent:

It is zero for a constant.

It is constant for any function which varies uniformly.

It is a function of the variable for any function which does not vary uniformly; and in such cases it has a differential.

Its value depends upon that of the differential coefficient and that assumed for the differential of the variable.

It may have values from $-\infty$ to $+\infty$.

It will be numerically greater or less than the differential coefficient depending upon whether the differential of the variable is assumed greater or less than unity.

It has the same sign as its differential coefficient, and therefore is positive for an increasing function and negative for a decreasing one.

Functions which are equal in all their successive states have their corresponding differentials equal.

73. *The differential coefficient of any function is equal to the reciprocal of the corresponding differential coefficient of its inverse function.*

Let $y = f(x)$ . . . (1) and $x = F(y)$ . . . (2) be any direct and inverse functions. Let Δx and Δy represent, respectively, any set of corresponding increments of x and y in (1). It follows (§ 4) that they will represent a set of the same in (2), and we have

$$\Delta y/\Delta x = 1/(\Delta x/\Delta y).$$

Hence (§ 70)

$$\frac{dy}{dx} = \underset{\Delta x \ggg\!\!\rightarrow 0}{\text{limit}} \frac{\Delta y}{\Delta x} = \lim \frac{1}{\Delta x/\Delta y} = \frac{1}{dx/dy}.$$

To illustrate, let the function y be represented by the

ordinate of the curve AM. Assume $dx = PR$, and from the figure we have, corresponding to the state PA,

$$\frac{QB}{PR} = \frac{dy}{dx} = \tan QAB.$$

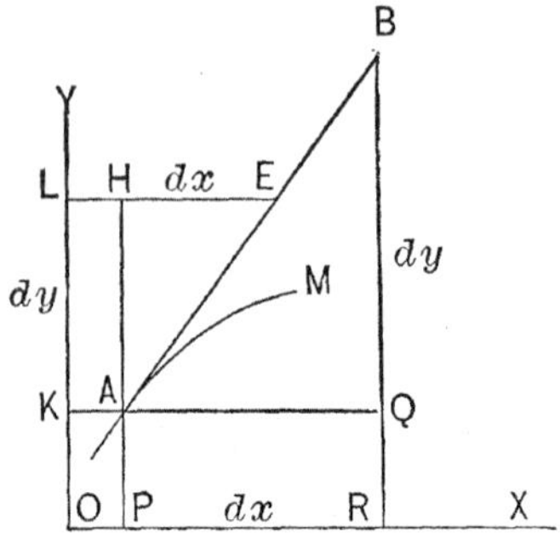

The inverse function will be represented by the abscissa of the curve AM regarded as a function of the ordinate; and assuming $dy = KL$, we have for the state KA, corresponding to A,

$$HE/AH = dx/dy = \tan EAH. \qquad EAH = 90^\circ - QAB.$$

Hence,

$$\tan QAB = \cot EAH = \frac{1}{\tan EAH}; \quad \text{or} \quad \frac{dy}{dx} = \frac{1}{dx/dy}.$$

It should be observed that, in general, dy in the first member of the above equation is not the same as dy in the second; for the first is the differential of y as a function, and the second is a differential of y as the independent variable. The same remarks apply to dx, in the two members, taken in reverse order.

The figure illustrates the differences referred to.

74. *The differential of the sum of any finite number of functions is equal to the sum of their differentials.*

Let $y = v - s + w +$ etc., in which v, s, w, etc., are functions of any variable, as x. Increasing x by Δx, we have

$$y + \Delta y = v + \Delta v - (s + \Delta s) + w + \Delta w + \text{etc.}$$

Whence

$$\Delta y = \Delta v - \Delta s + \Delta w + \text{etc.},$$

and (§ 70)

$$\frac{dy}{dx} = \underset{\Delta x \ggg\to 0}{\text{limit}} \left[\frac{\Delta v}{\Delta x} - \frac{\Delta s}{\Delta x} + \frac{\Delta w}{\Delta x} + \text{etc.}\right] = \frac{dv}{dx} - \frac{ds}{dx} + \frac{dw}{dx} + \text{etc.}$$

Therefore

$$\mathbf{d(v - s + w + etc.) = dv - ds + dw + etc.}$$

It follows that *the differential of the sum of any finite number of functions and constants is equal to the differential of the sum of the functions.* Thus, C being constant,

$$\mathbf{d[f(x) + C] = df(x).}$$

If corresponding differentials are equal it does not follow that the functions from which they were derived are equal.

75. *The differential of the product of any number of functions is equal to the sum of the products of the differential of each function by all of the other functions.*

Let $v = yz$ be the product of any two functions of any variable, as x, then

$$v + \Delta v = (y + \Delta y)(z + \Delta z) = yz + z \,.\, \Delta y + y \,.\, \Delta z + \Delta y \,.\, \Delta z,$$

whence $\Delta v = z \,.\, \Delta y + y \,.\, \Delta z + \Delta y \,.\, \Delta z$; and (§ 70)

$$\frac{dv}{dx} = \underset{\Delta x \ggg\to 0}{\text{limit}} \frac{\Delta v}{\Delta x} = \lim \left[z \frac{\Delta y}{\Delta x} + y \frac{\Delta z}{\Delta x} + \Delta y \frac{\Delta z}{\Delta x}\right].$$

$$= zdy/dx + ydz/dx.$$

Therefore, $\mathbf{dyz = zdy + ydz}$.

To illustrate, let $ONPM$ be a state of a rectangle with a variable diagonal represented by x. Two adjacent sides, denoted by z and y respectively, will be functions of x, and yz will be the variable area of the rectangle.

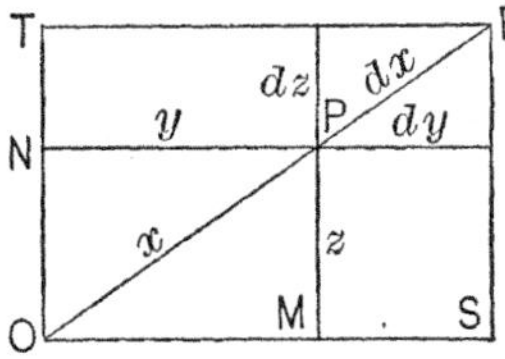

Assume $dx = PR$, and complete the rectangles $PT = ydz$

and $PS = zdy$. Then, since $dyz = ydz + zdy$, we have d (rect. OP) = rect. PT + rect. PS, which is the amount of change required by the definition of a differential. § 68.

It follows that *the differential of the product of a function and a constant is equal to the product of the constant and the differential of the function.* Thus, C being constant,

$$\mathbf{dCf(x) = Cdf(x).}$$

Let vsu be the product of any three functions of the same variable. Place $vs = r$, giving $vsu = ru$.

Differentiating, we have $dvsu = dru = rdu + udr$, in which

$dr = vds + sdv$. Hence, by substitution,

$$\mathbf{dvsu = vsdu + vuds + sudv.} \quad . \quad . \quad . \quad (1)$$

In a similar manner the principle may be established for any number of functions.

Dividing each member of (1) by vsu, we have

$$\mathbf{\frac{dvsu}{vsu} = \frac{du}{u} + \frac{ds}{s} + \frac{dv}{v}.}$$

Similarly, it may be shown that *the differential of the product of any number of functions divided by their product is equal to the sum of the quotients of the differential of each function by the function itself.*

EXAMPLES.

$d[(a+x)(b+x)] = (b+x)\,d(a+x) + (a+x)\,d(b+x) = (a+b+2x)dx.$

$a[3(c-x)] = -3dx. \qquad d[(a+x)x] = dx(a+x) + (a+x)dx.$

$$\frac{d[(a+x)x]}{(a+x)x} = \frac{d(a+x)}{a+x} + \frac{dx}{x} = \frac{dx}{a+x} + \frac{dx}{x}.$$

76. *The differential of a quotient of two functions is equal to the denominator into the differential of the numerator,* *minus the numerator into the differential of the denominator, divided by the square of the denominator.*

Let $y = v/s$, in which v and s are functions of any variable. Then $v = sy$, and

$$dv = sdy + yds = (s^2dy + vds)/s,$$

whence $\mathbf{dy = (sdv - vds)/s^2}$.

C being a constant, we have

$$\mathbf{d(C/s) = -Cds/s^2}, \quad \text{and} \quad \mathbf{d(v/C) = dv/C}.$$

EXAMPLES.

$d[x/(1+x)] = dx/(1+x)^2$. $\qquad d(3/x) = -3dx/x^2$.

$d[x(x+1)/(x-1)] = (x^2 - 2x - 1)dx/(x-1)^2$.

$d(x/3) = dx/3$. $\qquad d(2x/5a) = 2dx/5a$.

77. *The differential coefficient of y regarded as a function of x is equal to the product of the differential coefficient of y regarded as a function of u, by the differential coefficient of u regarded as a function of x.*

Having $y = f(u)$, and $u = \phi(x)$, let Δx, Δu and Δy be *corresponding increments* of x, u and y, respectively. Then (§ 4) Δu is the same in both cases, and

$$\Delta y/\Delta x = (\Delta y/\Delta u) \times (\Delta u/\Delta x). \quad \text{Hence,}$$

$$\lim_{\Delta x \to 0} [\Delta y/\Delta x] = \lim [\Delta y/\Delta u] \times \lim [\Delta u/\Delta x],$$

and (§ 70) $dy/dx = (dy/du) \times (du/dx)$.

Similarly, having $y = f(u)$, $u = \phi(x)$, $x = \psi(s)$, we find

$$dy/ds = (dy/du) \times (du/dx) \times (dx/ds);$$

and the same form holds true whatever be the number of the intermediate functions.

Having $y = f(u)$, and $x = \psi(u)$, we may write $u = \phi(x)$, and $dy/dx = (dy/du) \times (du/dx)$, but (§ 73) $du/dx = 1/(dx/du)$. Hence,

$$dy/dx = (dy/du)/(dx/du).$$

That is, *the differential coefficient of y regarded as a function of x is equal to the quotient of the differential coefficient of y regarded as a function of u, by the differential coefficient of x regarded as a function of u.*

EXAMPLES.

Given

1. $y = au^2, \quad u = bx.$ . . . . $dy/dx = 2ab^2x.$

2. $z = ay^2, \quad y^2 = 2px.$ . . . . $dz/dx = 2ap.$

3. $y = f(u), \quad x = \phi(u), \quad x = \psi(s).$ . . $\dfrac{dy}{ds} = \dfrac{dy/du}{dx/du} \times \dfrac{dx}{ds}.$

4. $y = u^2, \quad x = 3u, \quad x = 2s^2.$ . . . $dy/ds = 16s^3/9.$

5. $y = f(u), \quad u = F(s), \quad z = \psi(s).$ . . $\dfrac{dy}{dz} = \dfrac{du/ds}{dz/ds} \times \dfrac{dy}{du}.$

6. $y = f(u), \quad v = \phi(u), \quad v = \psi(s), \quad z = F(s), \quad z = F_1(x).$

$$\frac{dy}{dx} = \frac{dy/du}{dv/du} \times \frac{dv/ds}{dz/ds} \times \frac{dz}{dx}.$$

7. $y = f(s) = f(x + h)$, § 7. $\quad \therefore \quad s = x + h.$

Then $\quad \dfrac{dy}{dx} = \dfrac{dy}{ds} \times \dfrac{ds}{dx}, \quad$ and $\quad \dfrac{dy}{dh} = \dfrac{dy}{ds} \times \dfrac{ds}{dh}.$

But $\quad \dfrac{ds}{dx} = \dfrac{ds}{dh} = 1, \quad$ hence $\quad \dfrac{dy}{dx} = \dfrac{dy}{dh} = \dfrac{dy}{ds}.$

CHAPTER V.

DIFFERENTIATION OF FUNCTIONS.

FUNCTIONS OF A SINGLE VARIABLE.

78. The differential of any function of a single variable may be determined by applying the general rule, § 70, § 62, and multiplying the result by the differential of the variable; but by applying the general rule, § 70, § 62, to a general representative of any particular kind of function, there will result a *particular form*, or *rule, for differentiating such functions*, which is generally used in practice.

79. *The differential of any power of any function with a constant exponent is equal to the product of the exponent of the power, the function with its exponent diminished by unity, and the differential of the* **function.**

Let $y = x^n$, in which x is any variable and n is any constant. Then, increasing x by h, we have (§ 70)

$$\frac{dy}{dx} = \underset{h \ggg\!\rightarrow 0}{\text{limit}} \left[\frac{(x+h)^n - x^n}{h} \right].$$

Placing $x + h = s$, whence $h = s - x$, and as $h \ggg\!\rightarrow 0$, $s \ggg\!\rightarrow x$, we have (§ 40)

$$\frac{dy}{dx} = \underset{s \ggg\!\rightarrow x}{\text{limit}} \left[\frac{s^n - x^n}{s - x} \right] = nx^{n-1}, \text{ and } \mathbf{dx^n = nx^{n-1}dx}.$$

Having y^n, in which y is any function of any variable, as x, we have (§ 77)

$$dy^n/dx = (dy^n/dy) \times (dy/dx).$$

Hence,

$$dy^n/dx = ny^{n-1}(dy/dx), \quad \text{and} \quad \mathbf{dy^n = ny^{n-1}dy}. \qquad (1)$$

Substituting $1/n$ for n in (1), we have

$$\mathbf{dy^{1/n}} = \frac{1}{n}y^{1/n-1}dy = \frac{1}{n}y^{\frac{1-n}{n}}dy = \mathbf{dy/n\sqrt[n]{y^{n-1}}}.$$

Hence, *the differential of the nth root of any function is equal to the differential of the function divided by n times the nth root of the $n-1$ power of the function.*

EXAMPLES.

1. $d\sqrt{x} = dx/2\sqrt{x}$.
2. $dx^2 = 2xdx$.
3. $dx^3 = 3x^2dx$.
4. $d4x^4 = 16x^3dx$.
5. $dax^2 = 2axdx$.
6. $d(\log x)^2 = 2\log x\, d\log x$.
7. $d\sin^3 x = 3\sin^2 x\, d\sin x$.
8. $d3x^3 = 9x^2dx$.
9. $d(4\pi x^3/3) = 4\pi x^2dx$.
10. $d\sqrt[n]{a+x} = (a+x)^{\frac{1-n}{n}}dx/n$.
11. $d\sqrt{x^3-a^3} = 3x^2dx/2\sqrt{x^3-a^3}$.
12. $d\sqrt[3]{x} = dx/3\sqrt[3]{x^2}$.
13. $dx^{-n} = -nx^{-n-1}dx$.
14. $dx^{-4} = -4x^{-5}dx$.
15. $dx^{-\frac{1}{2}} = -\frac{1}{2}x^{-\frac{3}{2}}dx$.
16. $dx^{-\frac{1}{n}} = -\frac{1}{n}x^{-\frac{1}{n}-1}dx$.
17. $d(a^x)^2 = 2a^xda^x$.
18. $d\tan^n x = n\tan^{n-1}x\, d\tan x$.
19. $d[3(a+x^2)^3] = 18(a+x^2)^2xdx$.
20. $d(2\pi ax^2/3) = 4\pi axdx/3$.
21. $d(2x)^{10} = 20(2x)^9dx$.
22. $d[2x^{3/4}/7] = 3dx/14x^{1/4}$.
23. $d(a+x^2)^3 = 3(a+x^2)^2\,d(a+x^2) = 3(a+x^2)^2\,2xdx = 6x(a+x^2)^2dx$.
24. $d\sqrt[3]{a+x^2} = d(a+x^2)/3\sqrt[3]{(a+x^2)^2} = 2xdx/3\sqrt[3]{(a+x^2)^2}$.
25. $d(2x)^2 = 2(2x)d(2x) = 8xdx$.

26. $d(2x^2)^2 = 2(2x^2)d(2x^2) = 16x^3dx.$

27. $d(ax^2)^3 = 3(ax^2)^2d(ax^2) = 6a^3x^5dx.$

28. $d(3x)^{-2} = -2(3x)^{-3}d(3x) = -6(3x)^{-3}dx.$

29. $d(a^2-x^2)^{\frac{1}{2}} = \frac{1}{2}(a^2-x^2)^{-\frac{1}{2}}d(a^2-x^2) = -(a^2-x^2)^{-\frac{1}{2}}xdx.$

30. $d[(x^3+a)(3x^2+b)] = (15x^4+3bx^2+6ax)dx.$

31. $d\left[x^3/\sqrt{(1-x^2)^3}\right] = 3x^2dx/(1-x^2)^{5/2}.$

32. $d[a+x-3x^2+4x^3] = (1-6x+12x^2)dx.$

33. $d[(1+x^2)(1-x^3)] = (2x-3x^2-5x^4)dx.$

34. $d(a+bx^m)^n = bmn(a+bx^m)^{n-1}x^{m-1}dx.$

35. $d[(2x^2-1)/x\sqrt{1+x^2}] = (4x^2+1)dx/x^2(1+x^2)^{3/2}.$

36. $d[x/\sqrt{a^2-x^2}] = a^2dx/(a^2-x^2)^{3/2}.$

37. $d(1+x)^{-\frac{1}{2}} = -dx/2(1+x)^{\frac{3}{2}}.$

38. $d\sqrt{a^2+x^2} = xdx/\sqrt{a^2+x^2}.$

39. $d[x^3(a+x)^2] = (3a+5x)(a+x)x^2dx.$

40. $d[x^n/(1+x)^n] = nx^{n-1}dx/(1+x)^{n+1}.$

41. $d[1/\sqrt{1-x^2}] = xdx/(1-x^2)^{3/2}.$

42. $d[x/\sqrt{1-x}] = (2-x)dx/2(1-x)^{3/2}.$

43. $d[x/\sqrt{1-x^2}] = dx/(1-x^2)^{3/2}.$

44. $d[x^3/(1-x^2)^{3/2}] = 3x^2dx/(1-x^2)^{5/2}.$

45. $d[\sqrt{1+x}/\sqrt{1-x}] = dx/(1-x)\sqrt{1-x^2}.$

46. $d[x^{2n}/(1+x^2)^n] = 2nx^{2n-1}dx/(1+x^2)^{n+1}.$

47. $d[(1-x)/\sqrt{1+x^2}] = -(1+x)dx/(1+x^2)^{3/2}.$

48. $d[x/(x-\sqrt{1-x^2})] = -dx/\sqrt{1-x^2}(x-\sqrt{1-x^2})^2.$

49. $d\sqrt{x^3/(2a-x)} = (3a-x)\sqrt{x}dx/(2a-x)^{3/2}.$

50. $d(-\sqrt{a^2-x^2}/a^2x) = dx/x^2\sqrt{a^2-x^2}.$

51. $d[ax/(x+\sqrt{a+x^2})] = a^2dx/(x+\sqrt{a+x^2})^2\sqrt{a+x^2}.$

52. $d(x/a^2\sqrt{x^2+a^2}) = dx/\sqrt{(x^2+a^2)^3}$.

53. $d[x/(1+x)]^n = nx^{n-1}dx/(1+x)^{n+1}$.

54. $d[(a+bx^{3/2})/c\sqrt[4]{x^5}] = (bx^{3/2}-5a)dx/4c\sqrt[4]{x^9}$.

55. $d[x^2/(a+x^3)^2] = 2x(a-2x^3)dx/(a+x^3)^3$.

56. $dx(a^2+x^2)\sqrt{a^2-x^2} = (a^4+a^2x^2-4x^4)dx/(a^2-x^2)^{1/2}$.

57. $d[x^{3/2}(1+x)^{1/2}/(1-x)^{1/2}] = (3+2x-3x^2)\sqrt{x}dx/2(1-x)^{3/2}$.

58. $d(1+x)\sqrt{1-x} = (1-3x)dx/2\sqrt{1-x}$.

59. $d[\sqrt{1-\sqrt{x}}/\sqrt{1+\sqrt{x}}] = -dx/2(1+\sqrt{x})\sqrt{x-x^2}$.

60. $d\sqrt{x+\sqrt{1+x^2}} = dx\sqrt{x+\sqrt{1+x^2}}/2\sqrt{1+x^2}$.

61. $u = [a - b/\sqrt{x} + (c^2-x^2)^{2/3}]^{3/4}$.

1°. $d(a-b/\sqrt{x}+(c^2-x^2)^{2/3}) = [b/2x\sqrt{x}+2(c^2-x^2)^{-\frac{1}{3}}(-2x)/3]dx$

$$= [b/2x\sqrt{x} - 4x/3(c^2-x^2)^{1/3}]dx.$$

Hence, $$du = \frac{3}{4}\left[a - \frac{b}{\sqrt{x}} + (c^2-x^2)^{2/3}\right]^{-\frac{1}{4}}\left[\frac{b}{2x\sqrt{x}} - \frac{4x}{3(c^2-x^2)^{1/3}}\right]dx$$

$$= \left(\frac{3b}{2x\sqrt{x}} - \frac{4x}{\sqrt[3]{c^2-x^2}}\right)dx \Big/ 4\sqrt[4]{a - \frac{b}{\sqrt{x}} + \sqrt[3]{(c^2-x^2)^2}}.$$

62. In the parabola $y^2 = 9x$, find the rate of y with respect to x when $x = 4$. What value will x have when rate of y equals that of x? When rate of y is the greatest? When the least?

$$dy/dx = 9/2y = \pm 3/2\sqrt{x}. \quad \therefore (dy/dx)_{x=4} = \pm 3/4.$$

$$dy/dx = \pm 3/2\sqrt{x} = 1 \quad \text{gives} \quad x = 9/4.$$

dy/dx is the greatest when $x = 0$, and the least when $x = \infty$.

63. Find the slope of the curve $y = \pm\sqrt{9-x^2}$ when $x = 2$. Find values of x and y when the slope is 1.

Ans. ± 0.894. $x' = \pm\sqrt{4.5}$, $y' = \mp\sqrt{4.5}$

64. Find the angle which the curve $y = x/(1+x^2)$ makes with X at their point of intersection. Ans. $\pi/4$.

65 Find the angle at which the curves $y^2 = 10x$ and $x^2 + y^2 = 144$ intersect. Ans. $71°\ 0'\ 58''$.

66. Find the value of x at the point where the slope of $y^2 = 4x^3$ is unity. Ans. $1/9$.

67. At what rate does the volume of a cube change with respect to the length of an edge? Ans. $3\,(\text{edge})^2$.

68. Find the angles that a tangent to the curve $x^2 = 6y^2 + 3y + 1$, at the point (8, 3), makes with the axis X. Ans. $\tan^{-1}(16/39)$.

69. Find the rate of change of $(\sqrt{x} + 3/ax^2)$ when $x = 3$.

Ans. $1/2\sqrt{3} - 2/9a$.

70. Find the rate of change of the ordinate of a circle with respect to the abscissa. Ans. $\mp x/\sqrt{R^2 - x^2}$.

71. Find the rate of change of the ordinate of an ellipse with respect to the abscissa. Ans. $\mp bx/a\sqrt{a^2 - x^2}$.

80. *The differential of the logarithm of any function is equal to the differential of the* **function** *divided by the* **function.**

Let $y = \log x$, then (§ 70, § 42)

$$\frac{dy}{dx} = \underset{h \to 0}{\text{limit}} \frac{\log(x+h) - \log x}{h} = \lim \frac{\log \dfrac{x+h}{x}}{h}$$

$$= \lim \frac{\log\left(1 + \dfrac{h}{x}\right)}{h} = \lim \frac{\dfrac{h}{x}\log\left(1 + \dfrac{h}{x}\right)^{\frac{x}{h}}}{h}$$

$$= \frac{1}{x} \lim \log\left(1 + \frac{h}{x}\right)^{\frac{x}{h}} = \frac{1}{x}\log e = \frac{1}{x}.$$

Hence, $\mathbf{d \log x = dx/x}$.

Since $\log_a e = M_a$, it follows that $d \log_a x = \boldsymbol{M_a\, dx/x}$.

Having $y = f(x)$, we write (§ 77)

$$\frac{d \log y}{dx} = \frac{d \log y}{dy} \times \frac{dy}{dx} = \frac{1}{y} \times \frac{dy}{dx}.$$

Hence, $\mathbf{d \log y = dy/y}$.

EXAMPLES.

1. $d \log e^x = (de^x)/e^x$. 2. $d \log \sin x = (d \sin x)/\sin x$.

3. $d \log x^2 = dx^2/x^2 = 2xdx/x^2 = 2dx/x$.

4. $d \log \sqrt{x} = d\sqrt{x}/\sqrt{x} = (dx/2\sqrt{x})/\sqrt{x} = dx/2x$.

5. $d \log x^n = dx^n/x^n = nx^{n-1}dx/x^n = ndx/x$.

6. $d \log \sqrt{x^2 - 1} = d\sqrt{x^2 - 1}/\sqrt{x^2 - 1} = xdx/(x^2 - 1)$.

7. $d \log[(1 + x)/(1 - x)] = d[(1 + x)/(1 - x)]/[(1 + x)/(1 - x)]$
$= 2dx/(1 - x^2)$.

8. $d[\log \log x] = d \log x/\log x = (dx/x)/\log x = dx/x \log x$.

9. $d[x \log x] = xd \log x + \log xdx = (1 + \log x)dx$.

10. $d \log (x^2 + x^3) = d(x^2 + x^3)/(x^2 + x^3) = (2x + 3x^2)dx/(x^2 + x^3)$.

11. $d(\log x)^m = m(\log x)^{m-1} d \log x = m(\log x)^{m-1}dx/x$

12. $d(1/\log x^n) = - d \log x^n/(\log x^n)^2 = - ndx/x(\log x^n)^2$.

13. $d \log \sqrt{(1 + x)/(1 - x)} = dx/(1 - x^2)$.

14. $d \log [(1 + x^{\frac{1}{3}})/(1 - x^{\frac{1}{3}})] = 2dx/3x^{\frac{2}{3}}(1 - x^{\frac{2}{3}})$.

15. $d\, x^m (\log x)^n = [m(\log x)^n + n(\log x)^{n-1}]x^{m-1}dx$.

16. $d \log [\sqrt{x^3 + 1}/\sqrt{x^3 - 1}] = - 3x^2dx/(x^6 - 1)$.

17. $d \log [(1 + \sqrt{1 - x^2})/x] = - dx/x\sqrt{1 - x^2}$.

18. $d \log [(1 + \sqrt{x})/(1 - \sqrt{x})] = dx/(1 - x)\sqrt{x}$.

19. $d \log [x\sqrt{-1} + \sqrt{1 - x^2}] = - dx/\sqrt{x^2 - 1}$.

20. $d \log \dfrac{\sqrt{x^2 + 1} - 1}{\sqrt{x^2 + 1} + 1} = \dfrac{2dx}{x\sqrt{x^2 + 1}}$.

21. $d \log \dfrac{\sqrt{x^2+a^2}-x}{\sqrt{x^2+a^2}+x} = \dfrac{-2dx}{\sqrt{x^2+a^2}}$.

22. $d \log \dfrac{\sqrt{1+x}+\sqrt{1-x}}{\sqrt{1+x}-\sqrt{1-x}} = \dfrac{-dx}{x\sqrt{1-x^2}}$.

23. $d \log \sqrt{\dfrac{\sqrt{1+x^2}+x}{\sqrt{1+x^2}-x}} = \dfrac{dx}{\sqrt{1+x^2}}$.

24. $d \log \sqrt{a^2-x^2} = -xdx/(a^2-x^2)$.

25. $d \log \sqrt{a^2+x^2} = xdx/(a^2+x^2)$.

26. $d\left(\log [(a-x)/(a+x)]/2a\right) = dx/(x^2-a^2)$.

27. $d \log (x/\sqrt{1+x^2}) = dx/x(1+x^2)$.

28. $d \log (x \pm \sqrt{x^2 \pm a^2}) = \pm dx/\sqrt{x^2 \pm a^2}$.

29. d[log log...(repeated n times) of x] $= dx/x \log x (\log)^2 x \ldots (\log)^{n-1} x$.

30. $d \log [\log (a+bx^n)] = nbx^{n-1}dx/(a+bx^n) \log (a+bx^n)$.

31. $u = (\log x^n)^m$. Put $\log x^n = y$, then (§ 77)

$$du = d(\log x^n)^m = my^{m-1}(n/x)dx = mn(\log x^n)^{m-1}dx/x.$$

32. Find the rate of change of a logarithm in the common system with respect to the number. Ans. M_{10}/number.

33. Find the slope of the curve $y = \log_a x$ at the point (1, 0).

Ans. M_a.

81. *The differential of any exponential function with a constant base is equal to the product of the function, the logarithm of the base, and the differential of the* **exponent.**

Let $y = a^x$, then (§ 70, § 43)

$$\frac{dy}{dx} = \underset{h \to 0}{\text{limit}} \frac{a^{x+h}-a^x}{h} = a^x \lim \frac{a^h - 1}{h} = a^x \log a.$$

Hence, $$da^x = a^x \log a dx.$$

Having a^y, in which $y = f(x)$, we have (§ 77)

$$\frac{da^y}{dx} = \frac{da^y}{dy} \times \frac{dy}{dx} = a^y \log a \times \frac{dy}{dx}.$$

Therefore $\quad \mathbf{da^y = a^y \log a dy.}$

It follows that $de^y = e^y dy$.

EXAMPLES.

1. $da^{x^2} = a^{x^2} \log a\, dx^2 = 2a^{x^2} x \log a\, dx$.

2. $da^{\log x} = a^{\log x} \log a\, d \log x = a^{\log x} \log a\, dx/x$.

3. $da^{\sqrt{x}} = a^{\sqrt{x}} \log a\, d\sqrt{x} = a^{\sqrt{x}} \log a\, dx/2\sqrt{x}$.

4. $da^{1/x} = a^{1/x} \log a\, d(1/x) = -a^{1/x} \log a\, dx/x^2$.

5. $da^x x^n = a^x dx^n + x^n da^x = a^x x^{n-1}(x \log a + n)dx$.

6. $de^{-\frac{1}{x}} = e^{-\frac{1}{x}} dx/x^2$.

7. $d[(e^x - e^{-x})/2] = (e^x + e^{-x})dx/2$.

8. $d\left[a(e^{\frac{x}{a}} + e^{-\frac{x}{a}})/2\right] = (e^{\frac{x}{a}} - e^{-\frac{x}{a}})dx/2$.

9. $de^x \log x = (1/x + \log x)e^x dx$.

10. $d[(e^x - 1)/(e^x + 1)] = 2e^x dx/(e^x + 1)^2$.

11. $d \log [(e^x - 1)/(e^x + 1)] = 2e^x dx/(e^{2x} - 1)$.

12. $d \log (e^x + e^{-x}) = (e^x - e^{-x})dx/(e^x + e^{-x})$.

13. $d \log (e^x - e^{-x}) = (e^x + e^{-x})dx/(e^x - e^{-x})$.

14. $d[(a^x - 1)/(a^x + 1)] = 2a^x \log a\, dx/(a^x + 1)^2$.

15. $d(a^x + x)^2 = 2(a^x + x)(a^x \log a + 1)dx$.

16. $de^x(1 - x^3) = e^x(1 - 3x^2 - x^3)dx$.

17. $a[(e^x - e^{-x})/(e^x + e^{-x})] = 4dx/(e^x + e^{-x})^2$.

18. $a[x/(e^x - 1)] = [e^x(1 - x) - 1]dx/(e^x - 1)^2$.

19. $dx^n(1 + x)^n = nx^{n-1}(1 + x)^{n-1}(1 + 2x)dx$.

20. When $x = 0$, find the inclination of the curve $y = 10^x$ to X.

Ans. $66° 31' 30''$.

82. Logarithmic Differentiation. — The differentiation of an exponential function, or one involving a product or quotient, is frequently simplified by first taking the Napierian logarithm of the function.

Thus, let $u = y^z$, in which y and z are functions of the same variable.

$$\therefore \log u = z \log y$$

and (§ 80, § 75)

$$du/u = z dy/y + \log y dz.$$

Hence, $$du = dy^z = zy^{z-1}dy + y^z \log y dz,$$

which is the sum of the differentials obtained by applying first the rule in § 79, then that in § 81.

EXAMPLES.

1. $u = \dfrac{\sqrt{(x-1)^5}}{\sqrt[4]{(x-2)^3}\,\sqrt[3]{(x-3)^7}}$ $\quad \log u = \frac{5}{2}\log(x-1)$
$\qquad - \frac{3}{4}\log(x-2) - \frac{7}{3}\log(x-3),$

$$\frac{du}{u} = \frac{5}{2}\frac{dx}{x-1} - \frac{3}{4}\frac{dx}{x-2} - \frac{7}{3}\frac{dx}{x-3} = -\frac{7x^2+30x-97}{12(x-1)(x-2)(x-3)}dx,$$

$$du = -\frac{(x-1)^{3/2}(7x^2+30x-97)}{12(x-2)^{7/4}(x-3)^{10/3}}dx.$$

2. $dx^x = x^x(1 + \log x)dx.$

3. $dx^{\frac{1}{x}} = x^{\frac{1-2x}{x}}(1 - \log x)dx.$

4. $dx^{x^x} = x^{x^x} x^x(\log^2 x + \log x + 1/x)dx.$

5. $dx\sqrt{1-x}(1+x) = (2 + x - 5x^2)dx/2\sqrt{1-x}.$

6. $d\dfrac{(x-1)^{5/2}}{(x-2)^{3/4}(x-3)^{7/3}} = -\dfrac{(x-1)^{3/2}(7x^2+30x-97)dx}{12(x-2)^{7/4}(x-3)^{10/3}}.$

7. $d(x/n)^{nx} = n(x/n)^{nx}[1 + \log(x/n)]dx.$

8. $dx^{1/x} = x^{1/x}\log(e/x)dx/x^2.$

9. $de^{e^x} = e^{e^x}e^x dx.$ 10. $dx^{e^x} = x^{e^x}e^x dx(1 + x\log x)/x.$

11. $de^{x^x} = e^{x^x}x^x(1 + \log x)dx.$

12. $d(x^x + x^{1/x}) = [x^x \log ex - x^{\frac{1-2x}{x}}\log(x/e)]dx.$

13. $u = x^{\log x}$. Put $\log x = z$, then

$$du = dx^z = (\log x\, x^{\log x - 1} + x^{\log x}\log x/x)dx = 2x^{\log x - 1}\log x\, dx,$$

$$d(\log x)^x = [(\log x)^{x-1} + (\log x)^x \log\log x]dx.$$

14. $$d\frac{\sqrt{2x}(1 - x^2)^{3/4}}{(x-2)^{2/3}} = \frac{(-8x^3 + 24x^2 - x - 6)dx}{3(2x)^{1/2}(1 - x^2)^{1/4}(x-2)^{5/3}}.$$

Trigonometric Functions.

83. d sin x = cos x dx.

Let h be the increment of x, then (§ 70)

$$\frac{d\sin x}{dx} = \lim_{h \to 0}\left[\frac{\sin(x+h) - \sin x}{h}\right]$$

$$= \lim\left[\frac{2\sin\frac{h}{2}\cos\left(x + \frac{h}{2}\right)}{h}\right]$$

$$= \lim\left[\frac{\sin(h/2)}{h/2}\cos\left(x + \frac{h}{2}\right)\right] = \cos x.$$

Having $\sin y$, in which $y = f(x)$, we have (§ 77)

$$\frac{d\sin y}{dx} = \frac{d\sin y}{dy} \times \frac{dy}{dx} = \cos y \times \frac{dy}{dx}.$$

Hence, $$d\sin y = \cos y\, dy.$$

Similarly, by applying the general rule, § 70, and the principle in § 77, the differential of any trigonometric function may be determined; but it is perhaps simpler to make use of the relations existing between the functions.

$$\mathbf{d}\cos\mathbf{x} = d\sin(\pi/2 - x) = \cos(\pi/2 - x)d(\pi/2 - x)$$
$$= -\sin\mathbf{x}\,\mathbf{dx}.$$

$$\mathbf{d}\tan\mathbf{x} = d\frac{\sin x}{\cos x} = \frac{\mathbf{dx}}{\cos^2\mathbf{x}} = \sec^2\mathbf{x}\,\mathbf{dx} = (1 + \tan^2\mathbf{x})\mathbf{dx}.$$

$$\mathbf{d}\cot\mathbf{x} = d\tan(\pi/2 - x) = -\mathbf{dx}/\sin^2\mathbf{x} = -\operatorname{cosec}^2\mathbf{x}\,\mathbf{dx}$$
$$= -(1 + \cot^2\mathbf{x})\mathbf{dx}.$$

$$\mathbf{d}\sec\mathbf{x} = d(1/\cos x) = \sin x\,dx/\cos^2 x = \tan\mathbf{x}\sec\mathbf{x}\,\mathbf{dx}.$$

$$\mathbf{d}\operatorname{cosec}\mathbf{x} = d\sec(\pi/2 - x) = -\cot\mathbf{x}\operatorname{cosec}\mathbf{x}\,\mathbf{dx}.$$

$$\mathbf{d}\operatorname{vers}\mathbf{x} = d(1 - \cos x) = \sin\mathbf{x}\,\mathbf{dx}.$$

$$\mathbf{d}\operatorname{covers}\mathbf{x} = d\operatorname{vers}(\pi/2 - x) = -\cos\mathbf{x}\,\mathbf{dx}.$$

In order to illustrate the formulas for the differentials of the sine and cosine of any angle, let $ACB = \phi$ be any given angle. Assume $BCN = d\phi$, and with any radius, as $CO = R$, describe an arc, as OMN. Then

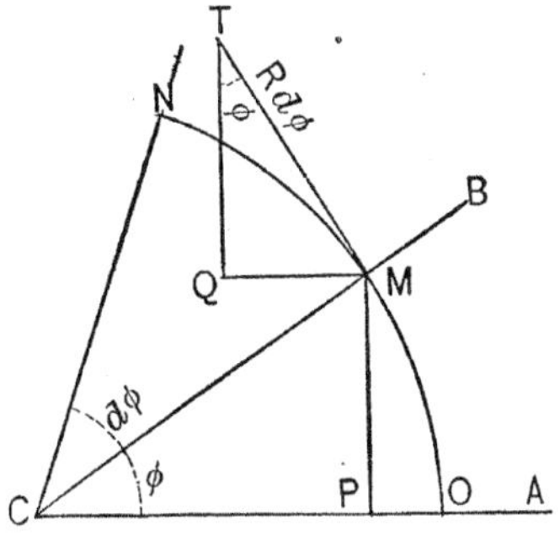

$$PM/R = \sin\phi,$$
$$CP/R = \cos\phi, \quad \text{arc } MN = Rd\phi.$$

The definition of a differential (§ 63), in this case, requires that $\sin\phi$ and $\cos\phi$, retaining their rates at the states corresponding to $\phi = ACB$, shall continue to change from those states while ϕ increases by the angle $BCN = d\phi$.

Draw the tangent line to the arc at M, and lay off MT equal to the arc $MN = Rd\phi$. Through T draw TQ parallel to MP, and through M draw MQ parallel to OC.

Then QT and $-MQ$ are, respectively, the changes that the lines PM and CP would undergo were they to continue to change with the rates they have when $\phi = ACB$, while ϕ increases by $d\phi$.

Hence, QT/R, and $-MQ/R$ are, respectively, the changes that sin ϕ and cos ϕ would undergo under the same requirements.

The angle $MTQ = \phi$. Hence,

$$QT = MT\cos\phi = R\cos\phi d\phi, \text{ and } QT/R = d\sin\phi = \cos\phi d\phi.$$

$$-MQ = MT\sin\phi = R\sin\phi d\phi, \text{ and } MQ/R = d\cos\phi = -\sin\phi d\phi.$$

Similarly the formulas for the differentials of the other trigonometric functions may be illustrated.

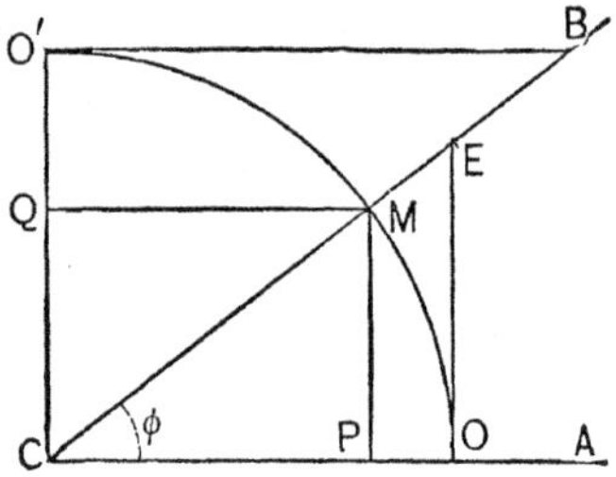

Regarding the right lines PM, CP, OE, $O'B$, etc., as functions of the variable angle ϕ, we have

$$dPM = dR\sin\phi = R\cos\phi d\phi. \qquad dCP = dR\cos\phi = -R\sin\phi d\phi.$$

$$dOE = dR\tan\phi = Rd\phi/\cos^2\phi. \qquad dO'B = dR\cot\phi = -Rd\phi/\sin^2\phi.$$

$$dCE = dR\sec\phi \qquad dCB = dR\operatorname{cosec}\phi$$

$$= R\tan\phi\sec\phi d\phi. \qquad = -R\cot\phi\operatorname{cosec}\phi d\phi.$$

$$dOP = dR\operatorname{vers}\phi = R\sin\phi d\phi. \qquad dO'Q = dR\operatorname{covers}\phi = -R\cos\phi d\phi.$$

It is important to notice the difference between the differentials of the above lines, which depend upon the radius of the circle used, and the differentials of the trigonometric functions which do not depend upon any radius or circle.

EXAMPLES.

1. $d \sin x^2 = 2x \cos x^2\, dx.$
2. $d \sin^2 x = 2 \sin x \cos x\, dx.$
3. $d \cos^2 x = -2 \cos x \sin x\, dx.$
4. $d \tan^2 x = 2 \tan x\, dx/\cos^2 x.$
5. $d \cot^2 x = -2 \cot x\, dx/\sin^2 x.$
6. $d \sec^2 x = 2 \sec^2 x \tan x\, dx.$
7. $d \operatorname{cosec}^2 x = -2 \operatorname{cosec}^2 x \cot x\, dx.$
8. $d \operatorname{vers}^2 x = 2 \operatorname{vers} x \sin x\, dx.$
9. $d \operatorname{covers}^2 x = -2 \operatorname{covers} x \cos x\, dx.$
10. $d \tan x^2 = 2x dx/\cos^2 x^2.$
11. $d e^{\sin x} = e^{\sin x} \cos x\, dx.$
12. $d \cos x^2 = -2x \sin x^2\, dx.$
13. $d \sin^3 x = 3 \sin^2 x \cos x\, dx.$
14. $d \cos^3 x = -3 \cos^2 x \sin x\, dx.$
15. $d \tan^3 x = 3 \tan^2 x\, dx/\cos^2 x.$
16. $d \cot^3 x = -3 \cot^2 x\, dx/\sin^2 x.$
17. $d \sec^3 x = 3 \sec^3 x \tan x dx.$
18. $d \operatorname{cosec}^3 x = -3 \operatorname{cosec}^3 \cot x\, dx.$
19. $d \operatorname{vers}^3 x = 3 \operatorname{vers}^2 \sin x\, dx.$
20. $d \cos x^3 = -3x^2 \sin x^3\, dx.$
21. $d \operatorname{covers}^3 x = -3 \operatorname{covers}^2 x \cos x\, dx.$
22. $d \log \cos x = -\tan x\, dx.$
23. $d \sin^2 x^2 = 2 \sin x^2 d \sin x^2 = 4x \sin x^2 \cos x^2\, dx.$
24. $d\sqrt{\tan 2x} = \sec^2 2x\, dx/\sqrt{\tan 2x}.$
25. $d \cos(1/x) = (1/x^2) \sin (1/x) dx.$
26. $d \tan^2 x^2 = 4x \tan x^2\, dx/\cos^2 x^2.$
27. $d \cos mx = -\sin mx\, d(mx) = -m \sin mx\, dx.$
28. $d \sin 3x = \cos 3x\, d\, 3x = 3 \cos 3x\, dx.$
29. $d \sin^2 2x = 4 \sin 2x \cos 2x\, dx.$
30. $d \sin^n ax = na \sin^{n-1} ax \cos ax\, dx.$
31. $d \tan^n x = n \tan^{n-1} x\, dx/\cos^2 x.$
32. $d(\tan x - x) = \tan^2 x\, dx.$
33. $d[(x - \sin x \cos x)/2] = \sin^2 x\, dx.$
34. $d \tan x \sec x = (\sec^2 x + \tan^2 x) \sec x\, dx.$
35. $d \tan a^{1/x} = -a^{1/x} \sec^2 a^{1/x} \log a\, dx/x^2.$

36. $d[(1 - \tan x)/\sec x] = -(\cos x + \sin x)dx.$

37. $d[\sin nx/\cos^n x] = n \cos (n - 1)x\, dx/\cos^{n+1} x.$

38. $d \tan \sqrt{1 - x} = -(\sec \sqrt{1 - x})^2 dx/2\sqrt{1 - x}.$

39. $d\sqrt{\sin \sqrt{x}} = \cos \sqrt{x}\, dx/4\sqrt{x \sin \sqrt{x}}.$

40. $d(\cos x)^{\sin x} = (\cos x)^{\sin x}(\cos x \log \cos x - \sin^2 x/\cos x)dx.$

41. $d \sin (\sin x) = \cos x \cos (\sin x)dx.$

42. $d \cos (\sin x) = -\cos x \sin (\sin x)dx.$

43. $d \sin (\log nx) = \cos (\log nx)dx/x.$

44. $dx^{\sin x} = x^{\sin x}(\cos x \log x + \sin x/x)dx.$

45. $d\dfrac{(\sin ax)^b}{(\cos bx)^a} = \dfrac{ab (\sin ax)^{b-1} \cos (bx - ax)dx}{(\cos bx)^{a+1}}.$

46. $d \sin^3 x \cos x = \sin^2 x(3 - 4 \sin^2 x)dx.$

47. $d \cos^5 x \sin^2 x = (2 \cos^2 x - 5 \sin^2 x) \cos^4 x \sin x\, dx.$

48. $d[(\sin x + \cos x)/(\sin x - \cos x)] = -2dx/(\sin x - \cos x)^2.$

49. $d(\sin x)^{\tan x} = (\sin x)^{\tan x}(1 + \sec^2 x \log \sin x)dx.$

50. Determine the manner in which the sine of an angle varies with the angle.

The rate of change of $\sin x$ is (§ 70) $d \sin x/dx = \cos x$, from which we see that as ϕ increases from 0 to $\pi/2$, the rate is +, but diminishing; hence, the sine increases, but its increments decrease.

From $\pi/2$ to π the rate is −, and diminishing; hence, the sine diminishes and its decrements increase numerically.

From π to $3\pi/2$ the rate is −, and increasing. That is, the sine decreases, but its decrements diminish numerically.

From $3\pi/2$ to 2π the rate is +, and increasing. That is, the sine increases, and its increments increase.

In a similar manner the circumstances of change of each trigonometric function with respect to the angle may be determined.

51. Assuming $dx = \pi/4$, we have, corresponding to $x = \pi/6$,

$$d \sin x = \frac{\sqrt{3}}{8}\pi. \qquad d \cos x = -\frac{\pi}{8}. \qquad d \tan x = \frac{\pi}{3}.$$

$d \cot x = -\pi$. $d \sec x = \frac{\pi}{6}$. $d \operatorname{cosec} x = -\frac{\sqrt{3}}{2}\pi$.

52. Corresponding to $x = \pi/4$, we have

$$\frac{d \sin x}{dx} = \frac{1}{\sqrt{2}}. \quad \frac{d \cos x}{dx} = -\frac{1}{\sqrt{2}}. \quad \frac{d \tan x}{dx} = 2.$$

$$\frac{d \cot x}{dx} = -2. \quad \frac{d \sec x}{dx} = \sqrt{2}. \quad \frac{d \operatorname{cosec} x}{dx} = -\sqrt{2}.$$

53. What is the value of x when $\tan x$ is increasing twice as fast as x? Ans. $\pi/4$.

54. Find the rate of change of the tangent, regarded as a function of the sine of an angle.

Since $\tan x = fx$, and $\sin x = Fx$, we have (§ 77)

$$d \tan x/d \sin x = (d \tan x/dx)/(d \sin x/dx)$$
$$= (1/\cos^2 x)/\cos x = 1/\cos^3 x.$$

In a similar manner the rate of change of any trigonometric function regarded as a function of any other may be found.

55. Find the slope of the curve $y = \sin x$ when $x = 0$, $x = \pi/4$, $x = \pi/2$. Ans. 1, $\sqrt{1/2}$, 0.

56. When $x = \pi/3$, find the inclination of the curve $y = \tan x$ to X. Ans. $75^\circ\ 57'\ 50''$.

57. Find the angle which the curves $y = \sin x$ and $y = \cos x$ make with each other at their point of intersection. Ans. $\tan^{-1} 2\sqrt{2}$.

Inverse Trigonometric Functions.

84. $d \sin^{-1} x = dx/\sqrt{1 - x^2}$.

Let $\phi = \sin^{-1} x$; then $x = \sin \phi$ and $dx/d\phi = \cos \phi$. Hence (§ 73),

$$d \sin^{-1} x/dx = 1/\cos \phi = 1/\pm \sqrt{1 - x^2}.*$$

* The sign depends upon that of $\cos \phi$. Formulas involving the double sign in this article are generally written with the plus sign only, which corresponds to angles ending in the first quadrant.

By applying the principle in § 77, it may be shown that

$$d \sin^{-1} y = dy/\sqrt{1 - y^2}, \qquad \text{in which } y = fx.$$

$$\mathbf{d \cos^{-1} x} = d(\pi/2 - \sin^{-1} x) = -\mathbf{dx}/\sqrt{\mathbf{1 - x^2}}.$$

$$\mathbf{d \tan^{-1} x = dx/(1 + x^2).}$$

Let $\phi = \tan^{-1} x$; then

$$x = \tan\phi, \quad \text{and} \quad dx/d\phi = 1 + \tan^2\phi.$$

Hence (§ 73), $d \tan^{-1} x/dx = 1/(1 + x^2)$.

$$\mathbf{d \cot^{-1} x} = d(\pi/2 - \tan^{-1} x) = -\mathbf{dx/(1 + x^2)}.$$

$$\mathbf{d \sec^{-1} x = dx/x\sqrt{x^2 - 1}.}$$

Let $\phi = \sec^{-1} x$,
then $x = \sec\phi$, and $dx/d\phi = \sec\phi \tan\phi$. Hence (§ 73),

$$\frac{d \sec^{-1} x}{dx} = \frac{1}{\sec\phi \tan\phi} = \frac{1}{\sec\phi\sqrt{\sec^2\phi - 1}} = \frac{1}{x\sqrt{x^2 - 1}}.$$

$$\mathbf{d\, cosec^{-1} x} = d(\pi/2 - \sec^{-1} \mathbf{x}) = -\mathbf{dx/x\sqrt{x^2 - 1}}.$$

$$\mathbf{d\, vers^{-1} x = dx/\sqrt{2x - x^2}}.$$

Let $\phi = \text{vers}^{-1} x$; then $x = \text{vers}\,\phi$, and $\dfrac{dx}{d\phi} = \sin\phi$.
Hence (§ 73),

$$\frac{d\, \text{vers}^{-1} x}{dx} = \frac{1}{\sin\phi} = \frac{1}{\sqrt{1 - \cos^2\phi}} = \frac{1}{\sqrt{1 - (1 - \text{vers}\,\phi)^2}}$$

$$= \frac{1}{\sqrt{2\,\text{vers}\,\phi - \text{vers}^2\phi}} = \frac{1}{\sqrt{2x - x^2}}.$$

$$\mathbf{d}\ \text{covers}^{-1}\,\mathbf{x} = d(\pi/2 - \text{vers}^{-1}\,x) = -\,\mathbf{dx}/\sqrt{\mathbf{2x} - \mathbf{x}^2}.$$

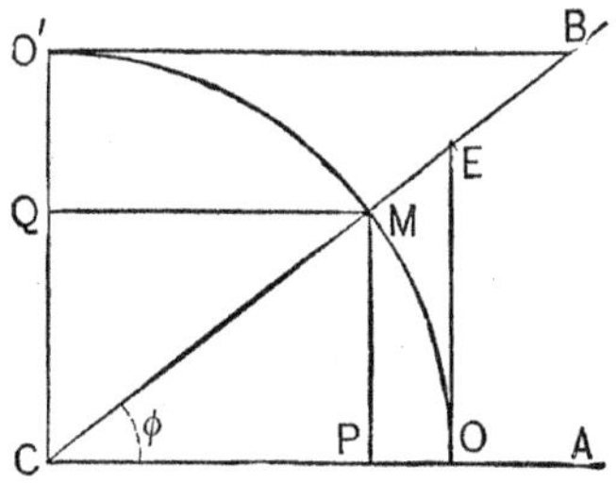

Regarding ϕ as a function of the line PM, denoted by y, we have $\phi = \sin^{-1}\dfrac{y}{R}$. Hence,

$$d\phi = \frac{d\dfrac{y}{R}}{\sqrt{1 - \dfrac{y^2}{R^2}}} = \frac{dy}{\sqrt{R^2 - y^2}}.$$

Similarly, having

$CP = y, \quad \therefore\ \phi = \cos^{-1}\dfrac{y}{R},$ we have $d\phi = \dfrac{-\,dy}{\sqrt{R^2 - y^2}}$

$OE = y, \quad \therefore\ \phi = \tan^{-1}\dfrac{y}{R},$ we have $d\phi = \dfrac{Rdy}{R^2 + y^2}.$

$O'B = y, \quad \therefore\ \phi = \cot^{-1}\dfrac{y}{R},$ we have $d\phi = \dfrac{-\,Rdy}{R^2 + y^2};$

$CE = y, \quad \therefore\ \phi = \sec^{-1}\dfrac{y}{R},$ we have $d\phi = \dfrac{Rdy}{y\sqrt{y^2 - R^2}};$

$CB = y, \quad \therefore\ \phi = \text{cosec}^{-1}\dfrac{y}{R},$ we have $d\phi = \dfrac{-\,Rdy}{y\sqrt{y^2 - R^2}};$

$PO = y, \quad \therefore\ \phi = \text{versin}^{-1}\dfrac{y}{R},$ we have $d\phi = \dfrac{dy}{\sqrt{2Ry - y^2}};$

$O'Q = y, \quad \therefore\ \phi = \text{coversin}^{-1}\dfrac{y}{R},$ we have $d\phi = \dfrac{-\,dy}{\sqrt{2Ry - y^2}}.$

EXAMPLES.

1. $d \sin^{-1} x^2 = 2xdx/\sqrt{1 - x^4}.$

2. $d \sin^{-1} 2x\sqrt{1 - x^2} = 2dx/\sqrt{1 - x^2}.$

3. $d \tan^{-1} [(a + x)/(1 - ax)] = dx/(1 + x^2).$

4. $d \sin^{-1} [(x+1)/\sqrt{2}] = dx/\sqrt{1-2x-x^2}$.

5. $d \operatorname{vers}^{-1} x^2 = 2dx/\sqrt{2-x^2}$.

6. $d \sqrt{\sin^{-1} x} = dx/2\sqrt{\sin^{-1} x(1-x^2)}$.

7. $d \tan^{-1} (-a/x) = adx/(a^2+x^2)$.

8. $d \cos^{-1} [(1-x^2)/(1+x^2)] = 2dx/(1+x^2)$.

9. $d \tan^{-1} (\sqrt{1-x}/\sqrt{1+x}) = -dx/2\sqrt{1-x^2}$.

10. $d \sec^{-1} [1/(2x^2-1)] = -2dx/\sqrt{1-x^2}$.

11. $d \tan^{-1} [x/\sqrt{1-x^2}] = dx/\sqrt{1-x^2}$.

12. $d \cos^{-1} [x/(a-x)] = -adx/(a-x)\sqrt{a^2-2ax}$.

13. $d \tan^{-1} [x\sqrt{3}/(2+x)] = \sqrt{3}\,dx/2(x^2+x+1)$.

14. $d \cos^{-1} \sqrt{1-x^2} = dx/\sqrt{1-x^2}$.

15. $d \sec^{-1} [a/\sqrt{a^2-x^2}] = dx/\sqrt{a^2-x^2}$.

16. $d \cos^{-1} [(4-3x^2)/x^3] = -3dx/x\sqrt{x^2-1}$.

17. $d - \cot^{-1} (x/a)/a = d \tan^{-1} (x/a)/a = dx/(a^2+x^2)$.

18. $d \tan^{-1} [2x/(1+x^2)] = 2(1-x^2)dx/(1+6x^2+x^4)$.

19. $d \sec^{-1} (x/a)/a = d - \operatorname{cosec}^{-1} (x/a)/a = dx/x\sqrt{x^2-a^2}$.

20. $d \sin^{-1} [(1-x^2)/(1+x^2)] = -2dx/(1+x^2)$.

21. $d \tan^{-1} [(1-x)/(1+x)] = dx/(1+x^2)$.

22. $d \sin^{-1} [x/\sqrt{1+x^2}] = dx/(1+x^2)$.

23. $d \cos^{-1} (2x-1) = -dx/\sqrt{x(1-x)}$.

24. $d \tan^{-1} [2x/(1-x^2)] = 2dx/(1+x^2)$.

25. $d \sin^{-1} (1/\sqrt{1+x^2}) = -dx/(1+x^2)$.

26. $d \sin^{-1} \sqrt{(1-x)/2} = -dx/2\sqrt{1-x^2}$.

27. $d \tan^{-1}(1/\sqrt{x^2-1}) = -dx/x\sqrt{x^2-1}$.

28. $d \cos^{-1}(4x^3-3x) = -3dx/\sqrt{1-x^2}$.

29. $d \tan^{-1} a^{1/x} = -a^{1/x} \log adx/x^2(1+a^{2/x})$.

30. $d \tan^{-1}\left(\sqrt{1+x^2}-x\right) = -dx/2(1+x^2)$.

31. $d \cos^{-1}(1-2x^2) = 2dx/\sqrt{1-x^2}$.

32. $d\, e^{\sin^{-1}x} = e^{\sin^{-1}x}\, dx/\sqrt{1-x^2}$.

33. $d \operatorname{vers}^{-1}(1/x) = -dx/x\sqrt{2x-1}$.

34. $d\left[r \operatorname{vers}^{-1}(y/r) - \sqrt{2ry-y^2}\right] = y\,dy/\sqrt{2ry-y^2}$.

35. $d\, x^{\sin^{-1}x} = x^{\sin^{-1}x}\left(\sin^{-1}x/x + \log x/\sqrt{1-x^2}\right)dx$.

36. $y = 2\tan^{-1}\sqrt{(1-x)/1+x}$. $(1-x)/(1+x) = \tan^2(y/2)$.

Hence, $x = \cos y$, and $dy = -dx/\sqrt{1-x^2}$.

37. $y = \cos^{-1}[(4-3x^2)/x^3]$. Put $(4-3x^2)/x^3 = z$.

Then $dy/dz = \left(-1/\sqrt{1-z}\right)(dz/dx) = -3/x\sqrt{x^2-1}$.

38. When $y = 0$, and $y = 2r$, find the slope of the curve

$$x = r \operatorname{vers}^{-1}(y/r) - \sqrt{2ry-y^2}.$$

Ans. ∞, and 0.

MISCELLANEOUS EXAMPLES.

1. $d \log \sin x = \dfrac{d \sin x}{\sin x} = \dfrac{\cos x\, dx}{\sin x} = \dfrac{dx}{\tan x}$.

2. $d \log \tan x = \dfrac{d \tan x}{\tan x} = \dfrac{dx}{\cos x \sin x} = \dfrac{2dx}{\sin 2x}$.

3. $d \log \tan \dfrac{x}{2} = \dfrac{dx}{\sin x} = \operatorname{cosec} x\, dx = d\left[\dfrac{1}{2}\log\dfrac{1-\cos x}{1+\cos x}\right]$.

4. $d \log \tan\left(\dfrac{\pi}{4}+\dfrac{x}{2}\right) = \dfrac{dx}{\sin(\pi/2+x)} = \dfrac{dx}{\cos x}$

$$= \sec x\, dx = d\left[\frac{1}{2}\log\frac{1+\sin x}{1-\sin x}\right].$$

5. $d - \log \cos x = d \log \sec x = \tan x\, dx$.

6. $de^x \cos x = e^x(\cos x - \sin x)dx$.

7. $dxe^{\sin x} = e^{\sin x}(1 + x\cos x)dx$.

8. $dxe^{\cos x} = e^{\cos x}(1 - x\sin x)dx.$

9. $d\sin(\log x) = \cos(\log x)dx/x.$

10. $u = \sin^m x/\cos^n x. \quad \therefore \quad \log u = m\log\sin x - n\log\cos x,$

and $$\frac{du}{u} = \left(m\frac{\cos x}{\sin x} + n\frac{\sin x}{\cos x}\right)dx.$$

$$\therefore du = \left(\frac{m\sin^{m-1}x}{\cos^{n-1}x} + \frac{n\sin^{m+1}x}{\cos^{n+1}x}\right)dx.$$

11. $u = xe^{\tan^{-1}x}. \quad \therefore \log u = \log x + \tan^{-1}x.$

$$du = u\big(1/x + 1/(1+x^2)\big)dx = e^{\tan^{-1}x}(1 + x + x^2)/(1 + x^2).$$

12. $dx^n e^{n\sin x} = nx^{n-1}e^{n\sin x}(1 + x\cos x)dx.$

13. $de^{\cos x}\sin x = e^{\cos x}(\cos x - \sin^2 x)dx.$

14. $d[\sin nx/\sin^n x] = -n\sin(n-1)x\,dx/\sin^{n+1}x$

15. $d\cos\log(1/x) = \sin\log(1/x)dx/x.$

16. $d\cos\sin x = -\cos x\sin\sin x\,dx.$

17. $de^x\log x = e^x\log(ex^x)dx/x.$

18. $d\log(xe^{\cos x}) = (1 - x\sin x)dx/x.$

19. $dx^m(\log x)^n = x^{m-1}(\log x)^{n-1}(m\log x + n)dx.$

20. $dx^n e^{\log x} = (n+1)x^{n-1}e^{\log x}dx.$ $\{e^{\log x} = x\}$

21. $d\log\sqrt{(1+\sin x)/(1-\sin x)} = dx/\cos x.$

22. $d\log\sqrt{(1-\cos x)/(1+\cos x)} = dx/\sin x.$

23. $d\sin\tan x = \cos\tan x\,dx/\cos^2 x.$

24. $$d\left[\frac{1}{6}\log\frac{(x+1)^3}{x^3+1} + \frac{1}{\sqrt{3}}\tan^{-1}\frac{2x-1}{\sqrt{3}}\right] = \frac{dx}{x^3+1}.$$

25. $d\cos\log\sin x = -\cot x\sin\log\sin x\,dx.$

26. $d\sin^{-1}\sqrt{\sin x} = dx/2\sqrt{1 + \operatorname{cosec} x}.$

27. $d\log(x/a^x) = \log(e^{1/x}/a)dx.$

28. $d\log\sin^{-1}x = dx/\sin^{-1}x\sqrt{1-x^2}.$

29. $d\ \sin^{-1}(\tan x) = \sec^2 x\, dx/\sqrt{1 - \tan^2 x}.$

30. $d\ \log \cos^{-1} x = -dx/\cos^{-1} x\, \sqrt{1 - x^2}.$

31. $d\ \log \tan^{-1} x = dx/(1 + x^2) \tan^{-1} x.$

32. $d \tan^{-1} \log x = dx/x[1 + (\log x)^2].$

33. $d \cos [a \sin^{-1}(1/x)] = a \sin (a\ \mathrm{cosec}^{-1} x) dx/x \sqrt{x^2 - 1}.$

34. $d \cot^{-1} (\mathrm{cosec}\ x) = \cos x\, dx/(1 + \sin^2 x).$

35. $d\ \sin^{-1}\left(e^{\tan^{-1} x}\right) = e^{\tan^{-1} x}\, dx/(1 + x^2) \sqrt{1 - e^{2 \tan^{-1} x}}.$

Hyperbolic Functions.

85. $$\mathbf{d \sinh x} = d\,\tfrac{1}{2}(e^x - e^{-x})$$

$$= \tfrac{1}{2}(e^x + e^{-x})dx = \mathbf{\cosh x\, dx}.$$

$$\mathbf{d \cosh x} = d\,\tfrac{1}{2}(e^x + e^{-x})$$

$$= \tfrac{1}{2}(e^x - e^{-x})dx = \mathbf{\sinh x\, dx}.$$

$$\mathbf{d \tanh x} = d(\sinh x/\cosh x) = \mathbf{sech^2 x\, dx}.$$

$$\mathbf{d \coth x} = d(\cosh x/\sinh x) = -\mathbf{cosech^2 x\, dx}.$$

$$\mathbf{d\ sech\ x} = d(1/\cosh x) = -\mathbf{sech\ x \tanh x\, dx}.$$

$$\mathbf{d\ cosech\ x} = d(1/\sinh x) = -\mathbf{cosech\ x \coth x\, dx}.$$

EXAMPLES.

1. $d\sqrt{\cosh x} = \sinh x\, dx/2\sqrt{\cosh x}.$

2. $d \log \cosh x = \tanh x\, dx.$

3. $d \log \sinh x = \coth x\, dx.$

4. $d(x - \tanh x) = \tanh^2 x\, dx.$

5. $d[(\sinh 2x)/4 + x/2] = \cosh^2 x\, dx.$

6. $d[(\sinh 2x)/4 - x/2] = \sinh^2 x\, dx.$

7. $d\left(x+\frac{x^3}{\underline{|3}}+\frac{x^5}{\underline{|5}}+\dots\right)=\cosh x\,dx.$

8. $d\left(1+\frac{x^2}{\underline{|2}}+\frac{x^4}{\underline{|4}}+\dots\right)=\sinh x\,dx.$

Inverse Hyperbolic Functions.

86. $\mathbf{d \sinh^{-1} x = d \log(x + \sqrt{1 + x^2}) = dx/\sqrt{1 + x^2}.}$

Let $y = \sinh^{-1} x$; then

$$x = \sinh y, \quad \text{and (§ 85)} \quad dx/dy = \cosh y.$$

Hence (§ 73),

$$\frac{d \sinh^{-1} x}{dx} = \frac{1}{\cosh y} = \frac{1}{\sqrt{1 + \sinh^2 y}} = \frac{1}{\sqrt{1 + x^2}}.$$

$$\mathbf{d \cosh^{-1} x = d \log(x + \sqrt{x^2 - 1}) = dx/\sqrt{x^2 - 1}.}$$

Let $y = \cosh^{-1} x$; then

$$x = \cosh y, \quad \text{and (§ 85)} \quad dx/dy = \sinh y.$$

Hence (§ 73),

$$\frac{d \cosh^{-1} x}{dx} = \frac{1}{\sinh y} = \frac{1}{\sqrt{\cosh^2 y - 1}} = \frac{1}{\sqrt{x^2 - 1}}.$$

$$\mathbf{d \tanh^{-1} x = d\,\frac{1}{2} \log \frac{1 + x}{1 - x} = \frac{dx}{1 - x^2}}\ (x < 1).$$

Let $y = \tanh^{-1} x$; then

$$x = \tanh y, \quad \text{and (§ 85)} \quad dx/dy = \operatorname{sech}^2 y.$$

Hence (§ 73),

$$\frac{d\tanh^{-1}x}{dx} = \frac{1}{\operatorname{sech}^2 y} = \frac{1}{1-\tanh^2 y} = \frac{1}{1-x^2}.$$

$$\mathbf{d\,coth^{-1}\,x = d\,\frac{1}{2}\,log\,\frac{x+1}{x-1} = \frac{-dx}{x^2-1}}(x > 1).$$

Let $y = \coth^{-1} x$; then

$$x = \coth y, \quad \text{and (§ 85)} \quad dx/dy = -\operatorname{cosech}^2 y.$$

Hence (§ 73),

$$\frac{d\coth^{-1}x}{dx} = \frac{-1}{\operatorname{cosech}^2 y} = \frac{-1}{\coth^2 y - 1} = \frac{-1}{x^2-1}.$$

$$\mathbf{d\,sech^{-1}\,x = d\,log\,\frac{1 \pm \sqrt{1-x^2}}{x} = \frac{-dx}{x\sqrt{1-x^2}}}.$$

Let $y = \operatorname{sech}^{-1} x$; then

$$x = \operatorname{sech} y, \quad \text{and (§ 85)} \quad dx/dy = -\operatorname{sech} y \tanh y.$$

Hence (§ 73),

$$\frac{d\operatorname{sech}^{-1}x}{dx} = \frac{-1}{\operatorname{sech} y \tanh y} = \frac{-1}{x\sqrt{1-x^2}}.$$

$$\mathbf{d\,cosech^{-1}\,x = d\,log\,\frac{1 \pm \sqrt{1+x^2}}{x} = \frac{-dx}{x\sqrt{x^2+1}}}.$$

Let $y = \operatorname{cosech}^{-1} x$; then

$$x = \operatorname{cosech} y, \quad \text{and (§ 85)} \quad dx/dy = -\operatorname{cosech} y \coth y.$$

Hence (§ 73),

$$\frac{d\,\text{cosech}^{-1}x}{dx} = \frac{-1}{\text{cosech}\,y\coth y} = \frac{-1}{x\sqrt{x^2+1}}.$$

EXAMPLES.

1. $d\cosh^{-1}(x/a) = dx/\sqrt{x^2-a^2}$.

2. $d\sinh^{-1}(x/a) = dx/\sqrt{a^2+x^2}$.

3. $d\tanh^{-1}(x/a) = a\,dx/(a^2-x^2)$, $(x < a.)$

4. $d\coth^{-1}(x/a) = a\,dx/(a^2-x^2)$, $(x > a.)$

5. $d\tan^{-1}(\tanh x) = \text{sech}\,2x\,dx$.

6. $d[\frac{1}{2}\tanh^{-1}x + \frac{1}{2}\tan^{-1}x] = d\left[\log\sqrt[4]{\frac{1+x}{1-x}} + \frac{1}{2}\tan^{-1}x\right]$
$= dx/(1-x^4)$.

7. $d[\frac{1}{2}\cot^{-1}x - \frac{1}{2}\coth^{-1}x] = d\left[\frac{1}{2}\cot^{-1}x - \log\sqrt[4]{\frac{x+1}{x-1}}\right]$
$= dx/(x^4-1)$.

8. $d[\frac{1}{2}x\sqrt{x^2-a^2} - \frac{1}{2}a^2\cosh^{-1}(x/a)] = \sqrt{x^2-a^2}\,dx$.

9. $d[\frac{1}{2}x\sqrt{a^2+x^2} + \frac{1}{2}a^2\sinh^{-1}(x/a)] = \sqrt{a^2+x^2}\,dx$.

Geometric Functions.

87. Differential of an Arc of a Plane Curve.—Let s represent the length of a varying portion of any plane curve in the plane XY. It will be a function of one independent variable only (§ 18), which we may take to be x.

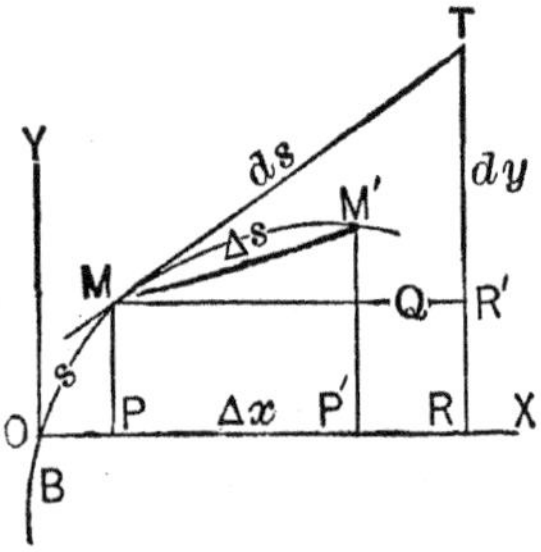

Assume any point of the curve, as M, and increase the corresponding value of $x = OP$, by $PP' = \Delta x$. $\Delta s = MM'$ will be the corresponding increment of s, and $\Delta y = QM'$.

$$\text{Then } (\S\ 70,\ \S\ 44)\ \frac{ds}{dx} = \underset{\Delta x \ggg\!\rightarrow 0}{\text{limit}} \frac{\Delta s}{\Delta x} = \lim \frac{\text{ch } MM'}{\Delta x}$$

$$= \lim \frac{\sqrt{\Delta x^2 + \Delta y^2}}{\Delta x} = \lim \sqrt{1 + \left(\frac{\Delta y}{\Delta x}\right)^2} = \sqrt{1 + \frac{dy^2}{dx^2}}$$

$$= \sqrt{dx^2 + dy^2}/dx. \quad \text{Similarly } \frac{ds}{dy} = \sqrt{dx^2 + dy^2}/dy.$$

Hence, $$\mathbf{ds} = \sqrt{\mathbf{dx}^2 + \mathbf{dy}^2}.*$$

The double sign is omitted because s may always be considered as an increasing function of x.

That is, *the differential of an arc of a plane curve is equal to the square root of the sum of the squares of the differentials of the coördinates of its extreme point.*

If s were to change from its state corresponding to any point, as M, with its rate at that state unchanged, the generatrix would move upon the tangent line at M; hence, $MT = \sqrt{dx^2 + dy^2}$ represents ds in direction and measure.

In order to express ds in terms of a single variable and its differential, find expressions for dy in terms of x and dx, or of dx in terms of y and dy, from the equation of the curve, and substitute them in the formula.

Thus, let s be an arc of the circle whose equation is $x^2 + y^2 = 4$. Solving with respect to y, and differentiating, we have

$$dy = \mp\, xdx/\sqrt{4 - x^2}.$$

---

* The square of the differential of a variable represented by a single letter is generally written as indicated in the above formula, and is similar in form to the symbol for the differential of the square of the variable. Similarly, the nth power of dx is generally written dx^n.

Hence, $$ds = \sqrt{dx^2 + \frac{x^2dx^2}{4 - x^2}} = \frac{2\,dx}{\sqrt{4 - x^2}},$$

88. Differential of any Arc. Let s represent the length of a varying portion of any curve in space. It will be a function of one independent variable only (§ 18), which we may assume to be x.

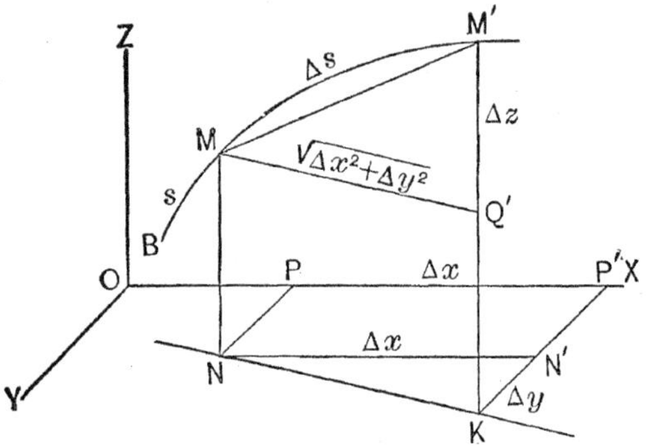

Through any assumed point of the curve, as M, draw the ordinate MN; and through N, the point where it pierces XY, draw NP parallel to Y. OP will be the value of x corresponding to M. Increase $x = OP$ by $PP' = \Delta x$, and through P' pass a plane parallel to YZ, intersecting the given curve at M'. $\Delta s =$ arc MM' will be the increment of s corresponding to the assumed increment of x.

Draw the chord MM' and the ordinate $M'K$. Through M draw MQ' parallel to a right line drawn through N and K; and through N draw NN' parallel to X. Then $N'K = \Delta y$ and $Q'M' = \Delta z$ will be the increments of y and z corresponding to Δx; and we have

$$\text{chord } MM' = \sqrt{(\Delta x)^2 + (\Delta y)^2 + (\Delta z)^2}.$$

Hence (§ 70, § 44),

$$\frac{ds}{dx} = \underset{\Delta x \ggg\!\!\rightarrow 0}{\text{limit}} \frac{\text{arc } MM'}{\Delta x} = \underset{\Delta x \ggg\!\!\rightarrow 0}{\text{limit}} \frac{\text{ch. } MM'}{\Delta x}$$

$$= \underset{\Delta x \ggg\!\!\rightarrow 0}{\text{limit}} \frac{\sqrt{(\Delta x)^2 + (\Delta y)^2 + (\Delta z)^2}}{\Delta x}$$

$$= \underset{\Delta x \ggg\!\!\rightarrow 0}{\text{limit}} \sqrt{1 + \left(\frac{\Delta y}{\Delta x}\right)^2 + \left(\frac{\Delta z}{\Delta x}\right)^2}$$

$$= \sqrt{1 + \left(\frac{dy}{dx}\right)^2 + \left(\frac{dz}{dx}\right)^2}; \text{ hence, } \mathbf{ds = \sqrt{dx^2 + dy^2 + dz^2}}.$$

s may be a curve of single or of double curvature. The increment Δs may or may not lie in the projecting plane of the chord MM'. If not, the projection of the chord MM' on the plane XY will change direction as Δx approaches zero, but the above relations will not be affected thereby.

Let α', β' and γ' represent the angles made by the chord MM' with X, Y and Z, respectively, then chord $MM'/\Delta x = 1/\cos \alpha'$. Let α, β and γ, respectively, represent the corresponding angles made by the tangent at M. Then

$$ds/dx = \underset{\Delta x \ggg\!\!\rightarrow 0}{\text{limit}} [1/\cos \alpha'] = 1/\cos \alpha, \text{ and } dx = ds \cos \alpha.$$

Similarly $dy = ds \cos \beta$, and $dz = ds \cos \gamma$, in which x, y, z or s may be considered as the independent variable.

89. Differential of a Plane Area.—Let u represent the area of the plane surface included between any varying portion of any plane curve, as AM, the ordinates of its extremities, and the axis X.

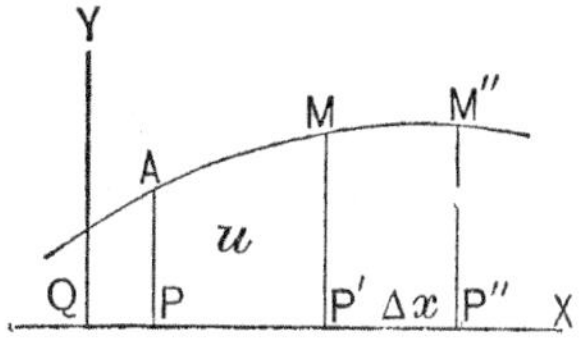

Regarding u as a function of x (§ 21), let $x = OP'$ be increased by $P'P'' = \Delta x$. $P'MM''P''$ will be the corresponding increment of u. Hence (§ 70, § 46),

$$du/dx = \underset{\Delta x \Rightarrow 0}{\text{limit}} [P'MM''P''/\Delta x] = y,$$

which gives **du = ydx.**

That is, *the differential of a plane area is equal to the ordinate of the extreme point of the bounding curve into the differential of the abscissa.*

To illustrate, let u represent the area $BAMP$, and $PR = dx$; then $du = ydx =$ rect. PQ, which fulfils the requirements of the definition of a differential (§ 68).

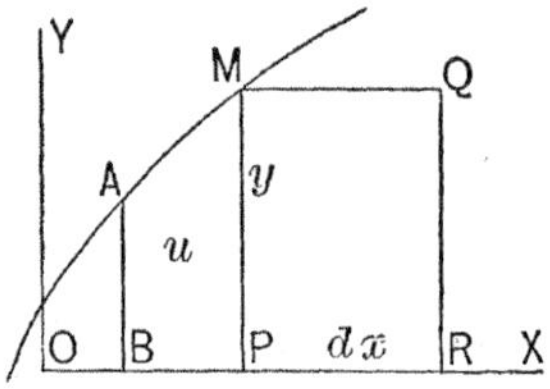

Similarly, it may be shown that xdy is the differential of the plane area included between any arc, the abscissas of its extremities, and the axis of Y.

In case the coördinate axes are inclined to each other by an angle θ, we have $du = y \sin \theta dx$, or $du = x \sin \theta dy$.

In order to express du in terms of x and dx, substitute for y, or dy, its expression determined from the equation of the bounding curve.

Thus, if $a^2y^2 + b^2x^2 = a^2b^2$ is the equation of the bounding curve, we have $y = \frac{b}{a}\sqrt{a^2 - x^2}$; and $du = \frac{b}{a}\sqrt{a^2 - x^2}dx$.

90. Differential of a Surface of Revolution.—Let the axis of X coincide with the axis of revolution; and let $BM = s$ be any varying portion of the meridian curve

* It is important to notice and remember that ydx is the differential of a plane area bounded as described; and that it is not, in general, the differential of a plane area otherwise bounded.

in the plane XY. Through M draw the tangent MT, the ordinate MP, and the right line MR' parallel to X. Let u represent the surface generated by s; and regarding it as a function of x (§ 24), let $x = OP$ be increased by $PP' = \Delta x$. $MM' = \Delta s$ will be the corresponding increment of s; and the surface generated by it will be the increment of the function u corresponding to Δx. Hence (§ 70, § 56),

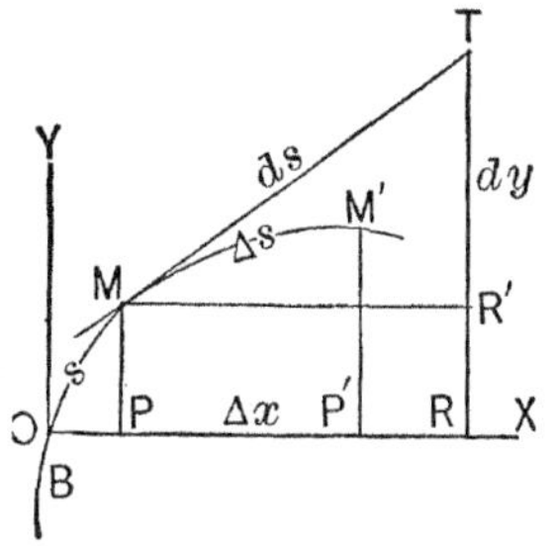

$$\frac{du}{dx} = \underset{\Delta x \ggg \to 0}{\text{limit}} \frac{\text{sur. gen. by arc } MM'}{\Delta x} = \frac{2\pi y}{\cos R'MT}.$$

Assume $PR = dx$; then $R'T = dy$, $MT = ds$, and $\cos R'MT = dx/ds$. Substituting this expression for $\cos R'MT$ in above, we have

$$\frac{du}{dx} = \frac{2\pi y\, ds}{dx}; \text{ and } \mathbf{du} = 2\pi \mathbf{y ds} = 2\pi \mathbf{y}\sqrt{\mathbf{dx}^2 + \mathbf{dy}^2}.$$

Hence, *the differential of a surface of revolution is equal to the product of the circum. of a circle perpendicular to the axis and the differential of the arc of the generating curve.*

Similarly, it may be shown that $2\pi x \sqrt{dx^2 + dy^2}$ is the differential of a surface of revolution generated by revolving a plane curve about the axis of Y.

In order to express du in terms of a single variable and its differential, find expressions for y and dy in terms of x and dx, or of dx in terms of y and dy, from the equation of the generating curve; and substitute them in the formula.

Thus, if $y^2 = 2px$ is the equation of the generating curve, we have $y = \sqrt{2px}$ and $dy = \dfrac{pdx}{\sqrt{2px}}$. Hence,

$$du = 2\pi\sqrt{2px}\sqrt{dx^2 + \frac{p^2dx^2}{2px}} = 2\pi(2px + p^2)^{\frac{1}{2}}dx.$$

91. Differential of a Volume of Revolution.—Let the axis of X coincide with the axis of revolution; and let BM be any varying portion of the meridian curve in the plane XY. Through M draw the ordinate MP, and the right line MQ' parallel to X. Let v represent the volume generated

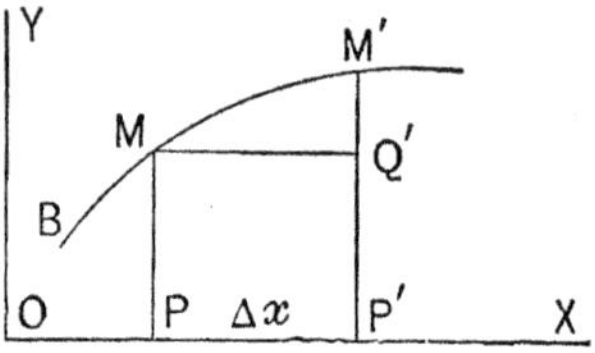

by the plane surface included between the arc BM, the ordinates of its extremities, and the axis of X. Regarding v as a function of x (§ 29), let x be increased by $PP' = \Delta x$. The volume generated by the plane surface $PMM'P'$ will be the corresponding increment of the function v. Then (§ 70, § 48)

$$\frac{dv}{dx} = \underset{\Delta x \Rightarrow 0}{\text{limit}} \frac{\text{vol. gen. by } PMM'P'}{\Delta x} = \pi y^2,$$

and $\quad \mathbf{dv = \pi y^2 dx.}$

Hence, *the differential of a volume of revolution is equal to the area of a circle perpendicular to the axis into the differential of the abscissa of the meridian curve.*

Similarly, it may be shown that $\pi x^2 dy$ is the differential of a volume of revolution generated by revolving a plane surface about the axis of Y.

In order to express dv in terms of a single variable and its differential, determine an expression for y in terms of x, or of dx in terms of y and dy, from the equation of the meridian curve, and substitute them in the formula.

Thus, if $x^2 + y^2 - 2Rx = 0$ is the equation of the meridian curve, we have $dv = \pi(2Rx - x^2)dx$; or since $dx = \mp ydy/\sqrt{R^2 - y^2}$, $dv = \mp \pi y^3 dy/\sqrt{R^2 - y^2}$.

92. Differential of an Arc of a Plane Curve in Terms of Polar Coördinates.—Let $r = f(v)$ be the polar equation

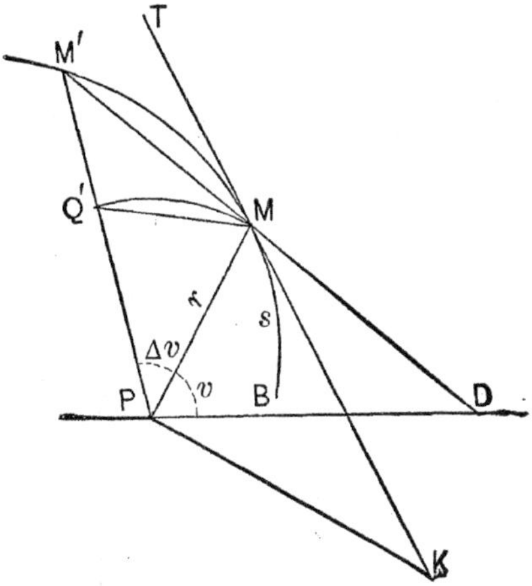

of any plane curve, as BMM', referred to the fixed right line PD, and the pole P. Let $BM = s$ be any varying portion of the curve, and $PM = r$ the radius vector corresponding to M. Regarding s as a function of v (§ 19), let v be increased by $MPM' = \Delta v$. The arc $MM' = \Delta s$ will be the corresponding increment of s. With P as a centre and PM as a radius, describe the arc MQ'. Denote PM' by r'; then $Q'M' = r' - r$ will be the increment of r corresponding to Δv. Through M draw the tangent MT, and the chords MM' and MQ'. Then (§ 70, § 55) we have

$$\frac{ds}{dv} = \underset{\Delta v \ggg 0}{\text{limit}} \frac{\text{arc } MM'}{\Delta v} = \underset{\Delta v \ggg 0}{\text{limit}} \sqrt{\left(\frac{r'-r}{\Delta v}\right)^2 + r^2} = \sqrt{\frac{dr^2}{dv^2} + r^2}.$$

Hence, $\quad \mathbf{ds = \sqrt{dr^2 + r^2 dv^2}}$.

Also,

$$\frac{dr}{dv} = \underset{\Delta v \ggg \to 0}{\text{limit}} \frac{Q'M'}{\Delta v} = \lim \frac{rQ'M'}{\text{arc } Q'M} = r \lim \frac{Q'M'}{\text{ch.} Q'M}.$$

If the radius vector PM coincides with the normal to the curve at M, the corresponding tangent to the arc MQ' will coincide with MT; and (§ 55)

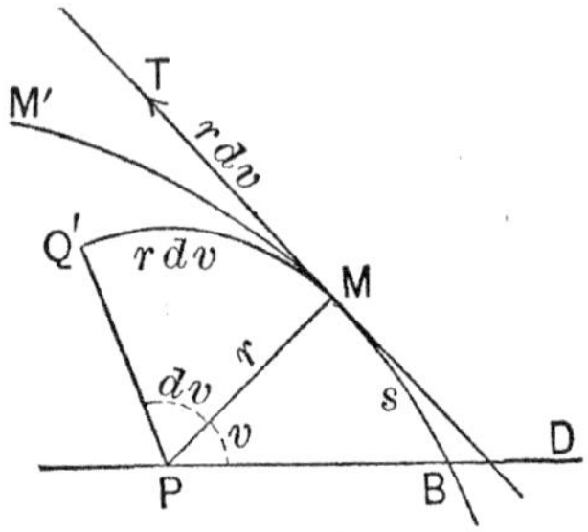

$$\frac{ds}{dv} = \underset{\Delta v \ggg \to 0}{\text{limit}} \frac{\text{arc } MM'}{\Delta v} = r,$$

giving $ds = rdv$.

In this case $dr = 0$, because the motion of the generatrix at the point considered is perpendicular to the radius vector.

An important example of this case is a circle with the pole at its centre.

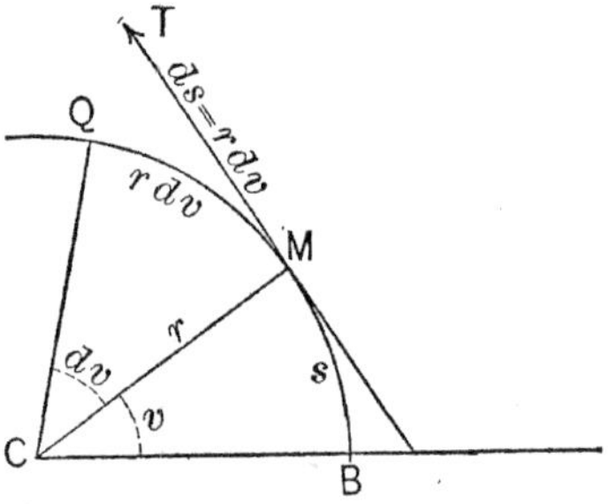

Let $BM = s$ be any arc of a circle, and $BCM = v$ the subtended angle. Then, since the radius is always normal to the arc, we have $ds = rdv$.

That is, *the differential of an arc of a circle regarded as a function of the corresponding angle at the centre, is equal to its radius into the differential of the angle.*

To illustrate, assume $MCQ = dv$; then will the arc $MQ = rdv$. The direction of the motion of the generatrix at any point is along the corresponding tangent to s; hence, by laying off from M upon the tangent at that point a distance $MT = ds = rdv$, we have ds represented in measure and direction.

In order to represent graphically the general case when $ds = \sqrt{dr^2 + r^2dv^2}$, let BM be the given curve, P the pole, M the assumed point, and $MPM' = dv$. If r were constant, as we have seen in the case of a circle, $MT' = rdv$

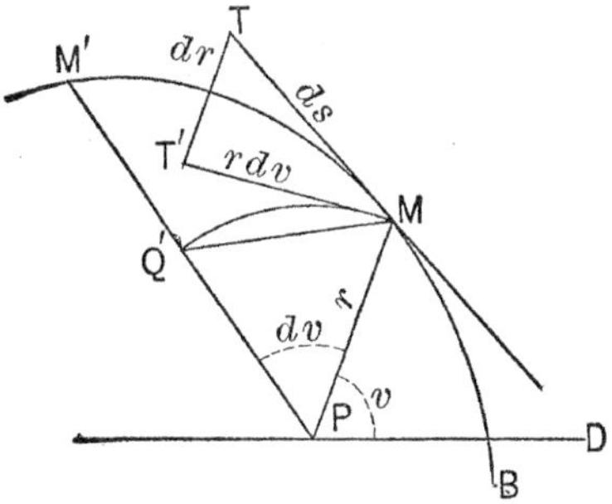

would be ds; but, in general, ds is affected by a uniform change in r, in the direction PM, equal to dr. To determine it we have

$$\frac{dr}{dv} = r \underset{\Delta v \to 0}{\text{limit}} \frac{Q'M'}{\text{ch. } Q'M} = r \tan T'MT.$$

At T' draw $T'T$ parallel to PM; then $T'T/T'M = \tan T'MT = T'T/rdv$. Hence, $dr/dv = rT'T/rdv = T'T/dv$, and $dr = T'T$. $MT = ds = \sqrt{dr^2 + r^2dv^2}$, therefore, represents ds in measure and direction.

In order to express ds in terms of a single variable and its differential, find expressions for r and dr in terms of v and dv, or an expression for dv in terms of r and dr, from the polar equation of the curve; and substitute in the formula.

93. Differential of a Plane Area in Terms of Polar Coördinates.—Let u represent the area of a varying portion of the surface generated by the radius vector PM revolving about the pole P. Regarding u as a function of v (§ 23), let $MPM' = \Delta v$. The area MPM', represented by Δu, will be the corresponding increment of u. Hence (§ 47),

$$\frac{du}{dv} = \underset{\Delta v \Rightarrow 0}{\text{limit}} \frac{\Delta u}{\Delta v} = \frac{r^2}{2}, \quad \text{and} \quad \mathbf{du} = \frac{\mathbf{r^2dv}}{\mathbf{2}}.$$

To illustrate, with $PM = r$, describe the arc of a circle $MQ = rdv$ corresponding to $MPQ = dv$; then $du = r^2dv/2$ = area of the circular sector MPQ.

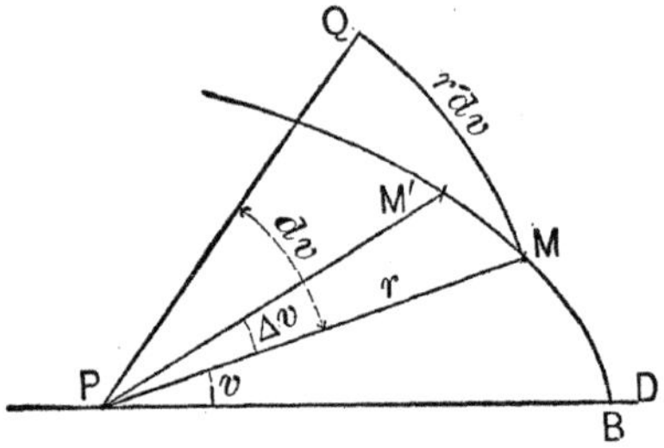

du may be expressed in terms of v and dv, by substituting for r its value in terms of v, determined from the polar equation of the bounding curve.

94. **Motion.**—When a point changes its position with respect to any origin it is said to be in motion with respect to that origin.

In general, the distance from any origin to a point in motion continually changes, and is a continuous function of the time during which the point moves.

When the distance changes so that *any* two increments of it whatever are proportional to the corresponding intervals of time, the distance changes uniformly with the time.

* Motion, without regard to cause, is generally discussed under the head of Kinematics, but many important applications of the Calculus involve motion, therefore some of the definitions and principles of Kinematics are here and elsewhere introduced.

The point is then said to be moving *uniformly*, or with *uniform* motion with respect to the origin.

If the distance does not change uniformly with the time the point is said to be moving with *varied* motion with respect to the origin.

A train of cars moves from a station with varied motion until it attains its greatest speed, after which its motion along the track is uniform while it maintains that speed.

With uniform motion equal distances are passed over in any equal portions of time, and with varied motion unequal distances are passed over in equal portions of time.

Let s in both figures represent the variable distance from any origin, as A, to a point moving on any line, as MNO;

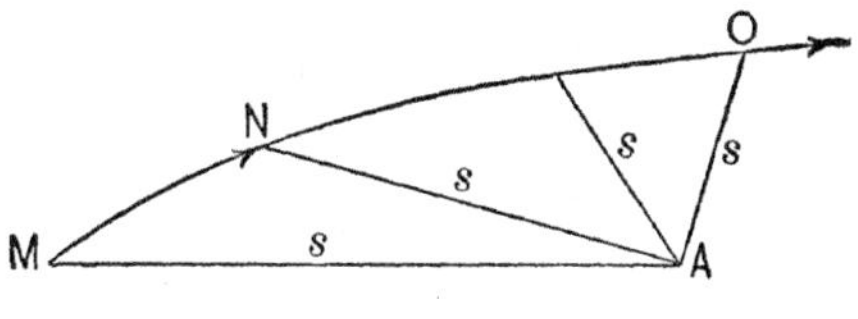

and let t denote the number of units of time during which the point moves; then $s = f(t)$.

If $f(t)$ is of the first degree with respect to t, the distance s will change uniformly; otherwise the point approaches or recedes from the origin with varied motion. § 57.

The rate of change of s, regarded as a function of t, corresponding to any position of the moving point, is called the *rate of motion* of the moving point with respect to the origin; and since uniform motion causes s to change uniformly with t, the rate of motion, in such cases, is constant. § 59.

In varied motion the rate varies with t, and is therefore a function of t.

95. If the differential of the variable is assumed equal to the unit of the variable, the differential of a function and the corresponding differential coefficient will have the same numerical value.

Thus, if $\frac{dy}{dx} = 2$, and $dx = 1$ inch, we have $\frac{dy}{dx}dx = 2$ inches. In such cases the differential of the function expresses the rate in terms of the unit of the variable; and since it is more definite, it is frequently used instead of the differential coefficient.

To illustrate, let s denote any variable distance regarded as a function of time, giving $s=f(t)$. Assuming any convenient length to represent the unit of t, we may, by substituting s for y and t for x (§ 20), determine a line, as AM, whose ordinate represents the given function.

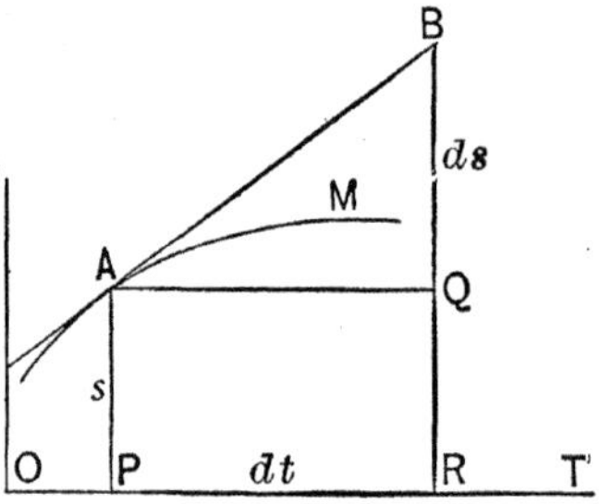

If $PR = dt$ represents one hour, $\frac{ds}{dt}dt = QB$ represents the change that s would undergo in one hour, from the state represented by PA, were it to retain its rate at that state; and is more definite than the corresponding abstract value of ds/dt.

96. Velocity.—The differential coefficient of the variable distance from any origin to a point in motion, regarded as a function of the time of the motion, is called the velocity of the moving point with respect to that origin.

Representing the variable distance by s, and the velocity by v, we have $v = ds/dt$.

For the reasons given above, velocity *is measured* by the product of ds/dt and the distance assumed to represent the unit of time.

That is, *the measure* of the velocity of a point in motion at any instant, in any required direction, *is the distance in that direction that the point would go in the next unit of time, were it to retain its rate at that instant.*

It should be noticed that the distance referred to above, and represented by s, may or may not be estimated along the line or path upon which the body moves. Thus, if a point moves from A towards B, and the velocity at any point, as C, in the direction AB is required, the distance s is estimated along the path described; but if the rate or velocity with which a point, moving from A to B, is approaching D is required, s must represent the variable distance from the moving point to D, in order that ds/dt shall be the required velocity.

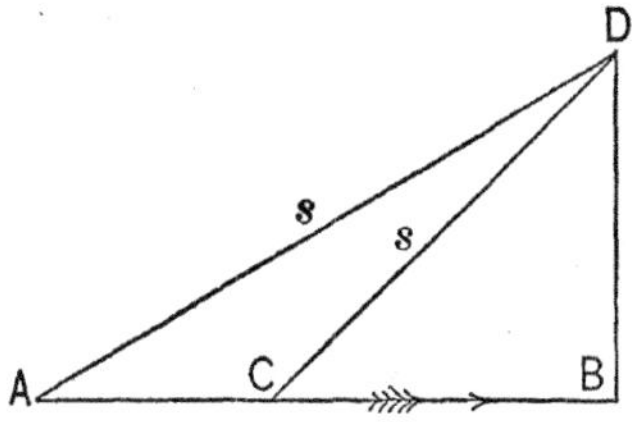

Since velocity is a rate of motion, it is constant in uniform motion, and a variable function of time in varied motion.

97. Acceleration.—The differential coefficient of velocity regarded as a function of time is called acceleration. It is denoted by dv/dt, in which v represents velocity. Since acceleration is the velocity of a velocity, it is generally expressed in terms of the distance which represents the unit of time.

98. Angular Motion.—Let C be a fixed point, CA a fixed right line, and B a point in motion so that the angle ACB,

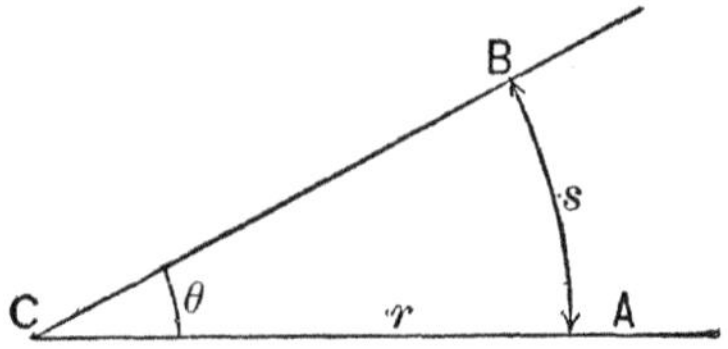

denoted by θ, is changing. Then the line CB is said to have an *angular motion* with respect to, or about, C.

Let s represent the length of the varying arc, of any convenient circle, subtending θ, giving $\theta = s/r$.

Both θ and s are functions of the time during which CB moves.

Angular motion is *uniform* when any two increments of the angle, or arc subtending the angle, are proportional to the corresponding intervals of time; otherwise it is *varied.*

99. Angular Velocity.—The differential coefficient of any varying angle regarded as a function of the time is called angular velocity.

Representing any varying angle by θ, and its angular velocity by ω, we have $\omega = d\theta/dt$.

If s denotes the varying arc of a circle whose radius is r, which subtends θ, we have

$$\theta = s/r; \quad \text{hence,} \quad \omega = d\theta/dt = ds/rdt.$$

That is, angular velocity is equal to the actual velocity of a point describing any convenient circle about the vertex of the angle as a centre, divided by its radius.

It is customary in applied mathematics to consider the radius equal to the unit of distance used in any particular case. Angular velocity will then be measured by the actual velocity of a point at *the* unit's distance from the vertex.

100. Angular Acceleration.—The differential coefficient of angular velocity regarded as a function of time is called angular acceleration. It is denoted by $d\omega/dt$, when ω represents angular velocity.

PROBLEMS.

1. The side of a square increases uniformly 3 in. a minute; find the rate per minute of its area when its side is 6 in.

Let x = side of square in inches, and $u = \text{area} = x^2$; then $dx/dt =$ 3 in. min. and $du/dx = 2x$. Hence (§ 77), $du/dt = (du/dx)(dx/dt) = 3 \times 2x = 6x$ sq. in. min. and $(du/dt)_{x=6} = 36$ sq. in. min.

2. The radius of a circle increases uniformly .01 in. per second; find the rate of its area when the radius is 1 in.

Let r = radius, and $u = \text{area} = \pi r^2$; then $dr/dt = .01$ in. sec. and $du/dr = 2\pi r$. Hence (§ 77), $du/dt = .01 \times 2\pi r = .02\pi r$ sq. in. sec.; and $(du/dt)_{r=1} = .02\pi$ sq. in. sec.

3. Find the rate of the radius when the area of a circle increases uniformly at the rate of $2\pi r$ sq. in. sec.

Ans. 1 in. sec.

4. The radius of a sphere increases uniformly .0491 in. sec.; find the rate of its volume when the radius is 1.5 ft.

Ans. 200 cu. in. sec.

5. The volume of a sphere increases uniformly 500 cu. in. sec.; when its radius increases at the rate of 2 in. sec., find the radius. Ans. 4.45 in.

6. The area of a rectangle increases uniformly 100 sq. in. min. Its base and altitude are increasing at the rates of 3 and 7 in. per min. respectively; find the area when the altitude is double the base. Ans. 118.34 sq. in.

7. The diameter of a circle increases uniformly 3 in. sec.; find the difference between the rates of the areas of the circle and its circumscribed square when the square is 1 sq. ft. Ans. 15.45 sq. in. sec.

8.* A man 6 feet in height walks away from a light 10 feet above the ground at the rate of 3 mi. per hour. At what rate is the end of his shadow moving, and at what rate does his shadow increase in length?

Let $x = AM$ = distance from foot of light to man, $y = AB$ = distance from foot of light to end of shadow, and $s = MB$ = length of shadow. Let t = number of hours.

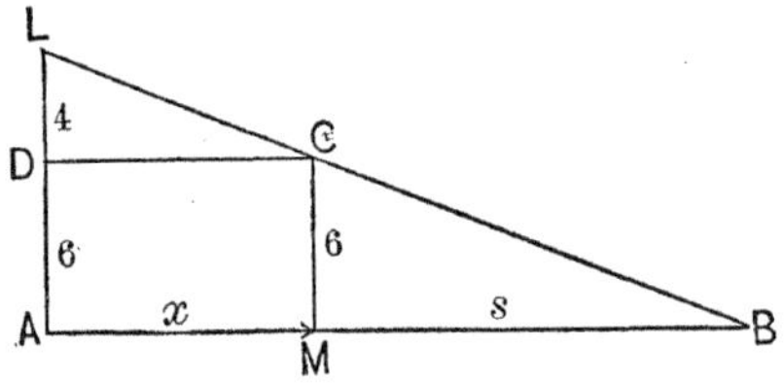

Then we have $dx/dt = 3$ mi. hr.; $AL = 10$ ft.; $MC = 6$ ft.; $DL = 4$ ft.; and it is required to find dy/dt and ds/dt.

The similar triangles ABL and DCL give $x : y :: 4 : 10$; hence, $y = 5x/2$, and $dy/dx = 5/2$. Therefore (§ 77)

$$dy/dt = (dy/dx)(dx/dt) = (5/2)(3) = 7.5 \text{ mi. hr.}$$

Also, $x : s :: 4 : 6$; hence, $s = 3x/2$, and $ds/dx = 3/2$,

and $$ds/dt = (ds/dx)(dx/dt) = 1.5 \times 3 = 4.5 \text{ mi. hr.}$$

9.* A vessel sailing south at the rate of 8 mi. per hour is 20 mi. north of a vessel sailing east at the rate of 10 mi. an hour. At what rate are they separating at the time? At the end of $1\frac{1}{2}$ hrs.? At the end of $2\frac{1}{2}$ hrs.? When are they neither separating from nor approaching each other?

---

* Rice and Johnson's Calculus.

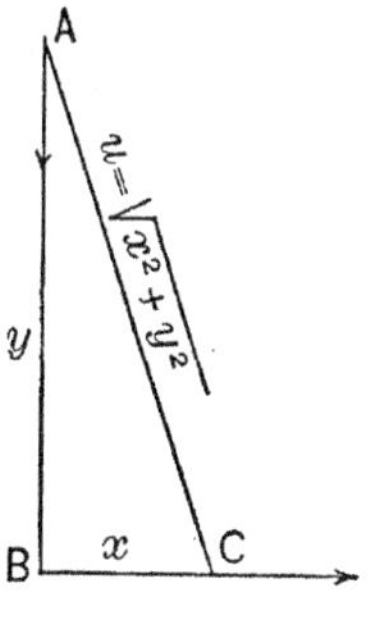

Let $t =$ time in hours from the given epoch.

Let $AB = y = 20 - 8t =$ distance of first ship from BC t hours after the given epoch.

Let $BC = x = 10t =$ distance of second ship from BA at the same time.

Let $u = AC = \sqrt{x^2 + y^2} = \sqrt{400 - 320t + 164t^2}$.

Given, $\dfrac{dy}{dt} = -8\dfrac{\text{mi.}}{\text{hr.}}$; $\quad \dfrac{dx}{dt} = 10\dfrac{\text{mi.}}{\text{hr.}}$.

Required, $\dfrac{du}{dt} = \dfrac{-160 + 164t}{\sqrt{400 - 320t + 164t^2}}$,

$$\left(\frac{du}{dt}\right)_{t=0} = -8\frac{\text{mi.}}{\text{hr.}} \qquad \left(\frac{du}{dt}\right)_{t=1\frac{1}{2}} = 5\frac{1}{17}\,\frac{\text{mi.}}{\text{hr.}} \qquad \left(\frac{du}{dt}\right)_{t=\frac{40}{41}} = 0.$$

The following general outline of steps may assist the student in solving similar problems :

1°. Draw a figure representing the magnitudes and directions under consideration ; and denote the variable parts by the final letters of the alphabet.

2°. Write, with the proper symbols, all known data ; and indicate the symbols for the required rates.

3°. From the relations between the magnitudes find an expression for the function whose rate is required, in terms of the variable.

4°. Differentiate and determine values or expressions for the required rates.

In case an explicit function of a variable cannot be found, make use of the principles in § 77.

10. $x^2 = 2pz$ is the equation of a parabola OM. A point starting from O moves along the curve in such a manner that $z = 16.1t^2$; in which z is expressed in feet, and t in seconds. Find the rate of x with respect to t.

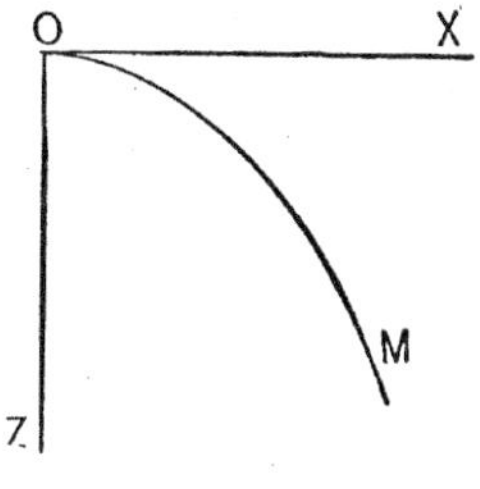

$$\frac{dx}{dz} = \frac{p}{\sqrt{2pz}} = \frac{p}{\sqrt{32.2pt^2}}. \qquad \frac{dz}{dt} = 32.2t.$$

Hence, $$\frac{dx}{dt} = \frac{dx}{dz} \times \frac{dz}{dt} = \frac{p}{\sqrt{32.2pt^2}} \times 32.2t = \sqrt{32.2p}.$$

11. One ship was sailing south 6 mi. per hour, another east 8 mi. per hour. At 4 P.M. the second crossed the track of the first at a point where the first was 2 hrs. before. How was the distance between the ships changing at 3 P.M.? When was the distance between them not changing? Ans. 2.8 mi. hr.; 3 hr. 16 min. 48 sec.

12. A ship is sailing south 60° east, 8 mi. per hour; find the rate of her latitude and longitude.

Ans. 4 mi. hr.; $4\sqrt{3}$ mi. hr.

13. A point P moves in a straight line away from a point B at the rate of 8 mi. hr.; find its velocity with respect to a point C situated upon the perpendicular to the line BP through B and at 100 ft. from B when $BP = 50$ ft.; when $BP = 150$ ft. Ans. $8/\sqrt{5}$ mi. hr.; $24/\sqrt{13}$ mi. hr.

14. If the diameter of a sphere increases uniformly at the rate of 1/10 inches per second, what is its diameter when the volume is increasing at the rate of 5 cubic inches per second? Ans. $10/\sqrt{\pi}$ in.

15. If the diameter D of the base of a cone increases uniformly at the rate of 1/10 inch per second, at what rate is its volume increasing when D becomes 10 inches, the height being constantly one foot?

Ans. 2π cu. in. sec.

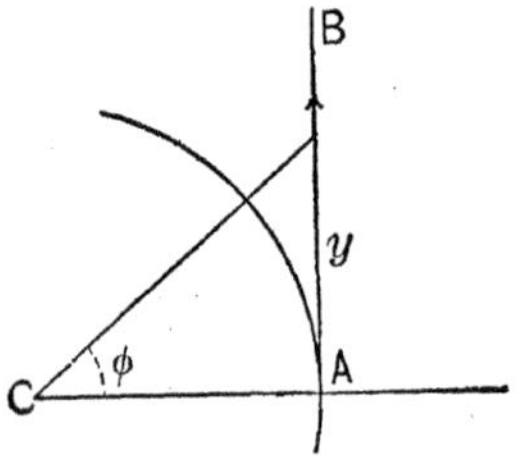

16. The base of a right triangle is 4 mi.; its altitude is variable and denoted by y, and ϕ is the variable angle opposite to y. Corresponding to $y = 2$ mi. find the

rate of ϕ, first as a function of y, then as a function of tan ϕ. Explain the difference between the two results.

Ans. $1/5$; $4/5$.

17. A train is running from A to B at the rate of 20 mi. an hour. The distance from A to C on a perpendicular to AB is 2 mi. Find the rate of the angle at C included between CA and a right line from C to the train.

Let $\phi =$ variable angle at C, and $y =$ mi. from A to train.
Then $CA = 2$ mi., and $dy/dt = 20$ mi. hr.

$$y = CA \tan\phi. \qquad \therefore \phi = \tan^{-1}\frac{y}{CA} \quad \text{and} \quad \frac{d\phi}{dy} = \frac{CA}{\overline{AC}^2 + y^2} = \frac{2}{4+y^2}.$$

$$\frac{d\phi}{dt} = (d\phi/dy) \times (dy/dt) = \frac{2}{4+y^2} \times 20 = \frac{40}{4+y^2}.$$

$$(d\phi/dt)_{y=0} = 10. \qquad (d\phi/dt)_{y=2} = 5. \qquad (d\phi/dt)_{y=\infty} = 0.$$

18. Find the rate of the surface and volume of a sphere when its radius decreases at the rate of 2 ft. per minute.

Ans. $-16\pi r$ sq. ft. min.; $-8\pi r^2$ cu. ft. min.

19. A ball of twine rolls along a floor in a right line at the rate of 4 mi. per hour. One end of it is 30 feet above the floor and is attached to the top of a pole. At what rate is the ball unwinding when it is 40 feet from the bottom of the pole on the floor? Ans. 3.2 mi. hour.

20. A ladder 20 ft. in length leans against a wall; if the bottom is drawn out at the rate of 2 ft. per second, at what rate will the top descend when the bottom is 8 ft. from the wall? Ans. 10.5 in. sec.

21. The side of an equilateral triangle increases at the rate of 2 in. per minute. Find the rate of its altitude, and the rate of its area when the side is 10 in.

Ans. $\sqrt{3}$ in. mi.; $10\sqrt{3}$ sq. in. min.

22. Two straight railways intersect at an angle of 60°. An engine approaches the intersection on one of the tracks at the rate of 25 mi. per hour, and on the other track an engine is leaving it at the rate of 30 mi. per hour. At what rate are the engines separating when each is 10 mi. from the intersection? Ans. 2.5 mi. hr.

23. A man walking on a horizontal plane approaches the foot of a pole 60 ft. in height, with a constant rate. When he is 40 feet from the foot how will the rate with which he approaches the top compare with that with which he approaches the bottom? How far will he be from the foot when he is approaching it twice as fast as he is the top?

Ans. $2\sqrt{1/13}$, $\sqrt{1200}$ ft.

FUNCTIONS OF TWO OR MORE VARIABLES.

101. **The Partial Differential of a Function of Two or more Variables,** with respect to one of the variables, is the change that the function would undergo from any state, were it to retain its rate at that state, with respect to that variable, while that variable changed by its differential.

The Total Differential of a Function of Two Variables is the change that the function would undergo from any state, were it to retain its rate at that state, with respect to each variable, while both variables changed by their differentials.

Any function of two variables which changes uniformly with each variable has a constant rate with respect to each, and its form must be some particular case of the general expression $Ax + By + C$ (§ 64).

Representing such a function by z, we have

$$z = Ax + By + C. \qquad (1)$$

Increasing x and y by their differentials, and denoting the corresponding new state of the function by z', we have

$$z' = A(x + dx) + B(y + dy) + C. \quad . \quad . \quad . \quad (2)$$

Subtracting (1) from (2), member from member, we have

$$z' - z = Adx + Bdy.$$

Since the function z changes uniformly with respect to each variable, the total differential of it, denoted by dz, is equal to the corresponding change in the function.

Therefore, $$dz = Adx + Bdy.$$

Adx is the corresponding partial differential of the function z with respect to x; and Bdy is the same with respect to y.

Hence, *the total differential of any function of two variables, which changes uniformly with respect to each, is equal to the sum of the corresponding partial differentials.*

The total differential of any function of two variables which does not vary uniformly with each variable is not, in in general, equal to the corresponding change in the function, but it is equal to the corresponding change of a function having a constant rate with respect to each variable, equal to that of the given function at the state considered. In other words, the total differential is equal to *that* of a function *which changes uniformly* with each variable, and which has at the state considered its partial differentials equal to the corresponding partial differentials of the given function.

Hence, *the total differential of any function of two variables is equal to the sum of the corresponding partial differentials.*

In a similar manner it may be shown that *the total differential of any function of any number of variables is equal to the sum of the corresponding partial differentials.*

In order to distinguish between a total and a partial differential the symbol ∂ is used to indicate a partial differential. Thus having $z = f(x, y)$, then

$$dz = (\partial z/dx)dx + (\partial z/dy)dy,$$

in which dz represents the total, ∂z a partial, differential.

EXAMPLES.

1. $d(xy) = xdy + ydx$.

2. $d(3ax^2y - 2y^2 + 3bx^3 - 5) = 6axydx + 9bx^2dx + 3ax^2dy - 4ydy$.

3. $d[(x+y)/(x-y)] = [2(xdy - ydx)]/(x-y)^2$.

4. $d(x^2y^2z^2) = 2y^2z^2xdx + 2x^2z^2ydy + 2x^2y^2zdz$.

5. $d\tan^{-1}(y/x) = (xdy - ydx)/(x^2+y^2)$.

6. $d[\sin(xy)] = \cos(xy)(ydx + xdy)$.

7. $d\log(x^y) = ydx/x + \log xdy$.

8 $dy^{\sin x} = y^{\sin x}[\log y \cos xdx + (\sin xdy/y)]$.

9. $d\operatorname{versin}^{-1}(x/y) = (ydx - xdy)/(y\sqrt{2xy - x^2})$.

10. $d\sin(x+y) = \cos(x+y)(dx+dy)$.

11. Deduce the formula $ds = \sqrt{dr^2 + r^2dv^2}$ (§ 92) from the formulas

$$\left.\begin{array}{l} x = a + r\cos v, \\ y = b + r\sin v, \end{array}\right\} \text{[Anal. Geo.,] and } ds = \sqrt{dx^2 + dy^2} \text{ (§87).}$$

12. One side of a rectangle increases at the rate of 3 in. per second and the other decreases at the rate of 2 in. per second. Find the rate of the area when the first side is 10 in. and the second 8 in. in length.

Ans. 4 sq. in. sec.

102. In order to represent $dz = (\partial z/dx)dx + (\partial z/dy)dy$...(1) graphically, let any state of the function z be represented by the ordinate NM of a surface. § 27.

Through NM pass the planes MNR and MNP parallel, respectively, to ZX and ZY. Let MC be the intersec-

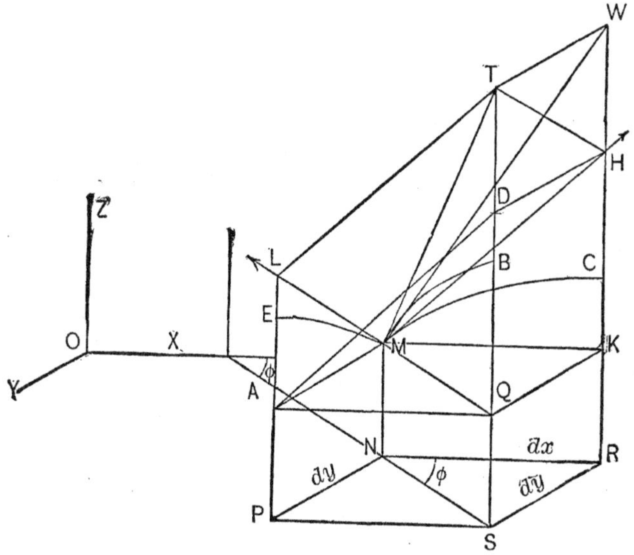

tion of the surface by the plane MNR, and let MH be the tangent to it at M. Let ME be the intersection of the surface by the plane MNP, and let ML be its tangent at M. MH and ML determine the tangent plane to the surface at M.

Assume $dx = NR$, and $dy = NP$. Complete the parallelogram NS and the parallelopipedon NQ. Produce RK, PA and SQ.

Then, $\tan KMH = \partial z/dx$, and $KH = (\partial z/dx)dx$.

$\tan AML = \partial z/dy$, and $AL = (\partial z/dy)dy$.

Draw HD parallel to KQ. Connect A and D by AD, and draw LT parallel to AD.

Then, $QD = KH$, and $DT = AL$.

Therefore, $QT = (\partial z/dx)dx + (\partial z/dy)dy = dz$.

LT is parallel to AD, which is parallel to MH. T therefore lies in the tangent plane, and MT is the tangent to MB, the line of intersection of the surface by the plane MNS.

While x and y are changing as assumed, the foot of the corresponding ordinate passes with uniform motion from N to S, and QT is the total differential required by definition.

As in the case of a function of a single variable, the expressions $\partial z/dx$ and $\partial z/dy$ are, respectively, independent of dx and dy, but for any state of a function of two variables there is no fixed *total differential coefficient.* In the figure,

$$QT/QM = \tan QMT = dz/\sqrt{dx^2 + dy^2}$$

is the total rate of change of z, corresponding to the values assumed for dx and dy, but any change in the relative value of dx and dy will change this total rate. For any values assumed for dx and dy, the total rate of change of the function, and therefore its total differential, will be the same as that of *the ordinate of the line* cut out of the tangent plane by the plane through NM, whose trace on XY makes with X an angle $\phi = \tan^{-1} dy/dx$.

It is now apparent that the function represented by the *ordinate of the tangent plane* at M is the one *with a constant rate with respect to each variable*, whose total differential for the same values of dx and dy is the same as that of the given function at the state corresponding to M.

$dz/\sqrt{dx^2 + dy^2} = \tan QMT$ is the tangent of the angle which a tangent to MB makes with XY, and (§ 65) its numerical value is the slope of the surface at M along MB.

From (1), we write

$$\left.\begin{aligned} dz/dx &= \partial z/dx + (\partial z/dy)(dy/dx), \\ dz/dy &= \partial z/dy + (\partial z/dx)(dx/dy), \end{aligned}\right\} \quad . \quad . \quad . \quad (2)$$

the first members of which depend upon the relative value of dx and dy.

Draw TW parallel to SR. Then MW is the projection of MT upon the plane MNR, $dz/dx = \tan KMW$, and dz/dy is the tangent of the angle which the projection of MT upon the plane MNP makes with Y.

$\partial z/dx = \tan KMH$ is the partial differential coefficient of the function with respect to x only, and is independent of dx and dy. It is equal to the tangent of the angle which the intersection of the tangent plane, and any plane parallel to ZX, makes with X or the plane XY. Whereas $dz/dx = \tan KMW$ is dependent upon the quotient dx/dy (Eq. 2), which has no definite value because x and y are independent, and their differentials arbitrary.

If, however, by means of a second equation $\phi(x,y) = 0$, x and y are related, thus making one of them dependent upon the other and determining dy/dx, then z becomes a function of a single independent variable; its graph becomes the line of intersection of the surface $z = f(x, y)$ by the cylinder $\phi(x, y) = 0$, dz/dx becomes the differential coefficient of z regarded as a function of x and is equal to $\tan \theta$, θ being the angle between the axis of X and the projection of the tangent to the graph on XZ.

Similarly both members of Eq. 1 may be divided by dt, t being any independent variable, giving

$$dz/dt = (\partial z/dx)(dx/dt) + (\partial z/dy)(dy/dt).$$

The values of both members will now depend upon the values assumed (§ 67) for dx, dy, and dt.

But if we also write $\phi(x, t) = 0$ and $\psi(y, t) = 0$, thus relating x and y to t, and determining dx/dt and dy/dt, then z becomes a function of one independent variable, and the first member becomes the differential coefficient of z regarded as a function of t.

CHAPTER VI.

SUCCESSIVE DIFFERENTIATION.

FUNCTIONS OF A SINGLE VARIABLE.

103. In general the differential coefficient, and therefore the differential of any function of a variable, are functions of the variable and may be differentiated.

Thus, having ax^3,

$$dax^3 = 3ax^2dx, \quad \text{and} \quad dax^3/dx = 3ax^2.$$

Differentiating again, denoting $d(dax^3)$ by d^2ax^3, read "*second differential of* ax^3," and representing $(dx)^2$ by dx^2, we have

$$d(dax^3) = d^2ax^3 = 6axdx^2, \quad \text{and} \quad d^2ax^3/dx^2 = 6ax.$$

$6axdx^2$ is the differential of $3ax^2dx$, which is the differential of ax^3. $6axdx^2$ is therefore called the *first* differential of $3ax^2dx$ and the *second* differential of ax^3.

Similarly, $6ax$ is the *first* differential coefficient of $3ax^2$ and the *second* differential coefficient of ax^3.

Differentiating again and extending the notation, we have

$$d(d^2ax^3) = d^3ax^3 = 6adx^3, \quad \text{and} \quad d^3ax^3/dx^3 = 6a.$$

$6adx^3$ is the *first* differential of $6axdx^2$, the *second* differential of $3ax^2dx$, and the *third* differential of ax^3. $6a$ is the *first differential coefficient of* $6ax$, the *second* of $3ax^2$, and the *third* of ax^3.

Representing the function x^n by y, we have $y = x^n$. Differentiating, we obtain

$$dy = nx^{n-1}dx, \quad \text{and} \quad dy/dx = nx^{n-1},$$

for the *first* differential and differential coefficient, respectively, of x^n.

Differentiating again, we have

$$d^2y = n(n-1)x^{n-2}dx^2, \quad \text{and} \quad d^2y/dx^2 = n(n-1)x^{n-2},$$

for the *second* differential and differential coefficient, respectively, of x^n.

Again, $\qquad d^3y = n(n-1)(n-2)x^{n-3}dx^3,$

and $\qquad d^3y/dx^3 = n(n-1)(n-2)x^{n-3},$

for the *third* differential and differential coefficient respectively, of x^n.

Similarly, the fourth, fifth, etc., differentials may be derived in succession.

It should be observed that the symbol d^2y/dx^2, which represents the *second* differential coefficient, is the differential coefficient of the symbol dy/dx, which represents the *first* differential coefficient.

Thus, $$\frac{d^2y}{dx^2} = d\left(\frac{dy}{dx}\right) \Big/ dx.$$

Similarly, $$\frac{d^3y}{dx^3} = d\left(\frac{d^2y}{dx^2}\right) \Big/ dx,$$

and $$\frac{d^ny}{dx^n} = d\left(\frac{d^{n-1}y}{dx^{n-1}}\right) \Big/ dx.$$

Successive differential coefficients are also called *derived functions* or *derivatives*, and are frequently represented in order by accents on the functional letter.

Thus, if $f(x)$ represents the primitive function, then

$$f'(x), \qquad f''(x), \qquad f'''(x), \qquad \text{etc.},$$

denote respectively the first, second, third, etc., derivatives of $f(x)$.

Other forms are also used. Thus, having $y = f(x)$,

$$D_x y, \qquad D_x^2 y, \qquad D_x^3 y, \qquad \text{etc.},$$

represent the successive derived functions in order.

Each successive differential coefficient or derivative in order is the rate of change of the immediately preceding one, and $d^{n-1}y/dx^{n-1}$ is an increasing or decreasing function of x, according as $d^n y/dx^n$ is positive or negative. § 71.

Let $y = f(x) = x^n$, n being entire and positive.

Then $dy/dx = f'(x) = D_x y = nx^{n-1}$.

$$d^2y/dx^2 = f''(x) = D_x^2 y = n(n-1)x^{n-2}.$$

etc. etc. etc.

$$d^ny/dx^n = f^n(x) = D_x^n y = n(n-1)(n-2)\ldots 2.1.$$

$$d^{n+1}y/dx^{n+1} = f^{n+1}(x) = D_x^{n+1}y = 0.$$

It should be observed that the symbols d^2y/dx^2, $f''(x)$, $D_x^2 y$, etc., serve only to represent expressions, and to indicate their relations to the primitive function. In the above d^2y/dx^2 denotes that $n(n-1)x^{n-2}$ is the *second* differential coefficient of x^n.

The differential of any order is the product of the corresponding derivative and power of the differential of the variable.

Thus, $$d^ny = (d^ny/dx^n)dx^n = f^n(x)dx^n \quad . \quad . \quad . \quad \text{(a)}$$

represents a differential of the nth order, and $d^ny/dx^n = f^n(x)$

represents a differential coefficient or derivative of the nth order.

Dividing both members of (a) by dx^{n-1}, dx^{n-2}, etc., in succession, we have, in order,

$$d^n y/dx^{n-1} = d(d^{n-1}y/dx^{n-1}) = f^n(x)dx.$$

$$d^n y/dx^{n-2} = d^2(d^{n-2}y/dx^{n-2}) = f^n(x)dx^2.$$

etc. etc. etc.

$$d^n y/dx^{n-r} = d^r(d^{n-r}y/dx^{n-r}) = f^n(x)dx^r.$$

etc. etc. etc.

$$d^n y/dx \quad = d^{n-1}(dy/dx) \quad = f^n(x)dx^{n-1}.$$

From which we see that the product of a derivative of any order by the *first* power of the differential of the variable is the *first* differential of the immediately preceding derivative, and its product by the *second* power of the differential of the variable is the *second* differential of the second preceding derivative, etc. The product of a derivative of the nth order by the $(n - 1)^{th}$ power of the differential of the variable is the $(n - 1)^{th}$ differential of the first derivative.

EXAMPLES.

1. $y = ax^4$. $\quad dy = 4ax^3dx, \quad d^2y = 12ax^2dx^2,$
$\quad d^3y = 24ax\,dx^3, \quad d^4y = 24a\,dx^4.$

2. $f(x) = (a - x)^{-1}$. $\quad f'(x) = (a - x)^{-2}, \quad f''(x) = 2(a - x)^{-3},$
$\quad f^n(x) = 2.3\ldots n(a - x)^{-n-1},$ etc.

3. $y = \sin x$. $\quad D_x y = \cos x, \quad D_x^2 y = -\sin x,$
$\quad D_x^3 y = -\cos x, \quad D_x^4 y = \sin x,$
etc. etc.

The exponent of the power of a function is diminished by unity at each differentiation (§ 79), and when entire and positive it will finally be reduced to zero. Hence, algebraic functions which do not contain fractional or negative exponents affecting the variable have a limited number of derivatives. All others, including transcendental functions, have an unlimited number.

4. $f(x) = ax^3 + bx^2$.

$$f'(x) = 3ax^2 + 2bx, \qquad f''(x) = 6ax + 2b, \qquad f'''(x) = 6a.$$

5. $y = ax^{\frac{1}{2}}$.

$$\frac{dy}{dx} = \tfrac{1}{2}ax^{-\frac{1}{2}}, \qquad \frac{d^2y}{dx^2} = -\tfrac{1}{4}ax^{-\frac{3}{2}}, \qquad \frac{d^3y}{dx^3} = \tfrac{3}{8}ax^{-\frac{5}{2}}, \quad \text{etc.}$$

6. $y = a^x$.

$$D_xy = a^x \log a, \qquad D_x^2y = a^x (\log a)^2, \qquad \text{etc.,} \qquad D_x^ny = a^x (\log a)^n.$$

7. $y = \cos x$.

$$\frac{dy}{dx} = -\sin x = \cos\left(x + \frac{\pi}{2}\right), \qquad \frac{d^2y}{dx^2} = -\cos x = \cos\left(x + \frac{2\pi}{2}\right),$$

$$\frac{d^3y}{dx^3} = \sin x = \cos\left(x + \frac{3\pi}{2}\right), \qquad \text{etc.,} \qquad \frac{d^ny}{dx^n} = \cos\left(x + \frac{n\pi}{2}\right).$$

8. $y = \log x$.

$$\frac{dy}{dx} = \frac{1}{x}, \quad \frac{d^2y}{dx^2} = -\frac{1}{x^2}, \quad \frac{d^3y}{dx^3} = \frac{2}{x^3}, \quad \frac{d^4y}{dx^4} = -\frac{2 \cdot 3}{x^4}, \quad \text{etc.}$$

$$d^ny/dx^n = (-1)^{n-1}\underline{|n-1}/x^n.$$

9. $\phi = \cos^{-1} u$.

$$d\phi/du = -1/\sqrt{1-u^2}, \quad d^2\phi/du^2 = -u/(1-u^2)^{3/2}, \quad \text{etc.}$$

10. $f(x) = \sin mx$.

$$f'(x) = m \cos mx, \qquad f''(x) = -m^2 \sin mx, \text{ etc.}$$

$$f^{2n}(x) = (-1)^n m^{2n} \sin mx, \quad f^{2n+1}(x) = (-1)^n m^{2n+1} \cos mx.$$

11. $y = e^{ax}$.

$$\frac{dy}{dx} = ae^{ax}, \quad \frac{d^2y}{dx^2} = a^2e^{ax}, \quad \text{etc.} \ldots \frac{d^ny}{dx^n} = a^ne^{ax}.$$

12. $y = \log (e^x + e^{-x})$. $\qquad d^3y/dx^3 = -8(e^x - e^{-x})/(e^x + e^{-x})^3$.

13. $y = \cos mx$. $\qquad d^ny/dx^n = m^n \cos (mx + n\pi/2)$.

14. $y = \cos^2 x$. $\qquad d^ny/dx^n = 2^{n-1} \cos (2x + n\pi/2)$.

15. $y = \sin x$. $\quad d^n y/dx^n = \sin(x + n\pi/2)$.

16. $y = (1 + x)/(1 - x)$. $\quad d^n y/dx^n = 2\lfloor n/(1 - x)^{n+1}$.

17. $y = \tan x$. $\quad d^3y/dx^3 = 6 \sec^4 x - 4 \sec^2 x$.

18. $y = \sqrt{2px}$. $\quad d^2y/dx^2 = -p^2/(2px)^{3/2} = -p^2/y^3$.

19. $f(x) = x^3/(1 - x)$. $\quad f^{iv}(x) = 24/(1 - x)^5$.

20. $y = e^x \cos x$. $\quad d^n y/dx^n = 2^{n/2} e^x \cos(x + n\pi/4)$.

21. $y = \pm\sqrt{R^2 - x^2}$. $\quad \dfrac{d^2y}{dx^2} = -\dfrac{R^2}{\pm(R^2 - x^2)^{3/2}} = -\dfrac{R^2}{\pm y^3}$.

22. $f(x) = \tan x + \sec x$. $\quad f''(x) = \cos x(1 - \sin x)^{-2}$.

23. $y = x^x$. $\quad d^2y/dx^2 = x^x(1 + \log x)^2 + x^{x-1}$.

24. $f(x) = x^3 \log x$. $\quad f^{iv}(x) = 6x^{-1}$.

25. $x = \sin^{-1} y$. $\quad d^4x/dy^4 = (9y + 6y^3)/(1 - y^2)^{7/2}$.

26. $y = \log \sin x$. $\quad D_x^{\,3} y = 2 \cos x/\sin^3 x$.

27. $y = \sqrt{\sec 2x}$. $\quad d^2y/dx^2 = 3(\sec 2x)^{5/2} - (\sec 2x)^{1/2}$.

28. $y = (x^2 + a^2)\tan^{-1}(x/a)$. $\quad d^3y/dx^3 = 4a^3/(a^2 + x^2)^2$.

29. $y = \sec x$. $\quad d^2y/dx^2 = 2 \sec^3 x - \sec x$,

$d^3y/dx^3 = \sec x \tan x(6 \sec^2 x - 1)$.

30. $y = x^{n-1} \log x$. $\quad d^n y/dx^n = \lfloor n - 1/x$.

31. $f(x) = (2ax)^{a/b}$. $\quad f''(2) = 4(a - b)a^{(a+b)/b}/b^2$.

32. $y = \tan^{-1}(1/x)$, $\therefore\ x = \cot y$.

$$\frac{dy}{dx} = \frac{-1}{1 + x^2} = -\sin^2 y, \quad \therefore \quad \frac{1}{(1 + x^2)^{n/2}} = \sin^n y.$$

$$\frac{d^2y}{dx^2} = -\frac{d \sin^2 y}{dx} = -\frac{2 \sin y \cos y\, dy}{dx} = \sin 2y \sin^2 y.$$

$$\frac{d^3y}{dx^3} = \frac{d(\sin 2y \sin^2 y)}{dx} = \frac{\sin 2y\ 2 \sin y \cos y\, dy + \sin^2 y \cos 2y\ 2dy}{dx}$$

$$= 2(\sin y)(\sin 2y \cos y + \cos 2y \sin y)(dy/dx)$$

$$= -2 \sin^3 y \sin 3y.$$

Similarly, $$d^4y/dx^4 = \underline{|3}\ \sin^4 y \ \sin 4y,$$

and $$d^ny/dx^n = (-1)^n \ \underline{|n-1}\ \sin^n y \ \sin ny.$$

Since $$\tan^{-1} x = \pi/2 - \tan^{-1}(1/x),$$

we have $$d^n(\tan^{-1} x)/dx^n = (-1)^{n-1}\underline{|n-1}\ \sin^n y \sin ny,$$

or $$d^n(\tan^{-1} x)/dx^n = (-1)^{n-1}\underline{|n-1}\ \sin\left(n \tan^{-1}\tfrac{1}{x}\right)/(1+x^2)^{n/2}.$$

33. $y = a\left(e^{x/a} + e^{-x/a}\right)/2.$ $dy/dx = \left(e^{x/a} - e^{-x/a}\right)/2,$ $d^2y/dx^2 = y/a^2.$

34. $y = a^2x/(a^2+x^2).$

$dy/dx = a^2(a^2-x^2)/(a^2+x^2)^2.$ $d^2y/dx^2 = 2a^2x(x^2-3a^2)/(a^2+x^2)^3.$

35. $y = Ct^2/2 + C't + C''.$ $dy/dx = Ct + C',$ $d^2y/dx^2 = C.$

36. $y = b + c(x-a)^{3/5},$

$dy/dx = 3c/5(x-a)^{2/5}.$ $d^2y/dx^2 = -6c/25(x-a)^{7/5}.$

37. $fx = x^2 \pm x^{5/2}.$ $f'x = 2x \pm 5x^{3/2}/2,$ $f''x = 2 \pm 15x^{1/2}/4.$

38. $fx = e^{1/x}.$ $f'x = -e^{1/x}/x^2,$ $f''x = (2x+1)e^{1/x}/x^4.$

39. $fx = e^{-1/x}.$ $f'x = e^{-1/x}/x^2.$ $f''x = e^{-1/x}(1-2x)/x^4.$

40. The relation between the time, denoted by t, and the distance, represented by s, through which a body, starting from rest, falls in a vacuum near the earth's surface, is expressed very nearly by the equation $s = 16.1t^2$; s being in feet and t in seconds. Construct a table giving the entire distance fallen through in 1 second; in 2 seconds; in 3 seconds; and in 4 seconds; the distance passed over during each of the above seconds; the velocity and acceleration at the end of each.

Time in Seconds.	Entire Distance in Feet.	Distance each Second.	Velocity.	Acceleration.
1	16.1	16.1	32.2	32.2
2	64.4	48.3	64.4	32.2
3	144.9	80.5	96.6	32.2
4	257.6	112.7	128.8	32.2

41. Having $s^2 = 5t^3$, find the velocity and acceleration when $t = 2$ seconds; $t = 3$ seconds. Ans. $V_{t=2} = 3\sqrt{10}/2$. $V_{t=3} = 3\sqrt{15}/2$.
$A_{t=2} = 3\sqrt{5/2}/4$. $A_{t=3} = 3\sqrt{5/3}/4$.

42. $y = \sin(m \sin^{-1} x)$. $dy/dx = m\cos(m \sin^{-1} x)/\sqrt{1 - x^2}$. Hence,

$$(1 - x^2)(dy/dx)^2 = m^2\cos^2(m \sin^{-1} x) = m^2(1 - y^2).$$

Differentiating again and dividing by $2dy$, we have

$$(1 - x^2)(d^2y/dx^2) - x(dy/dx) + m^2y = 0.$$

43. $fx = \sinh x$. $f''x = \sinh x$.

44. $fx = \cosh x$. $f''x = \cosh x$.

104. Leibnitz's Theorem.—Let $y = uv$, in which u and v are any functions of x; then (§ 75)

$$dy/dx = u\,dv/dx + v\,du/dx,$$

$$d^2y/dx^2 = u\,d^2v/dx^2 + 2(du/dx)(dv/dx) + v\,d^2u/dx^2,$$

$$\frac{d^3y}{dx^3} = u\frac{d^3v}{dx^3} + 3\frac{du}{dx}\frac{d^2v}{dx^2} + 3\frac{d^2u}{dx^2}\frac{dv}{dx} + v\frac{d^3u}{dx^3}$$

in which the numerical coefficients follow the law of those of the binomial formula. By a method similar to that used in deducing that formula for positive entire exponents it may be shown that

$$\frac{d^ny}{dx^n} = u\frac{d^nv}{dx^n} + n\frac{du}{dx}\frac{d^{n-1}v}{dx^{n-1}} + \frac{n(n-1)}{1\,.\,2}\frac{d^2u}{dx^2}\frac{d^{n-2}v}{dx^{n-2}} + \ldots$$

$$+ \frac{n(n-1)\ldots(n-r+1)}{\underline{|r}}\frac{d^ru}{dx^r}\frac{d^{n-r}v}{dx^{n-r}} + \ldots n\frac{d^{n-1}u}{dx^{n-1}}\frac{dv}{dx} + v\frac{d^nu}{dx^n}.$$

EXAMPLES.

1. $y = e^{ax}x$.

$$u = e^{ax},\quad du/dx = ae^{ax},\ \ldots\ d^nu/dx^n = a^ne^{ax}.$$

$$v = x,\quad dv/dx = 1,\quad \ldots\ d^nv/dx^n = 0.$$

$$d^ny/dx^n = na^{n-1}e^{ax} + a^ne^{ax}x.$$

2. $y = e^{ax}x^2$. $\quad d^ny/dx^n = a^{n-2}e^{ax}[a^2x^2 + 2nax + n(n - 1)]$.

3. $y = x^3 \tan x$.

$$d^3y/dx^3 = 2x^3 \sec^2 x(3 \tan^2 x + 1) + 18x^2 \sec^2 x \tan x + 18x \sec^2 x + 6 \tan x.$$

4. $y = e^{ax}v$.

$$\frac{d^ny}{dx^n} = e^{ax}\left(a^nv + na^{n-1}\frac{dv}{dx} + \frac{n(n-1)}{1 \cdot 2}a^{n-2}\frac{d^2v}{dx^2} + \ldots + \frac{d^nv}{dx^n}\right).$$

FUNCTIONS OF TWO OR MORE VARIABLES.

105. Successive Partial Differentiation.—A partial differential of a function of two or more variables is, in general, a function of each variable and may, therefore, be differentiated again with respect to each.

Thus, if $z = f(x, y)$ we write (§ 101) $dz = \frac{\partial z}{dx}dx + \frac{\partial z}{dy}dy$,

in which $\partial z/dx$ and $\partial z/dy$ are symbols for *the partial derivatives of the first order* with reference to x and y respectively.

Differentiating $\partial z/dx$ with respect to x, we write

$$\partial\left(\frac{\partial z}{dx}\right) \bigg/ dx = \frac{\partial}{dx}\left(\frac{\partial z}{dx}\right) = \left(\frac{\partial^2 z}{dx^2}\right)$$

for the partial derivative of the second order taken twice with respect to x.

Differentiating $\partial z/dx$ with respect to y, we have

$$\frac{\partial}{dy}\left(\frac{\partial z}{dx}\right) = \frac{\partial^2 z}{dx\,dy}$$

for the partial derivative of the second order taken once with respect to x, and then with respect to y.

Similarly, we write

$$\frac{\partial}{dx}\left(\frac{\partial z}{dy}\right) = \frac{\partial^2 z}{dy\, dx}, \quad \text{and} \quad \frac{\partial}{dy}\left(\frac{\partial z}{dy}\right) = \frac{\partial^2 z}{dy^2}$$

for the other partial derivatives of the second order.

Each partial derivative of the second order will, in general, admit of differentiation with respect to each variable, and the differentiation may be continued, in general, to any required order.

The notation adopted is as follows:

$$\frac{\partial}{dx}\left(\frac{\partial^2 z}{dx^2}\right) = \frac{\partial^3 z}{dx^3},$$

$$\frac{\partial}{dy}\left(\frac{\partial^2 z}{dx^2}\right) = \frac{\partial^3 z}{dx^2\, dy},$$

$$\frac{\partial}{dx}\left(\frac{\partial^2 z}{dx\, dy}\right) = \frac{\partial^3 z}{dx\, dy\, dx},$$

$$\frac{\partial}{dy}\left(\frac{\partial^2 z}{dx\, dy}\right) = \frac{\partial^3 z}{dx\, dy^2},$$

etc. etc.

$$\frac{\partial}{dy}\left(\frac{\partial^n z}{dx^{n-r}\, dy^r}\right) = \frac{\partial^{n+1} z}{dx^{n-r}\, dy^{r+1}},$$

which represents the partial derivative of the $(n+1)^{\text{th}}$ order, taken $(n-r)$ times with respect to x and $(r+1)$ with respect to y.

The numerator indicates the number of differentiations, and the denominator the order of the successive operations with respect to the variables.

By multiplying the symbol for any partial derivative by its denominator we obtain the symbol for the corresponding partial differential. Thus, $\partial^{m+n} z\, dx^m dy^n / dx^m dy^n$ represents a partial differential of the $(m+n)^{\text{th}}$ order, taken m times with respect to x, and n times with respect to y.

Having $u = f(x, y, z)$,

$$\partial u/dx, \qquad \partial u/dy, \qquad \partial u/dz,$$

represent, respectively, the partial derivatives of the first order, each being, in general, a function of x, y, and z. The differentiation may, therefore, be continued, and the successive operations indicated by extending the notation used above. Thus, $\partial^{m+n+r}u/dx^m dy^n dz^r$ represents the partial derivative of the $(m+n+r)^{\text{th}}$ order, taken m times with respect to x, n times with respect to y, and r times with respect to z.

In a similar manner, a partial derivative of any order of a function of any number of variables may be obtained and represented.

106. *Partial differentials and their corresponding derivatives are independent of the order of differentiation.*

Assume $z = f(x, y)$, and increase x by h, giving (§ 70)

$$\frac{\partial z}{dx} = \underset{h \ggg 0}{\text{limit}}\left[\frac{f(x+h, y) - f(x, y)}{h}\right].$$

In this expression increase y by k, and we have (§ 70)

$$\frac{\partial}{dy}\left(\frac{\partial z}{dx}\right) = \frac{\partial^2 z}{dx\,dy} =$$

$$\underset{k \ggg 0}{\text{limit}}\left[\frac{\underset{h \ggg 0}{\text{limit}}\left[\frac{f(x+h, y+k) - f(x, y+k) - [f(x+h, y) - f(x, y)]}{h}\right]}{k}\right]$$

$$= \underset{\substack{h \ggg 0 \\ k \ggg 0}}{\text{limit}}\left[\frac{f(x+h, y+k) - f(x, y+k) - [f(x+h, y) - f(x, y)]}{hk}\right]. \qquad (1)$$

Similarly, increasing y in the primitive function by k, we write

$$\frac{\partial z}{dy} = \underset{k \ggg \to 0}{\text{limit}} \left[\frac{f(x, y+k) - f(x, y)}{k} \right],$$

from which, increasing x by h, we obtain

$$\frac{\partial}{dx}\left(\frac{\partial z}{dy}\right) = \frac{\partial^2 z}{dy\,dx} =$$

$$\underset{h \ggg \to 0}{\text{limit}} \left[\frac{\underset{k \ggg \to 0}{\text{limit}} \left[\frac{f(x+h, y+k) - f(x+h, y) - [f(x, y+k) - f(x, y)]}{k} \right]}{h} \right]$$

$$= \underset{\substack{k \ggg \to 0 \\ h \ggg \to 0}}{\text{limit}} \left[\frac{f(x+h, y+k) - f(x+h, y) - [f(x, y+k) - f(x, y)]}{hk} \right],$$

which compared with (1) gives

$$\frac{\partial^2 z}{dx\,dy} = \frac{\partial}{dy}\left(\frac{\partial z}{dx}\right) = \frac{\partial}{dx}\left(\frac{\partial z}{dy}\right) = \frac{\partial^2 z}{dy\,dx}.$$

From which we have

$$\frac{\partial^3 z}{dx^2\,dy} = \frac{\partial}{dy}\left[\frac{\partial}{dx}\left(\frac{\partial z}{dx}\right)\right] = \frac{\partial}{dx}\left[\frac{\partial}{dy}\left(\frac{\partial z}{dx}\right)\right]$$

$$= \frac{\partial}{dx}\left[\frac{\partial}{dx}\left(\frac{\partial z}{dy}\right)\right] = \frac{\partial^3 z}{dy\,dx^2}.$$

Similarly, we derive

$$\frac{\partial^4 z}{dx^3\,dy} = \frac{\partial^4 z}{dx^2\,dy\,dx} = \frac{\partial^4 z}{dy\,dx^3},$$

and, in general,

$$\partial^{m+n} z / dx^m dy^n = \partial^{m+n} z / dy^n dx^m.$$

Similarly, having $u = f(x, y, z)$, it can be proved that

$$\partial^3 u / dz\,dx\,dy = \partial^3 u / dy\,dx\,dz, \text{ etc.}$$

Hence we infer that *the order* of differentiation in all cases does not affect the result.

EXAMPLES.

1 $z = x \sin y + y \sin x$. $\dfrac{\partial^2 z}{dx dy} = \cos y + \cos x = \dfrac{\partial^2 z}{dy dx}$.

2. $z = 2x^2y^3 + x^4y$. $\dfrac{\partial^3 z}{dx dy dx} = \dfrac{\partial^3 z}{dy dx^2} = 12(x^2 + y^2)$.

3. $z = x \log y$. $\dfrac{\partial^2 z}{dx dy} = \dfrac{1}{y} = \dfrac{\partial^2 z}{dy dx}$.

4. $z = x^3 y + 4y^2$. $\dfrac{\partial^2 z}{dx dy} = 3x^2 = \dfrac{\partial^2 z}{dy dx}$.

5. $z = \tan^{-1}\left(\dfrac{x}{y}\right)$. $\dfrac{\partial^2 z}{dx dy} = \dfrac{\partial^2 z}{dy dx} = \dfrac{x^2 - y^2}{(x^2 + y^2)^2}$.

6. $z = \sin (ax^n + by^n)$. $\dfrac{\partial^2 z}{dx dy} = - abn^2(xy)^{n-1} \sin (ax^n + by^n)$.

7. $z = y^x$. $\dfrac{\partial^2 z}{dx dy} = y^{x-1}(1 + x \log y) = \dfrac{\partial^2 z}{dy dx}$.

8. $u = e^{xyz}$. $\dfrac{\partial^3 u}{dx dy dz} = (1 + 3xyz + x^2y^2z^2)e^{xyz} = \dfrac{\partial^3 u}{dz dy dx}$.

9. $z = \dfrac{x^2 + y^2}{x^2 - y^2}$. $\dfrac{\partial^2 z}{dx dy} = - 8xy \dfrac{x^2 + y^2}{(x^2 - y^2)^3}$.

10. $z = \sin^{-1}\left(\dfrac{x}{y}\right)$. $\dfrac{\partial^2 z}{dx dy} = - \dfrac{y}{(y^2 - x^2)^{\frac{3}{2}}}$.

11. $u = \dfrac{x^2 y}{a^2 - z^2}$. $\dfrac{\partial^2 u}{dy dz} = \dfrac{2x^2 z}{(a^2 - z^2)^2}$. $\dfrac{\partial^2 u}{dx dy} = \dfrac{2x}{a^2 - z^2}$.

$\dfrac{\partial^2 u}{dx dz} = \dfrac{4xyz}{(a^2 - z^2)^2}$. $\dfrac{\partial^3 u}{dx dy dz} = \dfrac{4xz}{(a^2 - z^2)^2}$.

12. $z = \sin \dfrac{x}{y}$. $\dfrac{\partial^3 z}{dy dx^2} = \dfrac{2}{y^3} \sin \dfrac{x}{y} + \dfrac{x}{y^4} \cos \dfrac{x}{y}$.

13. $z = \sin (x + y)$. $\dfrac{\partial^2 z}{dy^2} = - \sin (x + y)$. $\dfrac{\partial^3 z}{dy^3} = -\cos (x+y)$.

$$\frac{\partial^4 z}{dx^4} = \sin(x+y). \qquad \frac{\partial^5 z}{dx^5} = \cos(x+y).$$

14. $z = \cos(x-y)$. $$\frac{\partial^2 z}{dx^2} = -\cos(x-y). \qquad \frac{\partial^3 z}{dx^3} = \sin(x-y).$$

$$\frac{\partial^2 z}{dy^2} = -\cos(x-y) \qquad \frac{\partial^3 z}{dy^3} = -\sin(x-y).$$

15. $z = \log(x+y)$. $$\frac{\partial^2 z}{dx^2} = \frac{-1}{(x+y)^2}. \qquad \frac{\partial^3 z}{dy^3} = \frac{2}{(x+y)^3}.$$

107. Partial Differentials of a Surface.—Let ATL be any surface, and $ABCD = u$ a portion of it included be-

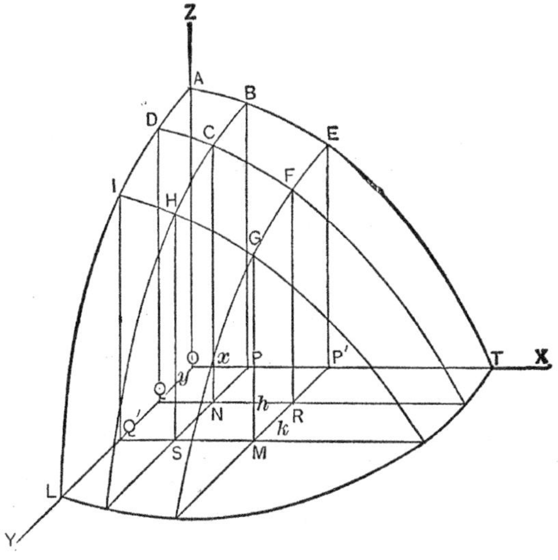

tween the coördinate planes XZ, YZ, and the planes DQR, BPS, parallel to them respectively. From § 26 we have $u = f(x, y)$.

Increase $OP = x$ by $PP' = h$, giving (§ 70)

$$\frac{\partial u}{dx} = \lim_{h \to 0} \left[\frac{f(x+h, y) - f(x, y)}{h}\right].$$

Now increase $OQ = y$ by $QQ' = k$, giving (§ 70)

$$\frac{\partial^2 u}{dx\,dy} = \underset{k \ggg\!\!\to 0}{\text{limit}} \left[\frac{\underset{h \ggg\!\!\to 0}{\text{limit}} \left[\frac{f(x+h, y+k) - f(x, y+k) - [f(x+h, y) - f(x, y)]}{h} \right]}{k} \right]$$

$$= \underset{\substack{h \ggg\!\!\to 0 \\ k \ggg\!\!\to 0}}{\text{limit}} \left[\frac{f(x+h, y+k) - f(x, y+k) - [f(x+h, y) - f(x, y)]}{hk} \right].$$

In which

$$f(x+h, y+k) = AEGI, \qquad f(x, y+k) = ABHI,$$

$$f(x+h, y) = AEFD, \qquad f(x, y) = ABCD.$$

Hence,

$$f(x+h, y+k) - f(x, y+k) = AEGI - ABHI = BEGH.$$

$$f(x+h, y) - f(x, y) = AEFD - ABCD = BEFC.$$

$$f(x+h, y+k) - f(x, y+k) - [f(x+h, y) - f(x, y)]$$

$$= BEGH - BEFC = CFGH.$$

Therefore (§ 50)

$$\frac{\partial^2 u}{dx\,dy} = \underset{\substack{h \ggg\!\!\to 0 \\ k \ggg\!\!\to 0}}{\text{limit}} \frac{CFGH}{hk} = \frac{1}{\cos \beta},$$

and $\qquad \partial^2 u\,dx\,dy / dx\,dy = \sec \beta\,dx\,dy.$

108. Partial Differentials of a Volume.—Let ATL (figure § 107) be any surface, and $ABCD\text{-}ON = v$ a volume limited by it, the three coördinate planes, and the planes DQR and BPS parallel, respectively, to XZ and YZ. From § 30 we have $v = f(x, y)$.

By the method used in the last Article, considering the corresponding volumes instead of the surfaces, we obtain

$$\frac{\partial^2 v}{dx\,dy} = \underset{\substack{h \ggg\!\!\to 0 \\ k \ggg\!\!\to 0}}{\text{limit}} \left[\frac{f(x+h, y+k) - f(x, y+k) - [f(x+h, y) - f(x, y)]}{hk} \right].$$

In which

$f(x+h, y+k)-f(x, y+k)-[f(x+h, y)-f(x, y)]=$vol. $CFGH$-NM.

Hence (§ 49),

$$\frac{\partial^2 v}{dx\,dy} = \lim_{\substack{h \to 0 \\ k \to 0}} \frac{CFGH - NM}{hk} = NC = z,$$

and $$\partial^2 v\,dx\,dy/dx\,dy = z\,dx\,dy.$$

109. Successive Total Differentiation.—Having $z = f(x, y)$, § 101 gives

$$dz = \frac{\partial z}{dx}dx + \frac{\partial z}{dy}dy, \quad . \quad . \quad . \quad . \quad . \quad . \quad (1)$$

in which the total differential of the first order dz, and the corresponding partial differentials $\frac{\partial z}{dx}dx$ and $\frac{\partial z}{dy}dy$, are, in general, functions of x and y.

Hence, the total differential of the second order, denoted by d^2z, is obtained by differentiating each term in the second member of (1) with respect to each variable, and taking the sum of the partial differentials of the second order.

Differentiating $\frac{\partial z}{dx}dx$ with respect to each variable, we have

$$d\left(\frac{\partial z}{dx}dx\right) = \frac{\partial^2 z}{dx^2}dx^2 + \frac{\partial^2 z}{dx\,dy}dx\,dy.$$

Similarly, $$d\left(\frac{\partial z}{dy}dy\right) = \frac{\partial^2 z}{dy\,dx}dy\,dx + \frac{\partial^2 z}{dy^2}dy^2.$$

Therefore

$$d^2z = \frac{\partial^2 z}{dx^2}dx^2 + 2\frac{\partial^2 z}{dx\,dy}dx\,dy + \frac{\partial^2 z}{dy^2}dy^2. \quad . \quad . \quad . \quad (2)$$

Differentiating again, since

$$d\left(\frac{\partial^2 z}{dx^2}dx^2\right) = \frac{\partial^3 z}{dx^3}dx^3 + \frac{\partial^3 z}{dx^2\,dy}dx^2\,dy,$$

$$d\left(2\frac{\partial^2 z}{dx\,dy}dx\,dy\right) = 2\frac{\partial^3 z}{dx^2\,dy}dx^2\,dy + 2\frac{\partial^3 z}{dx\,dy^2}dx\,dy^2,$$

$$d\left(\frac{\partial^2 z}{dy^2}dy^2\right) = \frac{\partial^3 z}{dy^2\,dx}dy^2\,dx + \frac{\partial^3 z}{dy^3}dy^3,$$

we obtain

$$\mathbf{d^3z = \frac{\partial^3 z}{dx^3}dx^3 + 3\frac{\partial^3 z}{dx^2\,dy}dx^2\,dy + 3\frac{\partial^3 z}{dy^2\,dx}dy^2\,dx + \frac{\partial^3 z}{dy^3}dy^3.}$$

Similarly, formulas for the total differentials of the higher orders may be obtained, the numerical coefficients of which will be found to follow the same law as those of the binomial formula. Thus the formula for the nth differential is

$$d^n z = \frac{\partial^n z}{dx^n}dx^n + n\frac{\partial^n z}{dx^{n-1}\,dy}dx^{n-1}dy + \frac{n(n-1)}{1\,.\,2}\,\frac{\partial^n z}{dx^{n-2}dy^2}dx^{n-2}dy^2$$

$$+ \ldots + n\frac{\partial^n z}{dx\,dy^{n-1}}dx\,dy^{n-1} + \frac{\partial^n z}{dy^n}dy^n,$$

otherwise written for abbreviation

$$d^n z = \left(\frac{\partial}{dx}dx + \frac{\partial}{dy}dy\right)^n z, \quad . \quad . \quad . \quad . \quad (a)$$

which form is not to be interpreted as usual, but as follows: Expand as indicated, regarding each term as a single quantity, and in the result replace each term, as $\left(\frac{\partial}{dx}dx\right)^n$, by $(\partial^n z/dx^n)dx^n$, and each combination, as

$$\left(\frac{\partial}{dx}dx\right)^{n-s}\left(\frac{\partial}{dy}dy\right)^{s}, \text{ by } \frac{\partial^n z}{dx^{n-s}dy^s}dx^{n-s}dy^s, \text{ etc.}$$

It is important to notice that in deriving eq. (2) from eq. (1) we write, in accordance with § 106,

$$\partial^2 z dx\, dy/dx\, dy = \partial^2 z dy\, dx/dy\, dx;$$

and it follows that having any expression in the form $Pdx + Qdy$, in order that it may be a *total* differential, it is necessary and sufficient that

$$\frac{\partial P}{dy} = \frac{\partial^2 z}{dx\, dy} = \frac{\partial^2 z}{dy\, dx} = \frac{\partial Q}{dx}.$$

This is known as **Euler's Test.** It determines whether or not the given expression is the result of the complete differentiation with respect to *both* variables of some other expression. Thus, having

$$2x\, dx + ydx + xdy + 2y\, dy,$$

we obtain

$$\frac{\partial P}{dy} = \frac{\partial}{dy}(2x + y) = \frac{\partial}{dx}(x + 2y) = \frac{\partial Q}{dx},$$

which shows that the given expression is a *total* differential of some function of x and y.

Having $2x\, dx + ydx + 2y\, dy$, the test fails and the expression is not a *total* differential of any function. By differentiating $x^2 + xy + y^2$, the student may confirm both results.

The successive total differentials of any function of any number of variables may be determined in a similar manner.

Thus, having $u = f(x, y, z)$, then

$$du = \frac{\partial u}{dx}dx + \frac{\partial u}{dy}dy + \frac{\partial u}{dz}dz.$$

$$d^2u = \frac{\partial^2 u}{dx^2}dx^2 + \frac{\partial^2 u}{dy^2}dy^2 + \frac{\partial^2 u}{dz^2}dz^2$$

$$+ 2\frac{\partial^2 u}{dx\,dy}dx\,dy + 2\frac{\partial^2 u}{dy\,dz}dy\,dz + 2\frac{\partial^2 u}{dx\,dz}dx\,dz.$$

etc. etc. etc.

$$d^n u = \left(\frac{\partial}{dx}dx + \frac{\partial}{dy}dy + \frac{\partial}{dz}dz\right)^n u.*$$

Rule. *Differentiate the function and obtain expressions for the several partial coefficients to the desired order. Substitute these in the proper formula for their respective symbols.*

EXAMPLES.

1. $z = x^3y^2$. $\qquad d^2z = 6xy^2dx^2 + 12x^2y\,dx\,dy + 2x^3dy^2$.

2. $z = (x^2 + y^2)/(x^2 - y^2)$.

$$d^2z = \frac{1}{(x^2 - y^2)^3}[4(y^4 + 3x^2y^2)dx^2 - 16xy(x^2 + y^2)dx\,dy$$
$$+ 4(x^4 + 3x^2y^2)dy^2].$$

3. $z = x^2y^2$. $\qquad d^2z = 2y^2dx^2 + 8xy\,dx\,dy + 2x^2dy^2$.

$$d^3z = 12y\,dx^2\,dy + 12x\,dx\,dy^2.$$

---

* Extension of the symbolic form (*a*), § 109.

4. $z = (x^2 + y^2)^{1/2}$.

$$d^2z = \frac{1}{(x^2 + y^2)^{3/2}}(y^2dx^2 - 2xy\,dx\,dy + x^2dy^2).$$

$$d^3z = \frac{1}{(x^2 + y^2)^{5/2}}(- 3xy^2dx^3 + 3y(2x^2 - y^2)dx^2dy$$

$$+ 3x(2y^2 - x^2)dx\,dy^2 - 3yx^2dy^3).$$

5. $z = e^{(ax+by)}$. $\quad d^2z = e^{(ax+by)}[a^2dx^2 + 2ab\,dx\,dy + b^2dy^2]$.

6. $z = x^3y^2 + y^3x^2$.

$$d^2z = (6xy^2 + 2y^3)dx^2 + 12(x^2y + xy^2)dx\,dy + (6x^2y + 2x^3)dy^2.$$

CHAPTER VII.

IMPLICIT FUNCTIONS AND DIFFERENTIAL EQUATIONS

IMPLICIT FUNCTIONS.

110. With the exceptions considered in § 73 and § 77, differentiation hitherto has been limited to *explicit* functions.

Let $y^2 = ax$, and assume x to be independent, then (§ 14) y is an *implicit* function of x.

Solving with respect to y, we have $y = \pm \sqrt{ax}$, which expresses y as an *explicit* function of x.

Differentiating, we have

$$dy = \pm\, adx/2\sqrt{ax}. \quad . \quad . \quad . \quad . \quad . \quad (1)$$

Otherwise, we may write

$$d(y^2) = adx, \quad \therefore \quad d(y^2)/dx = a, \quad . \quad . \quad . \quad (2)$$

in which $(y^2) = f(y)$, and $y = \phi(x)$. Therefore (§ 77)

$$\frac{d(y^2)}{dx} = \frac{d(y^2)}{dy} \cdot \frac{dy}{dx} = 2y\,\frac{dy}{dx}.$$

Substituting in first member of (2), we obtain

$$2y\,dy = adx, \quad . \quad . \quad (3) \qquad \therefore \qquad dy = \pm\, adx/2\sqrt{ax}, \quad . \quad (4)$$

which result corresponds with (1).

Examining (3) we see that it, and therefore (4), may be derived directly from $y^2 = ax$, by differentiating (y^2) as an *explicit* function of y, y, and in general dy, being functions of x.

That is, the equation may be solved with respect to y, and y differentiated as an *explicit* function, or we may differentiate, regarding y as an *implicit* function, and then solve the resulting *differential equation* with respect to dy.

This principle is general; for, having any equation containing two variables, x and y, it may be written

$$f(x, y) = \phi(x, y);$$

and regarding y as an *implicit* function of x, each member may be regarded as an *explicit* function of x, equal to the other for all values of x; therefore (§ 72) their differentials are equal. It is not essential that the value of the dependent variable shall be expressed in terms of the other, but it is necessary to remember which is assumed as the dependent and which the independent variable.

The advantage of differentiating without solving with respect to the implicit function beforehand increases, in general, with the degree of that function.

EXAMPLES.

1. $ay^2 - x^3 + bx = 0$. $dy/dx = \pm(3x^2 - b)/2\sqrt{a(x^3 - bx)}$.

2. $a^2y^2 + b^2x^2 = a^2b^2$. $dy/dx = -b^2x/a^2y$.

3. $(y + a)^2 = 4bx$. $dy/dx = \pm(b/x)^{1/2}$.

4. $\cos y = a \cos x$. $dy/dx = \tan x/\tan y$.

5. $\cos(x + y) = 0$. $dy/dx = -1$.

6. $x^y - y^x = 0$. $dy/dx = (y^2 - xy \log y)/(x^2 - xy \log x)$.

7. $(y - x^2)^2 = x^5$. $dy/dx = 2x \pm 5x^{3/2}/2$.

8. $ay^2 - x^3 + bx^2 = 0$. $dy/dx = \pm(3x - 2b)/2\sqrt{a(x - b)}$.

9. $ay^2 - x^3 + (b - c)x^2 + bcx = 0.$ $\quad \dfrac{dy}{dx} = \pm \dfrac{3x^2 - 2x(b-c) - bc}{2\sqrt{ax(x-b)(x+c)}}.$

10. $x^3 - 2x^2y - 2x^2 - 8y = 0.$ $\quad \dfrac{dy}{dx} = \dfrac{x(x^3 + 12x - 16)}{2(x^2+4)^2}.$

11. $y^2 - 2y\sqrt{a^2 + x^2} + x^2 = 0.$ $\quad dy/dx = x/\sqrt{a^2 + x^2}.$

12. $y \sin x - x \sin y + 1 = 0.$ $\quad \dfrac{dy}{dx} = \dfrac{\sin y - y \cos x}{\sin x - x \cos y}.$

13. $y^3 - 3yx^2 + 2x^3 = 0.$ $\quad dy/dx = 2x/(y + x).$

14. $xe^y - y + 1 = 0.$ $\quad dy/dx = e^y/(2 - y).$

15. $\log \dfrac{a + \sqrt{a^2 - y^2}}{y} - \dfrac{x + \sqrt{a^2 - y^2}}{a} = 0.$ $\quad \dfrac{dy}{dx} = -\dfrac{y}{\sqrt{a^2 - y^2}}.$

By continuing the differentiation in a similar manner, expressions for d^2y/dx^2, and derivatives of the higher orders, may be obtained.

16. $x^{2/3} + y^{2/3} = a^{2/3}.$ $\quad dy/dx = -y^{1/3}/x^{1/3},$

$$d^2y/dx^2 = (1/3x^{2/3})(1/y^{1/3} + y^{1/3}/x^{2/3}).$$

17. $x = r \operatorname{vers}^{-1}(y/r) - \sqrt{2ry - y^2}.$

$$dy/dx = (2r/y - 1)^{1/2}. \qquad d^2y/dx^2 = -r/y^2.$$

18. $a^2y^2 - b^2x^2 = -a^2b^2.$

$$\frac{dy}{dx} = \frac{b^2x}{a^2y} = \pm \frac{bx}{a\sqrt{x^2 - a^2}}. \qquad \frac{d^2y}{dx^2} = \frac{-b^4}{a^2y^3} = \frac{-ab}{(x^2 - a^2)^{3/2}}$$

111. Having $u = f(x, y) = 0$, and regarding y as an implicit function of x, we write (§ 102)*

---

* Also u may be differentiated with respect to x by differentiating it as a function of x and y (§ 101), and since y is a function of x $\partial u/dy$ must be multiplied by dy/dx (§ 77).

$$\frac{du}{dx} = \frac{\partial u}{dx} + \frac{\partial u}{\partial y}\frac{dy}{dx} = 0, \quad . \quad . \quad . \quad . \quad (1)$$

whence
$$\frac{dy}{dx} = -\frac{\partial u}{dx} \Big/ \frac{\partial u}{\partial y}. \quad . \quad . \quad . \quad . \quad . \quad (2)$$

EXAMPLES.

1. $u = y^3 - 3yx^2 + 2x^3 = 0.$

$$\frac{\partial u}{dx} = -6yx + 6x^2, \quad \frac{\partial u}{\partial y} = 3y^2 - 3x^2, \quad \frac{dy}{dx} = -\frac{-6yx + 6x^2}{3y^2 - 3x^2} = \frac{2x}{x+y}.$$

2. $y^2 - 2xy + a^2 = 0.$ $\qquad dy/dx = y/(y-x).$

3. $y^4 + 3a^2y^2 - 4a^2xy - a^2x^2 = 0.$ $\qquad \dfrac{dy}{dx} = \dfrac{(2y+x)a^2}{2y^3 + 3a^2y - 2a^2x}.$

4. $x^5 - ax^3y + bx^2y^2 - y^5 = 0.$ $\qquad \dfrac{dy}{dx} = \dfrac{5x^4 - 3ax^2y + 2bxy^2}{5y^4 - 2bx^2y + ax^3}.$

5. $y^2 - x^3 - 2ax^2 - a^2x = 0.$ $\qquad \dfrac{dy}{dx} = \dfrac{3x^2 + 4ax + a^2}{2y}.$

6. $y^2 - x^2(1 - x^2) = 0.$ $\qquad dy/dx = (x - 2x^3)/y.$

7. $x^4 - 2ay^3 - 3a^2y^2 - 2a^2x^2 + a^4 = 0.$

$$dy/dx = 2x(x^2 - a^2)/3ay(y+a).$$

8. $(4y - x^3)^2 - (x-4)^5(x-3)^6 = 0.$

$$\frac{dy}{dx} = \frac{3x^2}{4} + \frac{5(x-4)^4(x-3)^6}{8(4y - x^3)} + \frac{3(x-4)^5(x-3)^5}{4(4y-x^3)}.$$

Differentiating (1), remembering that

$$\frac{d}{dx}\left(\frac{\partial u}{dx}\right) = \frac{\partial^2 u}{dx^2} + \frac{\partial^2 u}{dx\,\partial y}\frac{dy}{dx},$$

$$\frac{d}{dx}\left(\frac{\partial u}{\partial y}\right) = \frac{\partial^2 u}{\partial y\,dx} + \frac{\partial^2 u}{\partial y^2}\frac{dy}{dx},$$

we have

$$\frac{d^2u}{dx^2} = \frac{\partial^2 u}{dx^2} + \frac{\partial^2 u}{dx\,\partial y}\frac{dy}{dx} + \left(\frac{\partial^2 u}{\partial y\,dx} + \frac{\partial^2 u}{\partial y^2}\frac{dy}{dx}\right)\frac{dy}{dx} + \frac{\partial u}{\partial y}\frac{d^2y}{dx^2} = 0,$$

or $$\frac{\partial^2 u}{dx^2} + 2\frac{\partial^2 u}{dx\,\partial y}\frac{dy}{dx} + \frac{\partial^2 u}{\partial y^2}\left(\frac{dy}{dx}\right)^2 + \frac{\partial u}{\partial y}\frac{d^2y}{dx^2} = 0. \quad . \quad (3)$$

Hence,

$$\frac{d^2y}{dx^2} = -\left[\frac{\partial^2 u}{dx^2} + 2\frac{\partial^2 u}{dx\,\partial y}\frac{dy}{dx} + \frac{\partial^2 u}{\partial y^2}\left(\frac{dy}{dx}\right)^2\right] \Big/ \frac{\partial u}{\partial y}. \quad . \quad . \quad . \quad (4)$$

Substituting expression for dy/dx from (2), and simplifying, we have

$$\frac{d^2y}{dx^2} = -\left[\frac{\partial^2 u}{dx^2}\left(\frac{\partial u}{\partial y}\right)^2 - 2\frac{\partial^2 u}{dx\,\partial y}\frac{\partial u}{dx}\frac{\partial u}{\partial y} + \frac{\partial^2 u}{\partial y^2}\left(\frac{\partial u}{dx}\right)^2\right] \Big/ \left(\frac{\partial u}{\partial y}\right)^3. \quad (5)$$

If $\dfrac{dy}{dx} = 0$, (4) gives $\dfrac{d^2y}{dx^2} = -\dfrac{\partial^2 u}{dx^2} \Big/ \dfrac{\partial u}{\partial y}$. . . . . (6)

EXAMPLES.

1. $x^3 - 3axy + y^3 = 0$.

$$\frac{\partial u}{dx} = 3(x^2 - ay), \qquad \frac{\partial u}{\partial y} = 3(-ax + y^2).$$

$$\frac{\partial^2 u}{dx^2} = 6x. \qquad \frac{\partial^2 u}{dx\,\partial y} = -3a, \qquad \frac{\partial^2 u}{\partial y^2} = 6y.$$

$$\frac{dy}{dx} = \frac{ay - x^2}{y^2 - ax}, \quad \frac{d^2y}{dx^2} = -\frac{2a^3xy}{(y^2 - ax)^3}.$$

If $dy/dx = 0$, (6) gives

$$\frac{d^2y}{dx^2} = -\frac{6x}{3(-ax + y^2)} = \frac{2x}{ax - y^2}.$$

2. $x^2 + y^2 - 2bxy - a^2 = 0.$

$$dy/dx = (x - by)/(bx - y),$$

$$d^2y/dx^2 = (b^2 - 1)a^2/(y - bx)^3.$$

If $dy/dx = 0$, $d^2y/dx^2 = -1/a\sqrt{1 - b^2}$.

3. $2xy - y^2 - a^2 = 0.$

$$dy/dx = y/(y - x),$$

$$d^2y/dx^2 = y(y - 2x)/(y - x)^3.$$

4. $x^3 + 3axy + y^3 = 0.$

$$dy/dx = -(x^2 + ay)/(y^2 + ax),$$

$$d^2y/dx^2 = 2a^3xy/(y^2 + ax)^3.$$

5. $y^2 - x^3/(2a - x) = 0.$

$$dy/dx = \pm(3a - x)\sqrt{x}/(2a - x)^{\frac{3}{2}},$$

$$d^2y/dx^2 = \pm 3a^2/(2a - x)^{\frac{5}{2}}\sqrt{x}.$$

Differentiating (3), we obtain

$$\frac{d^3u}{dx^3} = \frac{\partial^3 u}{dx^3} + 3\frac{\partial^3 u}{dx^2\partial y}\frac{dy}{dx} + 3\frac{\partial^3 u}{dx\,\partial y^2}\left(\frac{dy}{dx}\right)^2 + \frac{\partial^3 u}{\partial y^3}\left(\frac{dy}{dx}\right)^3$$

$$+ 3\left[\frac{\partial^2 u}{dx\partial y} + \frac{\partial^2 u}{\partial y^2}\frac{dy}{dx}\right]\frac{d^2y}{dx^2} + \frac{\partial u}{\partial y}\frac{d^3y}{dx^3} = 0. \quad . \; . \quad (7)$$

Similarly, it may be shown that

$$\frac{d^n u}{dx^n} = \frac{\partial^n u}{dx^n} + \ldots + \frac{\partial u}{\partial y}\frac{d^n y}{dx^n} = 0, \quad . \; . \quad (8)$$

in which the intermediate terms involve differential coefficients of y with respect to x of orders less than the nth.

112. Having any equation containing three variables

x, y and z, one must be an implicit function of the other two (§ 9). Each member may, therefore, be regarded as a function of but two independent variables.

The differential of each member is equal to the sum of its partial differentials; and since the partial differentials of the two members are respectively equal to each other, the total differentials are equal.

It is not necessary to express the implicit function or dependent variable in terms of the others, but it is always important to distinguish it.

In a similar manner it may be shown that the total differentials of the members of any equation are equal, remembering that the number of *independent* variables is one less than the number of variables.

DIFFERENTIAL EQUATIONS.

113. An equation which contains differential coefficients is called a **differential equation.**

A differential equation obtained by one differentiation may, in general, be differentiated again, giving a differential equation of the *second* order, and so on to differential equations of the *third*, etc., orders.

Thus, having $y^2 - 2mxy + x^2 = a^2$, regarding x as the independent variable, differentiating, equating the results, and reducing, we have

$$y\,dy - mx\,dy - my\,dx + x\,dx = 0,$$

or $$dy/dx = (my - x)/(y - mx).$$

Differentiating again,* we have

* The importance of distinguishing between the independent and dependent variables becomes apparent at the second operation of differentiation—as in above dx is a constant, whereas $dy = f(x)$.

$$dy^2 + y\,d^2y - mx\,d^2y - 2m\,dx\,dy + dx^2 = 0,$$

or, dividing by dx^2,

$$(y - mx)\,d^2y/dx^2 + dy^2/dx^2 - 2m\,dy/dx + 1 = 0.$$

The *order* of a differential equation is the same as that of the highest derivative it contains, and its *degree* is denoted by the greatest exponent of the derivative of the highest order in any term; provided that all such exponents are entire and positive. Thus,

$(dy/dx)^2 - a/x = 0$ is of the 1st order and 2d degree.

$d^2y/dx^2 + a^2y \quad = 0$ is of the 2d order and 1st degree.

$(d^ny)/dx^n)^m + M(d^{n-1}y/dx^{n-1})^{m-1} + \ldots = 0$ is of the nth order and m^{th} degree.

EXAMPLES.

1. $y^2 + x^2 = r^2$. $\quad dy/dx = -x/y, \quad d^2y/dx^2 = -r^2/y^3$.

2. $y^2 = 2px$. $\quad dy/dx = p/y, \quad d^2y/dx^2 = -p^2/y^3$.

3. $a^2y^2 + b^2x^2 = a^2b^2$. $\quad dy/dx = -b^2x/a^2y. \quad d^2y/dx^2 = -b^4/a^2y^3$.

4. $y^3 = a^2x$. $\quad dy/dx = a^2/3y^2. \quad d^2y/dx^2 = -2a^4/9y^5$.

5. $a^2x = y(x-a)^2$. $\quad \dfrac{dy}{dx} = -\dfrac{a^2(a+x)}{(x-a)^3}, \quad \dfrac{d^2y}{dx^2} = \dfrac{2a^2(x+2a)}{(x-a)^4}$.

6. $y^2 = bx^3$. $\quad dy/dx = 3bx^2/2y. \quad d^2y/dx^2 = 3b^2x^4/4y^3$.

7. $y^2 = 2px + r^2x^2$. $\quad \dfrac{dy}{dx} = \dfrac{(p + r^2x)}{y}, \quad \dfrac{d^2y}{dx^2} = -\dfrac{p^2}{y^3}$.

8. $x^2y = 4a^2(2a - y)$. $\quad dy/dx = -2xy/(x^2 + 4a^2),$

$d^2y/dx^2 = 2y(3x^2 - 4a^2)/(x^2 + 4a^2)^2$.

9. $y^3 - 3axy + x^3 = 0.$ $(y^2 - ax)\dfrac{d^2y}{dx^2} + 2y\left(\dfrac{dy}{dx}\right)^2 - 2a\dfrac{dy}{dx} + 2x = 0.$

10. $\log(xy) + x - y = a.$ $(x - xy)\dfrac{dy}{dx} + y + xy = 0.$

11. $y^3 - 3y + x = a.$ $(y^2 - 1)\dfrac{d^2y}{dx^2} + 2y\left(\dfrac{dy}{dx}\right)^2 = 0.$

12. $y^2 - 2xy + a^2 = 0.$ $d^2y/dx^2 = y(y - 2x)/(y - x)^3.$

13. $x^3 + 3axy + y^3 = 0.$ $d^2y/dx^2 = 2a^3 xy/(y^2 + ax)^3.$

14. $x^2y = 4a^2(2a - y).$ $\dfrac{d^2y}{dx^2} = \dfrac{16a^3(3x^2 - 4a^2)}{(x^2 + 4a)^3}.$

15. $y^3 = 2ax^2 - x^3.$ $d^2y/dx^2 = -8a^2/9x^{4/3}(2a - x)^{5/3}.$

16. $x^3 - xy + 1 = 0.$ $d^2y/dx^2 = 2(1 + 1/x^3).$

17. $\log(x + y) = x - y.$ $d^2y/dx^2 = 4(x + y)/(x + y + 1)^3.$

18. $x^3 - 2x^2y - 2x^2 - 8y = 0.$ $\dfrac{d^2y}{dx^2} = 4\dfrac{-x^3 + 6x^2 + 12x - 8}{(x^2 + 4)^3}.$

It is important to notice the difference between the successive differentials of an independent expression which contains variables and those of the same expression limited to a constant value. Thus, suppose we have $(x^2 + y^2)$ unlimited, and $(x^2 + y^2) = a^2$.

Then,

$d(x^2+y^2) = 2x\,dx + 2y\,dy$, and $d(x^2+y^2) = 2x\,dx + 2y\,dy = 0.$

$d^2(x^2+y^2) = 2dx^2 + 2dy^2$, and $d^2(x^2+y^2) = 2dx^2 + 2dy^2 + 2yd^2y = 0.$

$d^3(x^2+y^2) = 0$, and $d^3(x^2+y^2) = 4dy\,d^2y + 2dy\,d^2y + 2y\,d^3y = 0.$

etc. etc. etc.

In the first case both variables are independent, but in the second only one. The difference becomes apparent at the second differentiation.

Equations derived by differentiating primitive equations or other differential equations are called *immediate differential equations.*

114. Differential equations also arise by combining successive immediate differential equations with each other and the primitive equation in such a manner as to eliminate certain *constants*, or *particular functions* which enter the primitive equation. Thus, from

$$y = ax + b \quad \text{we obtain} \quad dy/dx = a.$$

Eliminating a, $\quad dy/dx = (y - b)/x,$

which is independent of a.

Differentiating again, we have $d^2y/dx^2 = 0$, which is independent of both a and b.

As another example, take the equation of a circle

$$(x - a)^2 + (y - b)^2 = R^2. \quad . \quad . \quad . \quad . \quad (1)$$

Differentiating three times, we have

$$2(x - a)dx + 2(y - b)dy = 0. \quad . \quad . \quad . \quad (2)$$

$$dx^2 + dy^2 + (y - b)d^2y = 0. \quad . \quad . \quad . \quad . \quad (3)$$

$$2dy\, d^2y + dy\, d^2y + (y - b)d^3y = 0. \quad . \quad . \quad (4)$$

Dividing (3) by (4), member by member, we have

$$(dx^2 + dy^2)/3dy\, d^2y = d^2y/d^3y, \quad \text{which gives}$$

$$\frac{dx^2+dy^2}{dx^2}\bigg/\frac{d^2y}{dx^2}=3\frac{dy}{dx}\frac{d^2y}{dx^2}\bigg/\frac{d^3y}{dx^3}, \quad \text{and}$$

$$\frac{d^3y}{dx^3}\left[\left(\frac{dy}{dx}\right)^2+1\right]-3\left(\frac{d^2y}{dx^2}\right)^2\frac{dy}{dx}=0,$$

which is independent of a, b and R.

And, in general, by differentiating an equation n times we obtain n differential equations between which and the primitive equation n constants or particular functions may be eliminated, giving a differential equation of the nth order.

EXAMPLES.

1. Eliminate e^x and $\sin x$ from $y = e^x \sin x$.

$$dy/dx = e^x \sin x + e^x \cos x = y + e^x \cos x,$$

$$d^2y/dx^2 = dy/dx + e^x \cos x - e^x \sin x;$$

and since $e^x \cos x = dy/dx - y$,

$$d^2y/dx^2 = 2dy/dx - 2y.$$

2. $y = a \sin x + b \sin 2x$.

$$dy/dx = a \cos x + 2b \cos 2x,$$

$$d^2y/dx^2 = - a \sin x - 4b \sin 2x,$$

$$d^3y/dx^3 = - a \cos x - 8b \cos 2x,$$

$$d^4y/dx^4 = a \sin x + 16b \sin 2x.$$

Hence, $\quad d^4y/dx^4 + 5d^2y/dx^2 + 4y = 0.$

3. $y^3 - 2ax^2 = - a^2$. $\quad 9y^4(dy/dx)^2 - 24x^3\, dy/dx = - 16yx^2$.

4. $xy - ax^4 - b = 0$. $\quad x^2d^2y/dx^2 - xdy/dx = 3y$.

5. $(1 + x^2)(1 + y^2) = ax^2$. $\quad (1 + x^2)xy\, dy/dx = 1 + y^2$.

6. $y = a \sin x - b \cos x.$ $dy^2/dx^2 = -y.$

7. $y = ax + a - a^3.$ $(x + 1)dy/dx - (dy/dx)^3 = y.$

8. $y = (a + bx)e^{cx}.$ $d^2y/dx^2 - 2c\,dy/dx = -c^2y.$

9. $y = \sin x.$ $(dy/dx)^2 + y^2 = 1.$

10. $y = e^x \cos x.$ $d^2y/dx^2 = 2dy/dx - 2y.$

11. $y = \sin \log x.$ $x^2d^2y/dx^2 + x\,dy/dx + y = 0.$

12. $y = \log \sin x.$ $d^2y/dx^2 + (dy/dx)^2 = -1.$

13. $y^2 = 2px + r^2x^2.$ $yx^2\dfrac{d^2y}{dx^2} + x^2\left(\dfrac{dy}{dx}\right)^2 - 2xy\dfrac{dy}{dx} + y^2 = 0.$

14. $y = a \cos (bx + c).$ $d^2y/dx^2 = -b^2y.$

15. $y = \sin^{-1} x.$ $(1 - x^2)d^2y/dx^2 - x\,dy/dx = 0.$

16. $y = cx + \sqrt{1 + c^2}.$ $y = xdy/dx + \sqrt{1 + (dy/dx)^2}.$

17. $y = (x + c)e^{ax}.$ $dy/dx - ay = e^{ax}.$

18. $y = ce^{-\tan^{-1}x} + \tan^{-1} x - 1.$ $(1 + x^2)dy/dx + y = \tan^{-1}x.$

19. $y = (cx + \log x + 1)^{-1}.$ $xdy/dx + y = y^2 \log x.$

20. $y^2 - 2cx - c^2 = 0.$ $y(dy/dx)^2 + 2xdy/dx = y.$

21. $\begin{cases} y = c \cos mx + c' \sin mx. \\ y = c \cos (mx + c'). \end{cases}$ $d^2y/dx^2 + m^2y = 0.$

22. $y = x \log [(c + c'x)/x].$ $x^3d^2y/dx^2 + (y - xdy/dx)^2 = 0.$

23. $y = x \sin nx/2n + c \cos nx + c' \sin nx.$

$$d^2y/dx^2 + n^2y = \cos nx$$

24. $y^2 \sin^2 x + 2ay + a^2 = 0.$ $(dy/dx)^2 + 2y \cot xdy/dx = y^2.$

25. $y = (e^x + e^{-x})/(e^x - e^{-x}).$ $y^2 - 1 + dy/dx = 0.$

26. $e^{2y} + 2axe^y + a^2 = 0.$ $(x^2 - 1)(dy/dx)^2 = 1.$

27. $y = ae^{cx} + be^{-cx}$. $d^2y/dx^2 = c^2y$.

28. $y = ae^{2x} + be^{-3x} + ce^{x}$. $d^3y/dx^3 - 7dy/dx = -6y$.

29. $y = (a + bx + x^2/2)e^{x} + c$. $d^3y/dx^3 - 2d^2y/dx^2 + dy/dx = e^{x}$.

30. $y = (a + bx + cx^2)e^{x} + d$. $d^4y/dx^4 - 3d^3y/dx^3 + 3d^2y/dx^2 - dy/dx = 0$.

CHAPTER VIII.

CHANGE OF THE INDEPENDENT VARIABLE.

115. Having any expression or equation containing differentials, or derivatives, of y regarded as a function of x, it is sometimes desirable to obtain a corresponding expression or equation, in which y, or some other variable upon which y and x depend, is the independent one. This operation is called *changing the independent variable.*

The principle deduced in § 73 enables us to make the change in cases involving the first derivative only.

When differentials of a higher order are involved it must be remembered that we have written

$$\frac{d}{dx}\left(\frac{dy}{dx}\right) = \frac{d^2y}{dx^2}, \quad \text{and} \quad \frac{d}{dx}\left(\frac{d^2y}{dx^2}\right) = \frac{d^3y}{dx^3},$$

in which x is the independent variable, and dx is a constant.

Regarding both dy and dx as variable, we have

$$\frac{d}{dx}\left(\frac{dy}{dx}\right) = \frac{dx\,d^2y - dy\,d^2x}{dx^3}, \quad . \quad . \quad . \quad . \quad (1)$$

and

$$\frac{d}{dx}\cdot\frac{d}{dx}\left(\frac{dy}{dx}\right) = \frac{d}{dx}\left(\frac{dx\,d^2y - dy\,d^2x}{dx^3}\right)$$

$$= \frac{(dx\,d^3y - dy\,d^3x)dx + 3(dy\,d^2x - dx\,d^2y)d^2x}{dx^6}, \quad (2)$$

which should be substituted for d^2y/dx^2 and d^3y/dx^3, respectively, in order that the results may be general, that is, in which neither x nor y is independent.

If then y is made independent, we have $dy =$ constant, $d^2y = 0$, $d^3y = 0$, and (1) and (2) reduce to

$$\frac{d^2y}{dx^2} = \frac{-dy\,d^2x}{dx^3} = -\frac{d^2x}{dy^2}\Big/\left(\frac{dx}{dy}\right)^3, \quad . \; . \; . \quad (3)$$

$$\frac{d^3y}{dx^3} = \frac{3dy(d^2x)^2 - dy\,dx\,d^3x}{dx^5}$$

$$= \left[3\left(\frac{d^2x}{dy^2}\right)^2 - \frac{d^3x}{dy^3}\frac{dx}{dy}\right]\Big/\left(\frac{dx}{dy}\right)^5, \quad . \; . \quad (4)$$

which may be used when the independent variable is changed from x to y.

EXAMPLES.

Change the independent variable from x to y in the following:

1. $\dfrac{d^2y}{dx^2} + \left(\dfrac{dy}{dx}\right)^2 = 0.$ $\qquad -\dfrac{d^2x}{dy^2} + \dfrac{dx}{dy} = 0.$

2. $\dfrac{d^2y}{dx^2} + 2y\left(\dfrac{dy}{dx}\right)^2 = 0.$ $\qquad \dfrac{-d^2x}{dy^2} + 2y\dfrac{dx}{dy} = 0.$

3. $y\dfrac{d^2y}{dx^2} + \dfrac{dy^2}{dx^2} + 1 = 0.$ $\qquad y\dfrac{d^2x}{dy^2} - \dfrac{dx^3}{dy^3} - \dfrac{dx}{dy} = 0.$

4. $x\dfrac{d^2y}{dx^2} + \left(\dfrac{dy}{dx}\right)^3 - \dfrac{dy}{dx} = 0.$ $\qquad x\dfrac{d^2x}{dy^2} + \left(\dfrac{dx}{dy}\right)^2 - 1 = 0.$

5. $(dy^2 + dx^2)^{3/2} + a\,dx\,d^2y = 0.$ $\qquad (1 + dx^2/dy^2)^{3/2} - a\,d^2x/dy^2 = 0.$

If x or y is given as a function of a third variable, θ, which we wish to make the independent one, we first transform the given expression, by means of (1) and (2), into its general form in which neither x nor y is independent, and then substitute for x, dx, d^2x, d^3x, or y, dy, d^2y, d^3y, their values in terms of θ and its differential.

6. Having $t = x + x^2$, show that

$$\frac{d^2y}{dx^2} = (4t + 1)\frac{d^2y}{dt^2} + 2\frac{dy}{dt}.$$

Change the independent variable from x to θ in the following equations:

7. $\dfrac{d^2y}{dx^2} - \dfrac{x}{1 - x^2}\dfrac{dy}{dx} + \dfrac{y}{1 - x^2} = 0$, when $x = \cos\theta$.

$$\frac{d^2y}{d\theta^2} + y = 0.$$

8. $x^2\dfrac{d^2y}{dx^2} + 2x\dfrac{dy}{dx} + \dfrac{a^2}{x^2}y = 0$, when $x = \dfrac{1}{\theta}$.

$$\frac{d^2y}{d\theta^2} + a^2y = 0.$$

9. $\dfrac{d^2y}{dx^2} + \dfrac{1}{x}\dfrac{dy}{dx} + y = 0$, when $x^2 = 4\theta$.

$$\frac{d^2y}{d\theta^2}\theta + \frac{dy}{d\theta} + y = 0.$$

10. $x^2\dfrac{d^2y}{dx^2} + ax\dfrac{dy}{dx} + by = 0$, when $x = e^\theta$.

$$\frac{d^2y}{d\theta^2} + (a - 1)\frac{dy}{d\theta} + by = 0.$$

11. Having $x = r\cos\theta$, and $y = r\sin\theta$, change the independent variable from x to θ in the equation

$$R = \left(1 + \frac{dy^2}{dx^2}\right)^{3/2} \bigg/ \frac{d^2y}{dx^2}. \quad . \quad . \quad . \quad . \quad . \quad . \quad (a)$$

From (1) we have $\qquad R = \dfrac{(dx^2 + dy^2)^{3/2}}{dx\,d^2y - dy\,d^2x};$

and, since $d^2\theta = 0$,

$$dx = \cos\theta\,dr - r\sin\theta\,d\theta,$$

$$dy = \sin\theta\,dr + r\cos\theta\,d\theta,$$

$$d^2x = \cos\theta\,d^2r - 2\sin\theta\,dr\,d\theta - r\cos\theta\,d\theta^2,$$

$$d^2y = \sin\theta\,d^2r + 2\cos\theta\,dr\,d\theta - r\sin\theta\,d\theta^2,$$

$$dx^2 + dy^2 = dr^2 + r^2d\theta^2,$$

$$d^2x\,dy - d^2y\,dx = rd^2rd\theta - 2dr^2d\theta - r^2d\theta^3.$$

Substituting these values, we have

$$R = \left(\frac{dr^2}{d\theta^2} + r^2\right)^{3/2} \Big/ \left(- r\frac{d^2r}{d\theta^2} + 2\frac{dr^2}{d\theta^2} + r^2\right).$$

12. Having $x = a(\phi - \sin\phi)$ and $y = a(1 - \cos\phi)$, change the independent variable in eq. (a) from x to ϕ.

Ans. $R = - 4a \sin(\phi/2)$.

13. Having $x = r\cos\theta$, and $y = r\sin\theta$, transform

$$z = (xdy - ydx)/(ydy + xdx)$$

so that r will be independent. Ans. $z = rd\theta/dr$.

14. Having $dy/dx = 3(x - 2)^2$, show that $d^2x/dy^2 = - 2/9(x - 2)^5$.

15. $\dfrac{d^2y}{dx^2} - \dfrac{x}{1 - x^2}\dfrac{dy}{dx} = 0$, when $x = \sin\theta$;

$$\frac{d^2y}{d\theta^2} = 0.$$

16. $\dfrac{d^2y}{dx^2} + \dfrac{2x}{a^2 + x^2}\dfrac{dy}{dx} = 0$, when $x = a\tan\theta$;

$$\frac{d^2y}{d\theta^2} = 0.$$

PART II.

ANALYTIC APPLICATIONS.

CHAPTER IX.

LIMITS OF FUNCTIONS WHICH ASSUME INDETERMINATE FORMS.

116. The symbols

$$0/0, \quad \infty/\infty, \quad 0.\infty, \quad \infty - \infty, \quad 0^0, \quad \infty^0, \quad 1^\infty$$

are indeterminate forms.

When for a particular value of the variable a function of the variable assumes any one of the above forms its corresponding *value* is indeterminate.

The *limit* of such a function, under the law that the variable approaches the particular value, is determinate. (§ 39.)

In many cases this limit may be found by simple algebraic methods, otherwise a method of the Calculus is generally used.*

117. Form 0/0.

Let $fx/\phi x$ be a fraction such that both terms vanish when $x = a$. It is required to find $\underset{x \ggg\!\rightarrow a}{\text{limit}} \dfrac{fx}{\phi x}$.

---

* Any operation by which this limit is determined is generally called the evaluation of the corresponding indeterminate form.

From eq. (1), § 62, we write

$$f(x+h) = fx + hf'(x+\theta h),$$
$$\phi(x+h) = \phi x + h\phi'(x+\theta' h).$$

Hence

$$\frac{f(a+h)}{\phi(a+h)} = \frac{f'(a+\theta h)}{\phi'(a+\theta' h)},$$

from which, as $h \ggg\!\!\rightarrow 0$, we have

$$fa/\phi a = f'a/\phi' a,$$

or

$$\underset{x \ggg\!\!\rightarrow a}{\text{limit}} \frac{fx}{\phi x} = \lim \frac{f'x}{\phi' x} = \frac{f'a}{\phi' a}.$$

Similarly we may deduce

$$\underset{x \ggg\!\!\rightarrow a}{\text{limit}} \frac{f'x}{\phi' x} = \lim \frac{f''x}{\phi'' x} = \ldots = \lim \frac{f^n x}{\phi^n x} = \frac{f^n a}{\phi^n a},$$

and
$$\underset{x \ggg\!\!\rightarrow a}{\text{limit}} \frac{fx}{\phi x} = \frac{f^n a}{\phi^n a},$$

$f^n x$ and $\phi^n x$ representing, respectively, the derivatives of the numerator and denominator of the *same* and *lowest* order, both of which do not vanish when $x = a$.

If $f^n a$ and $\phi^n a$ are both finite, the limit is finite.

If $f^n a$ is zero, and $\phi^n a$ is not zero, the limit is zero.

If $f^n a$ is not zero, and $\phi^n a$ is, the limit is infinite.

If $f^n a$ and $\phi^n a$ are both infinite, the limit is undetermined.

Therefore the required limit is the ratio, corresponding to the particular value of the variable, of the derivatives of the numerator and denominator, of the *same* and *lowest* order, both of which do not vanish or become infinite.

Each ratio as obtained should be carefully inspected and factors common to both terms should be cancelled if possible before proceeding to the next ratio.

EXAMPLES.

1. $\underset{x \to a}{\text{limit}} \left[\dfrac{x^m - a^m}{x^n - a^n}\right] = \text{lim.} \left[\dfrac{mx^{m-1}}{nx^{n-1}}\right] = \dfrac{m}{n}a^{m-n}.$

2. $\underset{x \to a}{\text{limit}} \left[\dfrac{(x-a)^m}{(x-a)^n}\right] = \text{lim.} \left[\dfrac{m(x-a)^{m-1}}{n(x-a)^{n-1}}\right] = \begin{cases} 0, \text{ if } m > n. \\ 1, \text{ if } m = n \\ \infty, \text{ if } m < n. \end{cases}$

3. $(\sin x/x)_{x=0} = \cos x\Big]_0 = 1.$*

4. $(e^x - 1)/x\Big]_0 = e^x\Big]_0 = 1.$

5. $\tan x/x\Big]_0 = \sec^2 x\Big]_0 = 1.$

6. $(x^3 - a^3)/(x^2 - a^2)\Big]_a = 3x^2/2x\Big]_a = 3a/2.$

7. $(a^2 - x^2)/(a-x)^2\Big]_a = 2x/2(a-x)\Big]_a = \infty.$

8. $(x-a)^{3/4}/(a-x)^{2/3}\Big]_a = \frac{3}{4}(x-a)^{1/3}/\frac{2}{3}(x-a)^{1/4}\Big]_a = 0.$

9. $(1 - \sin x)/\cos x\Big]_{\pi/2} = \cos x/\sin x\Big]_{\pi/2} = 0.$

10. $(e^x - e^{-x})/\sin x\Big]_0 = (e^x + e^{-x})/\cos x\Big]_0 = 2.$

11. $x^2/\sin x\Big]_0 = 2x/\cos x\Big]_0 = 0.$

12. $\dfrac{x^2}{\sin x} \Big/ \dfrac{2x}{\cos x}\Bigg]_0 = \dfrac{x \cos x}{2 \sin x}\Bigg]_0 = \dfrac{\cos x - x \sin x}{2 \cos x}\Bigg]_0 = \dfrac{1}{2}.$

13. $\dfrac{x - \sin x}{x^3}\Bigg]_0 = \dfrac{1 - \cos x}{3x^2}\Bigg]_0 = \dfrac{\sin x}{6x}\Bigg]_0 = \dfrac{\cos x}{6}\Bigg]_0 = \dfrac{1}{6}.$

14. $\dfrac{1 - \cos x}{\sin^2 x}\Bigg]_0 = \dfrac{\sin x}{2 \sin x \cos x}\Bigg]_0 = \dfrac{1}{2}.$

---

* Hereafter, for abbreviation, the forms $f(x)_{x=a}$ and $fx\Big]_a$ will frequently be used to express $\underset{x \to a}{\text{limit}} (fx)$.

15. $$\left.\frac{e^x - 2\sin x - e^{-x}}{x - \sin x}\right]_0 = \left.\frac{e^x - 2\cos x + e^{-x}}{1 - \cos x}\right]_0 = \left.\frac{e^x + 2\sin x - e^{-x}}{\sin x}\right]_0$$
$$= \left.\frac{e^x + 2\cos x + e^{-x}}{\cos x}\right]_0 = 4.$$

16. $$\left.\frac{e^{mx} - e^{ma}}{(x-a)^r}\right]_a = \left.\frac{me^{mx}}{r(x-a)^{r-1}}\right]_a = \begin{cases} \infty \text{ when } r > 1. \\ 0 \text{ when } r < 1. \end{cases}$$

Limits of factors of the given or any derived ratio may be determined separately (§ 36).

17. $$\left.\frac{\sqrt{x}\tan x}{\sqrt{(e^x - 1)^3}}\right]_0 = \sqrt{\left.\frac{x}{e^x - 1}\right]_0} \left.\frac{\tan x}{x}\right]_0 \left.\frac{x}{e^x - 1}\right]_0 = 1.$$

The given or any derived ratio may be separated into parts (§ 35).

18. $$\left.\frac{\sqrt{x} - \sqrt{a} + \sqrt{x-a}}{\sqrt{x^2 - a^2}}\right]_a = \left.\left(\frac{1}{2\sqrt{x}} + \frac{1}{2\sqrt{x-a}}\right) \Big/ \frac{x}{\sqrt{x^2 - a^2}}\right]_a$$
$$= \left.\frac{\sqrt{x^2 - a^2}}{2x\sqrt{x}} + \frac{\sqrt{x^2 - a^2}}{2x\sqrt{x-a}}\right]_a$$
$$= \left.\frac{1}{2x}\sqrt{\frac{x^2 - a^2}{x - a}}\right]_a = \frac{1}{\sqrt{2a}}.$$

19. $$\left.\frac{\tan x - \sin x}{x^3}\right]_0 = \left.\left(\frac{\sin x}{x}\right)\left(\frac{\sec x - 1}{x^2}\right)\right]_0 = \left.\frac{\sec x - 1}{x^2}\right]_0$$
$$= \left.\frac{\sec x \tan x}{2x}\right]_0 = \left.\frac{\sec^3 x + \tan^2 x \sec x}{2}\right]_0 = \frac{1}{2}.$$

20. $$\left.\frac{\tan x - \sin x}{x^n}\right]_0 = \left.\left(\frac{\tan x}{x}\right)\left(\frac{1 - \cos x}{x^{n-1}}\right)\right]_0 = \left.\frac{1 - \cos x}{x^{n-1}}\right]_0$$
$$= \left.\frac{\sin x}{(n-1)x^{n-2}}\right]_0 = \left.\frac{\cos x}{(n-1)(n-2)x^{n-3}}\right]_0 = \infty.$$

21. $\log(1 + x)/x\big]_0 = 1/(1 + x)\big]_0 = 1.$

In some cases it is advisable to transform the terms before applying the above rule. Thus—

22. $$\left.\frac{\sin x}{1 - \cos x}\right]_0 = \left.\frac{2\sin(x/2)\cos(x/2)}{2\sin^2(x/2)}\right]_0 = \cot(x/2)\big]_0 = \infty.$$

118. Form ∞/∞.

If $fa = \infty = \phi a$, we may write

$$\frac{fa}{\phi a} = \frac{1}{\phi a} \Big/ \frac{1}{fa} = \frac{0}{0}, \quad \text{and (§ 117)}$$

$$\underset{x \Longrightarrow a}{\text{limit}} \left[\frac{1}{\phi x} \Big/ \frac{1}{fx}\right] = \lim \left[\frac{\phi' x}{\overline{\phi x}^2} \Big/ \frac{f' x}{\overline{fx}^2}\right]$$

$$= \lim \left[\left(\frac{fx}{\phi x}\right)^2 \left(\frac{\phi' x}{f' x}\right)\right].$$

Hence $$\underset{x \Longrightarrow a}{\text{limit}} \frac{fx}{\phi x} = \lim \left[\left(\frac{fx}{\phi x}\right)^2 \left(\frac{\phi' x}{f' x}\right)\right];$$

and since limits of equimultiples of two variables with equal limits are equal, we have, multiplying by

$$\phi x\, f' x / fx\, \phi' x,\text{*}$$

$$\underset{x \Longrightarrow a}{\text{limit}} \frac{fx}{\phi x} = \lim \frac{f' x}{\phi' x}.$$

Similarly, we may deduce

$$\underset{x \Longrightarrow a}{\text{limit}} \frac{f' x}{\phi' x} = \lim \frac{f'' x}{\phi'' x} = \ldots = \lim \frac{f^n x}{\phi^n x} = \frac{f^n a}{\phi^n a}.$$

The method is therefore the same as in the preceding article; but by § 71 we have $f^n a / \phi^n a = \infty/\infty$, when $fa/\phi a = \infty/\infty$ for a finite value of a. Hence for finite values of a this method will fail to determine the limit unless factors common to both terms of some derived ratio become apparent or the limit of $f^n x / \phi^n x$ becomes otherwise known.

---

* Taylor's Calculus, § 79.

EXAMPLES.

1. $$\left.\frac{\log \sin 2x}{\log \sin x}\right]_0 = \left.\frac{2 \cot 2x}{\cot x}\right]_0 = 2\left.\left(\frac{\cos 2x}{\cos x}\right)\left(\frac{\sin x}{\sin 2x}\right)\right]_0$$

$$= 2\left.\left(\frac{2\cos^2 x - 1}{\cos x}\right)\left(\frac{\sin x}{\sin 2x}\right)\right]_0 = 2\left.\frac{\sin x}{\sin 2x}\right]_0$$

$$= 2\left.\frac{\cos x}{2 \cos 2x}\right]_0 = 1.$$

2. $$\left.\frac{\log x}{1/x}\right]_0 = -\left.\frac{1/x}{1/x^2}\right]_0 = -x\Big]_0 = 0 = x \log x\Big]_0.$$

3. $$\left.\frac{1/x}{\cot x}\right]_0 = \left.\frac{1/x^2}{1/\sin^2 x}\right]_0 = \left.\frac{\sin^2 x}{x^2}\right]_0 = 1.$$

4. $$\left.\frac{\log x}{\operatorname{cosec} x}\right]_0 = -\left.\frac{1/x}{\cot x \operatorname{cosec} x}\right]_0 = -\left.\frac{\sin^2 x}{x \cos x}\right]_0$$

$$= -\left.\frac{2 \sin x \cos x}{\cos x - x \sin x}\right]_0 = 0.$$

5. $$\left.\frac{\log x}{\cot x}\right]_0 = -\left.\frac{\sin^2 x}{x}\right]_0 = -2 \sin x \cos x\Big]_0 = 0.$$

6. $$\left.\frac{x^n}{e^x}\right]_\infty = \left.\frac{nx^{n-1}}{e^x}\right]_\infty = \ldots = \left.\frac{\lfloor n}{e^x}\right]_\infty = 0,$$

in which n is a positive integer. If n is fractional, the exponent of x ultimately becomes negative, giving the same result.

7. Putting $y^2 = 1/x$, whence $y \ggg\!\rightarrow 0$ as $x \ggg\!\rightarrow \infty$, we have

$$e^{-1/y^2}/y^{2n}\Big]_0 = x^n/e^x\Big]_\infty = 0.$$

In some cases we may with advantage place $x = a + h$ and subsequently make $h = 0$. Thus $\sqrt[3]{x-a}/\sqrt[4]{x^2-a^2}$ reduces to $0/0$, and the ratios of all derivatives of both terms become ∞/∞ when $x = a$; but putting $x = a + h$, we obtain

$$\left.\frac{\sqrt[3]{x-a}}{\sqrt[4]{x^2-a^2}}\right]_{x=a} = \left.\frac{h^{1/3}}{h^{1/4}(2a+h)^{1/4}}\right]_{h=0} = \left.\frac{h^{1/12}}{(2a+h)^{1/4}}\right]_{h=0} = 0.$$

119. Form $0 . \infty$.

If $fa = 0$ and $\phi a = \infty$, we write

$$fx \times \phi x = \frac{fx}{1/\phi x} = \frac{\phi x}{1/fx},$$

which takes the form $0/0$ or ∞/∞ when $x = a$, and the limit may be determined by the method of §§ 117, 118 with the same limitations.

EXAMPLES.

1. $e^{1/x}x\Big]_0 = \frac{e^{1/x}}{x^{-1}}\Big]_0 = \frac{-e^{1/x}/x^2}{-x^{-2}}\Big]_0 = e^{1/x}\Big]_0 = \infty.$

2. $e^{1/x}x^2\Big]_0 = \frac{e^{1/x}}{x^{-2}}\Big]_0 = \frac{e^{1/x}x}{2}\Big]_0 = \infty.$

3. $e^{-1/x}x^2\Big]_0 = x^2/e^{1/x}\Big]_0 = 0.$

4. $e^{-x}\log x\Big]_\infty = \log x/e^x\Big]_\infty = (1/x)/e^x\Big]_\infty = 0.$

5. $\sin x \log \cot x\Big]_0 = (\sin x/x)\, x \log \cot x\Big]_0 = \log \cot x/(1/x)\Big]_0$

$= (x^2/\sin^2 x)(1/\cot x)\Big]_0 = 0.$

6. $x \tan^{-1}\frac{1}{x}\Big]_\infty = \tan^{-1}\frac{1}{x}\Big/\frac{1}{x}\Big]_\infty = \frac{1}{x^2+1}\Big/\frac{1}{x^2}\Big]_\infty$

$= 1 - \Big[1/(x^2+1)\Big]_\infty = 1,$ when $\tan^{-1}(1/\infty) = 0.$

7. $\sec x\,(x \sin x - \pi/2)\Big]_{\pi/2} = (x \sin x - \pi/2)/\cos x\Big]_{\pi/2}$

$= (x \cos x + \sin x)/-\sin x\Big]_{\pi/2} = -1.$

8. $\log\left(2 - \frac{x}{a}\right) . \tan\frac{\pi x}{2a}\Big]_a = \log\left(2 - \frac{x}{a}\right)\Big/\cot\frac{\pi x}{2a}\Big]_a$

$= \frac{-1}{a(2 - x/a)}\Big/\frac{-\pi/2a}{\sin^2(\pi x/2a)}\Big]_a = \frac{2}{\pi}.$

9. $x^n \log x\Big]_0 = \frac{\log x}{1/x^n}\Big]_0 = \frac{1}{x}\Big/\frac{-nx^{n-1}}{x^{2n}}\Big]_0 = \frac{x^n}{-n}\Big]_0 = 0.$

10. $x^m (\log x)^n\Big]_0 = \dfrac{x^m}{1/(\log x)^n}\Big]_0 = \dfrac{mx^{m-1}}{-n/x(\log x)^{n+1}}\Big]_0$,

which remains indeterminate under the method; but placing $x = e^{-y}$, whence $x^m(\log x)^n = (-1)^n y^n / e^{my}$ and $y \Rrightarrow \infty$ as $x \Rrightarrow 0$, we have (example 7, § 118) 0 for the limit.

120. Form $\infty - \infty$.

If $fa - \phi a = \infty - \infty$, we write

$$fx - \phi x = \left(\frac{1}{\phi x} - \frac{1}{fx}\right) \Big/ \left(\frac{1}{\phi x fx}\right),$$

which becomes 0/0 when $x = a$, and the rule of § 117 will apply after the given expression is placed under the form indicated in the second member.

EXAMPLES.

1. $\dfrac{1}{x} - \cot x\Big]_0 = \left(\dfrac{1}{\cot x} - x\right) \Big/ \dfrac{x}{\cot x}\Big]_0 = \dfrac{\tan x - x}{x \tan x}\Big]_0$

$$= \frac{\sin x - x\cos x}{x \sin x}\Big]_0 = \frac{x \sin x}{\sin x + x \cos x}\Big]_0 = \frac{\sin x + x\cos x}{2\cos x - x \sin x}\Big]_0 = 0.$$

In some cases the desired form may be obtained by a more simple transformation.

2. $x - \sqrt{x^2 - ax}\Big]_\infty = \dfrac{1 - \sqrt{1 - a/x}}{1/x}\Big]_\infty = \dfrac{a}{2\sqrt{1 - a/x}}\Big]_\infty = \dfrac{a}{2}$.

3. $\tan x - \sec x\Big]_{\pi/2} = (\sin x - 1)/\cos x\Big]_{\pi/2} = -\cos x/\sin x\Big]_{\pi/2} = 0$.

4. $\dfrac{2}{x^2 - 1} - \dfrac{1}{x - 1}\Big]_1 = \dfrac{2(x-1) - (x^2 - 1)}{(x^2 - 1)(x - 1)}\Big]_1 = \dfrac{2 - x - 1}{x^2 - 1}\Big]_1$

$$= \frac{-1}{2x}\Big]_1 = -\frac{1}{2}.$$

5. $x \tan x - \pi \sec x/2\Big]_{\pi/2} = (x \sin x - \pi/2)/\cos x\Big]_{\pi/2}$

$$= (x \cos x + \sin x)/- \sin x\Big]_{\pi/2} = -1.$$

121. Forms 0^0, ∞^0, 1^∞.

If $(fa)^{\phi a} = 0^0$ or ∞^0 or 1^∞, we write

$$\log (fx)^{\phi x} = \phi x \log fx, \text{ whence } (fx)^{\phi x} = e^{\phi x \log fx},$$

and

$$\underset{x \to a}{\text{limit}} (fx)^{\phi x} = \lim e^{\phi x \log fx} = e^{\lim [\phi x \log fx]},$$

in which $\phi a \log fa = 0 . \infty$ in each of the above cases, and the method of § 119 will apply after the given expression is placed in the form indicated.

EXAMPLES.

1. $x^x\Big]_0 = e^{x \log x}\Big]_0 = e^{-x}\Big]_0 = 1.$

2. $\left(\frac{1}{x}\right)^{\sin x}\Bigg]_0 = e^{-\sin x \log x}\Big]_0 = 1.$

3. $x^{1/x}\Big]_\infty = e^{\log x/x}\Big]_\infty = e^{1/x}\Big]_\infty = 1.$

4. $(1 + x)^{1/x}\Big]_0 = e^{[\log (1+x)]/x}\Big]_0 = e^{1/(1+x)}\Big]_0 = e.$

5. $(2 - x/a)^{\tan\frac{\pi x}{2a}}\Bigg]_a = e^{\tan\frac{\pi x}{2a} \log \left(2 - \frac{x}{a}\right)}\Bigg]_a = e^{\frac{2}{\pi}}.$

6. $\cot x^{\sin x}\Big]_0 = e^{\sin x \log \cot x}\Big]_0 = 1.$

7. $\left[\frac{a^{nx} + b^{nx} + \ldots + n^{nx}}{n}\right]_0^{1/x} = e^{\left[\log\frac{a^{nx} + b^{nx} + \ldots}{n}\right]/x}\Bigg]_0$

$$= e^{\frac{n(a^{nx} \log a + b^{nx} \log b + \ldots}{a^{nx} + b^{nx} + \ldots}}\Bigg]_0 = e^{\log (ab \ldots n)} = ab \ldots n.$$

122. Evaluation of Derivatives of Implicit Functions. Derivatives which for particular values of the variables assume indeterminate forms may be evaluated as in the preceding cases.

EXAMPLES.

1. $y^3 + x^3 = ax^2$.

Whence $$\left.\frac{dy}{dx}\right]_{0.0} = \left.\frac{2ax - 3x^2}{3y^2}\right]_{0.0} = \left.\frac{a - 3x}{3y(dy/dx)}\right]_{0.0}.$$

Hence, $(dy/dx)^2_{0.0} = \infty$, and $dy/dx\Big]_{0.0} = \pm\infty$.*

2. $x^3 + y^3 = 3axy$. $dy/dx\Big]_{0.0} = 0$ or ∞.

3. $y^3 + 3ay^2 - 2axy = ax^2$. $dy/dx\Big]_{0.0} = 1$ or $-1/3$.

4. $(x^2 + y^2)^2 = a^2(x^2 - y^2)$. $dy/dx\Big]_{0.0} = \pm 1$.

5. $u = y^4 + 3a^2y^2 - 4a^2xy - a^2x^2 = 0$.

$$\partial u/dx = -4a^2y - 2a^2x. \quad \partial u/dy = 4y^3 + 6a^2y - 4a^2x.$$

$$\left.\frac{dy}{dx}\right]_{0.0} = \left.\frac{2a^2y + a^2x}{2y^3 + 3a^2y - 2a^2x}\right]_{0.0} = \left.\frac{2a^2(dy/dx) + a^2}{(6y^2 + 3a^2)(dy/dx) - 2a^2}\right]_{0.0}$$

$$= \left.\frac{2(dy/dx) + 1}{3(dy/dx) - 2}\right]_{0.0}. \quad \text{Hence} \quad \left.\left(\frac{dy}{dx}\right)^2 - \frac{4}{3}\frac{dy}{dx}\right]_{0.0} = \frac{1}{3}$$

and $$dy/dx\Big]_{0.0} = (2 \pm \sqrt{7})/3.$$

6. $u = a^2y^2 - a^2x^2 - x^4 = 0$. $dy/dx\Big]_{0.0} = \pm 1$.

7. $u = x^4 + ax^2y - ay^3 = 0$. $dy/dx\Big]_{0.0} = 0$ or ± 1.

8. $u = ay^3 - bx^2y + x^4 = 0$. $dy/dx\Big]_{0.0} = 0$ or $\pm\sqrt{b/a}$.

9. $u = x^4 - a^2xy + y^2 = 0$. $dy/dx\Big]_{0.0} = 0$ or a^2.

10. $u = x^3 + y^3 + a^3 - 3axy = 0$. $dy/dx\Big]_{a.a} = \frac{1}{2}(1 \pm \sqrt{-3})$.

11. $u = y^3 - 3axy + x^4 = 0$. $dy/dx\Big]_{0.0} = -0$ or ∞.

12. $u = x^4 + 2ax^2y - ay^3 = 0$. $dy/dx\Big]_{0.0} = 0$, or $\pm\sqrt{2}$.

---

* $dy/dx\Big]_{0.0}$ has two limiting values for the same value of the variable, hence it is discontinuous at the corresponding state.

MISCELLANEOUS EXAMPLES.

1. $\left.\dfrac{\log x}{x}\right]_{\infty} = 0.$

2. $\left.\dfrac{e^x - e^{-x} - 2x}{x - \sin x}\right]_0 = 2.$

3. $\left.\dfrac{\log x}{x - 1}\right]_1 = 1.$

4. $\left.\dfrac{(e^x - 1)\tan^2 x}{x^3}\right]_0 = 1.$

5. $\left.\dfrac{1 - \cos x}{x^2}\right]_0 = \dfrac{1}{2}.$

6. $\left.\sec x - \tan x\right]_{\pi/2} = 0.$

7. $\left.\dfrac{x^2}{e^{h^2x^2}}\right]_{\infty} = 0.$

8. $\left.\dfrac{\cos^{-1}(1 - x)}{\sqrt{2x - x^2}}\right]_0 = 1.$

9. $\left.2^x \sin \dfrac{a}{2^x}\right]_{\infty} = a.$

10. $\left.e^{\frac{\log(1 + nx)}{x}}\right]_0 = e^n.$

11. $\left.x^{-n}e^x\right]_{\infty} = \infty.$

12. $\left.\left(\dfrac{\tan x}{x}\right)^{1/x^2}\right]_0 = e^{\frac{1}{3}}.$

13. $\left.\dfrac{1 - \log x}{e^{1/x}}\right]_0 = 0.$

14. $\left.\dfrac{(1 - x)e^x - 1}{\tan^2 x}\right]_0 = -\dfrac{1}{2}.$

15. $\left.\dfrac{e^{-1/x}}{x}\right]_0 = 0.$

16. $\left.\dfrac{e^{1/x}}{-x}\right]_0 = \infty.$

17. $\left.(\sin x)^{\sin x}\right]_0 = 1.$

18. $\left.\dfrac{x - \sin^{-1} x}{\sin^3 x}\right]_0 = -\dfrac{1}{6}.$

19. $\left.(1 + ax)^{\frac{b}{x}}\right]_0 = e^{ab}.$

20. $\left.\log\left(\dfrac{a + x}{a}\right)\Big/x\right]_0 = \dfrac{1}{a}.$

21. $\left.(1 + ax)^{\frac{b}{x}}\right]_{\infty} = 1.$

22. $\left.\dfrac{a^{\log x} - x}{\log x}\right]_1 = \log\left(\dfrac{a}{e}\right).$

23. $\left.\dfrac{\log x}{x^a}\right]_{\infty} = 0.$

24. $\left.\dfrac{\pi/4x}{\cot \dfrac{\pi x}{2}}\right]_0 = \dfrac{\pi^2}{8}.$

25. $\left.\dfrac{a^x - b^x}{x}\right]_0 = \log \dfrac{a}{b}$

26. $\left.2^x \sin \dfrac{a}{2^x}\right]_{\infty} = a.$

27. $\left.(\cos ax)^{\operatorname{cosec}^2 cx}\right]_0 = e^{-\frac{a^2}{2c^2}}.$

28. $\left.\dfrac{\cot x + \operatorname{cosec} x - 1}{\cot x - \operatorname{cosec} x + 1}\right]_{\pi/2} = 1$

29. $x^{\sin x}\Big]_0 = 1.$

30. $\left.\dfrac{1}{\log x} - \dfrac{x}{\log x}\right]_1 = -1.$

31. $\left.\dfrac{e^x - e^{\sin x}}{x - \sin x}\right]_0 = 1.$

32. $\left.\dfrac{x}{x - 1} - \dfrac{1}{\log x}\right]_1 = \dfrac{1}{2}.$

33. $\left.\dfrac{x^2 - a^{\frac{3}{2}}x^{\frac{1}{2}}}{\sqrt{ax} - a}\right]_a = 3a.$

34. $\left.\dfrac{x^3 - 3x + 2}{x^4 - 6x^2 + 8x - 3}\right]_1 = \infty.$

35. $\left.\dfrac{x^3 - 1}{x^3 - 2x^2 + 2x - 1}\right]_1 = 3.$

36. $\left.\dfrac{\tan nx - n\tan x}{n\sin x - \sin nx}\right]_0 = 2.$

37. $\tan x/\log(x - \pi/2)\Big]_{\pi/2} = -\infty.$

38. $\left(x - a + \sqrt{2ax - 2a^2}\right)/\sqrt{x^2 - a^2}\Big]_a = 1.$

39. $x/\cot x - \pi/2\cos x\Big]_{\pi/2} = -1.$

40. $(1 - x)\tan(\pi x/2)\Big]_1 = 2/\pi.$

41. $(ax^2 - 2acx + ac^2)/(bx^2 - 2bcx + bc^2)\Big]_c = a/b.$

42. $(1 - \sin x + \cos x)/(\sin x + \cos x - 1)\Big]_{\pi/2} = 1.$

43. $(a - x - a\log a + a\log x)/\left(a - \sqrt{2ax - x^2}\right)\Big]_a = -1.$

44. $(x\sin x - \pi/2)/\cos x\Big]_{\pi/2} = -1.$

45. $(e^{2x} - e^{-2x} - 2)\sec x/x^2\Big]_0 = 4.$

46. $\left(\sqrt{2} - \sin x - \cos x\right)/\log\sin x\Big]_{\pi/4} = -1/2\sqrt{2}.$

47. $\left[(e^x - e^{-x})^2 - 2x^2(e^x + e^{-x})\right]/x^4\Big]_0 = -2/3.$

48. $\left[(x - a)^n/(e^x - e^a)\right]_a = 0$ or e^{-a} or ∞, according as $n > 1$, $n = 1$, $n < 1$.

49. $1/(1 + e^{1/x}) + e^{1/x}/x(1 + e^{1/x})^2\Big]_0 = 0.$

50. $1/(1 + e^{-1/x}) - e^{-1/x}x(1 + e^{-1/x})^2\Big]_0 = 1.$

51. $\left.\dfrac{x + \cos x}{x - \sin x}\right]_\infty = \left.\dfrac{1 + \cos x/x}{1 - \sin x/x}\right]_\infty = 1.$

CHAPTER X.

DEVELOPMENTS.

123. The **development** of a function is the operation of determining an equivalent *finite* or *convergent-infinite* series. When this can be done, the function will be the sum of the series, which, in the case of a convergent-infinite series, is, also, the *limit of the sum of n terms* as n increases without limit.

A convergent series having a given function as a limit may be used to determine approximate values of the function, the degree of approximation depending upon the rapidity of convergence and the number of terms considered. A divergent series should not be employed in finding approximate values of a function, or in the deduction of a general principle or formula.

Let S represent a function giving

$$S = u_1 + u_2 + u_3 + \ldots + u_n + \text{etc.}$$

Denote the sum of the first n terms by S_n, and the sum of the following terms by R, called the remainder; then $S = S_n + R$.

The series is convergent if R is an infinitesimal as n increases without limit, in which case S is the limit of S_n.

When S is the limit of S_n as n increases, it is also the limit of S_{n-1}, and we have

$$\underset{n \to \infty}{\text{limit}} \left[S_n - S_{n-1} \right] = \lim u_n = 0.$$

That is, in a convergent series the nth term is an infinitesimal as n increases, but the converse is not necessarily true unless S_n has a finite limit under the law; for $\lim u_n = 0 = \lim [S_n - S_{n-1}]$ may occur when $S = \infty$. Therefore a series is not necessarily convergent when the nth term is an infinitesimal as n increases.

124. **Taylor's Formula** *has for its object the development of a function of the sum of two variables into a series arranged according to the ascending powers of one of the variables with coefficients which are functions of the other.*

Assuming an expansion of the proposed form, we write

$$f(x+h) = X_1 + X_2h + X_3h^2 + \ldots + X_{n+1}h^n + R, \quad (b)$$

in which X_1, X_2, etc., are functions of x to be determined, and R the remainder after $n+1$ terms.

$$h = 0 \quad \text{gives} \quad fx = X_1.$$

Placing $x + h = s$, and differentiating, first with respect to x and then with respect to h, we have

$$df(x+h)/dx = (dfs/ds)(ds/dx), \ (\S\ 77).$$

$$df(x+h)/dh = (dfs/ds)(ds/dh).$$

But $ds/dx = ds/dh$, hence

$$df(x+h)/dx = df(x+h)/dh.$$

Hence, differentiating the second member of (b) first with respect to x and then with respect to h, we have

$$\frac{dX_1}{dx} + \frac{dX_2}{dx}h + \frac{dX_3}{dx}h^2 + \frac{dX_4}{dx}h^3 + \text{etc.} + \frac{dX_n}{dx}h^{n-1} + \text{etc.}$$

$$= X_2 + 2X_3h + 3X_4h^2 + 4X_5h^3 + \text{etc.} + nX_{n+1}h^{n-1} + \text{etc.},$$

which is an identical equation, and by the principle of indeterminate coefficients we have

$$\frac{dX_1}{dx} = X_2, \quad \frac{dX_2}{dx} = 2X_3, \quad \frac{dX_3}{dx} = 3X_4, \ldots \frac{dX_n}{dx} = nX_{n+1}, \text{ etc.}$$

Since $X_1 = fx$, $dX_1/dx = f'x$, $\therefore\ X_2 = f'x$.

Therefore
$$\left\{\begin{array}{ll} \dfrac{dX_2}{dx} = f''x = 2X_3, & \text{and } X_3 = \tfrac{1}{2}f''x. \\ \dfrac{dX_3}{dx} = \tfrac{1}{2}f'''x = 3X_4, & \text{and } X_4 = \dfrac{1}{2.3}f'''x. \\ \text{etc.} & \text{etc.} \\ \dfrac{dX_n}{dx} = \dfrac{1}{\lfloor n-1}f^n x = nX_{n+1}, & \text{and } X_{n+1} = \dfrac{1}{\lfloor n}f^n x. \\ \text{etc.} & \text{etc.} \end{array}\right.$$

Substituting these expressions for X_1, X_2, X_3, etc., in (b), we have **Taylor's formula** * :

$$\mathbf{f(x+h) = fx + f'x\,h + f''x\,h^2/2 + f'''x\,h^3/\lfloor 3 + \ldots}$$
$$\mathbf{+ f^nx\,h^n/\lfloor n + R}, \quad . \quad . \quad . \quad . \quad (c)$$

in which fx represents what the given function becomes when $h = 0$, $f'x$, $f''x$, etc., represent the first, second,

* Formula published in 1715 by Dr. Brook Taylor.

etc., derivatives of fx, and R the sum of all of the terms after the $(n+1)^{\text{th}}$.

The second member of (c) is also known as the development of the second state of a function of a single variable (§ 5).

Designating $f(x+h)$ by y', and fx by y, we have

$$y' = f(x+h) = y + \frac{dy}{dx}h + \frac{d^2y}{dx^2}\frac{h^2}{2} + \frac{d^3y}{dx^3}\frac{h^3}{\underline{|3}} + \ldots + \frac{d^ny}{dx^n}\frac{h^n}{\underline{|n}} + R,$$

which is another form of (c).

To apply Taylor's formula, *cause the variable with reference to which the development is to be arranged, to vanish. Differentiate the result and its derivatives in succession until one of the highest order desired is obtained, and substitute them, respectively, for their corresponding symbols in the formula.*

Thus, to develop $(x+y)^m$, place $y=0$, and differentiate x^m, whence

$$f(x) = x^m, \quad f'(x) = mx^{m-1}, \text{ etc.}$$
$$f^n(x) = m(m-1)\ldots(m-n+1)x^{m-n}.$$

Substituting in (c), we have

$$(x+y)^m = x^m + mx^{m-1}y + \frac{m(m-1)}{1.2}x^{m-2}y^2 + \ldots$$
$$+ \frac{m(m-1)\ldots(m-n+1)}{\underline{|n}}x^{m-n}y^n + R.$$

125. Lagrange's Expression for the Remainder.—In Taylor's formula put $x+h=X$, whence $h=X-x$, giving

$$fX = fx + f'x(X-x) + f''x(X-x)^2/2 + \ldots$$
$$+ f^nx(X-x)^n/\underline{|n} + R. \quad . \quad . \quad (1)$$

Assume $R = P(X-x)^{n+1}/\underline{\lfloor n+1}$, in which P is an unknown function of X and x, which will make (1) exact for all values of x and X, giving

$$fX - fx - f'x(X-x) - \dots \\ - f^n x(X-x)^n/\underline{\lfloor n} - P(X-x)^{n+1}/\underline{\lfloor n+1} = 0. \quad (2)$$

Substitute z for x (except in P), and let Fz represent the result which in general will not be equal to zero, giving

$$Fz = fX - fz - f'z(X-z) - \dots \\ - f^n z(X-z)^n/\underline{\lfloor n} - P(X-z)^{n+1}/\underline{\lfloor n+1}. \quad (3)$$

From (2) and (3) we see that $Fx = 0$, and from (3) we have $FX = 0$. As z varies from x to X, Fz increases and then decreases or the reverse, and $F'z$, if continuous, must change its sign by vanishing for some value of z between x and X. (§ 16, § 63.)

Differentiating (3) with respect to z and reducing, we have, since the terms with the exception of the last two cancel in pairs,

$$F'z = -f^{n+1}z(X-z)^n/\underline{\lfloor n} + P(X-z)^n/\underline{\lfloor n}.$$

Let $x+\theta_n(X-x)$, in which θ_n is a positive number less than unity, represent the value of z for which $F'z = 0$. Then

$$P = f^{n+1}(x+\theta_n(X-x)) = f^{n+1}(x+\theta_p h),$$

and $$R = f^{n+1}(x + \theta_n h)h^{n+1}/\underline{|n+1}\,*,$$

in which $0 < \theta_n < 1$.

This expression for R enables us to determine the conditions of applicability of Taylor's formula.

When as $n \ggg\!\!\rightarrow \infty$, R is an infinitesimal, the formula gives a finite or convergent-infinite series for the function.

This condition is fulfilled when as $n \ggg\!\!\rightarrow \infty$, fx and $f^n x$ remain continuous between states corresponding to all values of x from any assumed value of x to $x + h$, for then $f^{n+1}(x+\theta_n h)$ is always real and finite, and (§ 41) $h^{n+1}/\underline{|n+1}$ approaches zero as a limit.

With any assumed value of x which fulfils the above condition, h will frequently have limiting values. They are the numerically least positive and negative values of h that cause fx or any of its derivatives to become discontinuous when $x + h$ is substituted for x.

If, for any assumed value of x, fx or any of its derivatives becomes imaginary or infinite, the corresponding limiting value of h must be zero, and the formula is inapplicable. It may, however, develop the function for other values of x.

To illustrate the use of R, take the example

$$(x+y)^m = x^m + mx^{m-1}y + \frac{m(m-1)}{1.2}x^{m-2}y^2 + \ldots$$
$$+ \frac{m(m-1)\ldots(m-n+1)}{\underline{|n}}x^{m-n}y^n + R,$$

in which $f^n x = m(m-1)\ldots.(m-n+1)x^{m-n}$,

and $$R = \frac{m(m-1)\ldots(m-n)}{\underline{|n+1}}\,\frac{y^{n+1}}{(x+\theta_n y)^{n-m+1}}.$$

* Known as Lagrange's expression for the remainder.

When m is fractional or negative, n may increase without limit, the series will have an unlimited number of terms, and $m - n$ will become negative when $n > m$ numerically, giving

$$f^n x = m(m-1)\dots(m-n+1)/x^{n-m}$$

If $x < 0$, the development fails when the exponent $n - m$ is a fraction with an even denominator.

If $x = 0$, $f^n x$ ultimately becomes ∞ and the formula is inapplicable.

If $x > 0$, we have the ratio of R to the $(n+2)$th term equal to

$$(x/(x+\theta_n y))^{n-m+1},$$

which vanishes as $n \ggg \infty$; and R, therefore, diminishes indefinitely when the successive terms in order likewise decrease.

The ratio of the $(n+1)$th term to the nth is

$$\frac{m-n+1}{n}\frac{y}{x} = \frac{m/n - 1 + 1/n}{1}\frac{y}{x},$$

the limit of which, as $n \ggg \infty$, is $-(y/x)$.

Hence, when x is numerically greater than y the successive terms in order will ultimately decrease indefinitely.

In which case R will be an infinitesimal; and we conclude that when m is fractional or negative, the binomial formula develops $(x+y)^m$ for all positive values of x numerically greater than y.

EXAMPLES.

1. Develop $(x+y)^{1/2}$.

$$f(x) = \sqrt{x},\ f'(x) = 1/2\sqrt{x},\ f''(x) = -1/4\sqrt{x^3}, \text{ etc.}$$

$$(x+y)^{1/2} = \sqrt{x} + y/2\sqrt{x} - y^2/8\sqrt{x^3} + R,$$

which fails for $x =$ or < 0.

2. Develop cos $(x+y)$.

$$f(x) = \cos x, \ f'(x) = -\sin x, \ f''(x) = -\cos x,$$
$$f'''(x) = \sin x, \text{ etc.}$$

$$\cos(x+y) = \cos x - y \sin x - y^2 \cos x/2 + y^3 \sin x/\underline{|3} + y^4 \cos x/\underline{|4} + R,$$

which is true for all values of x and y.

Making $x = 0$, we have

$$\cos y = 1 - y^2/2 + y^4/\underline{|4} - y^6\underline{|6} + R.$$

3. $$\sin(x+y) = \sin x + y \cos x - y^2 \sin x/2 - y^3 \cos x/\underline{|3} + y^4 \sin x/\underline{|4} + y^5 \cos x/\underline{|5} + R.$$

Whence $\quad \sin y = y - y^3/\underline{|3} + y^5/\underline{|5} - y^7/\underline{|7} + R.$

4. $$\sin^{-1}(x+y) = \sin^{-1}x + \frac{y}{\sqrt{(1-x^2)}} + \frac{x}{\sqrt{(1-x^2)^3}}\frac{y^2}{2} + \frac{1+2x^2}{\sqrt{(1-x^2)^5}}\frac{y^3}{\underline{|3}} + \frac{3x(3+2x^2)}{\sqrt{(1-x^2)^7}}\frac{y^4}{\underline{|4}} + R,$$

which fails when $x = 1$ or > 1 numerically.

For values of $x < 1$ numerically, limiting value of $y = 1 - x$.

Making $x = 0$, we have

$$\sin^{-1}y = y + y^3/\underline{|3} + 3^2y^5/\underline{|5} + R.$$

5. $$\tan^{-1}(x+y) = \tan^{-1}x + \frac{y}{1+x^2} - \frac{2x}{(1+x^2)^2}\frac{y^2}{2} + \frac{(3x^2-1)}{(1+x^2)^3}\frac{y^3}{3} + R,$$

which is true for all values of x and y.

Whence $\quad \tan^{-1}y = y - y^3/3 + y^5/5 + R.$

6. Develop $\log_a(x+y)$.

$$fx = \log_a x, \quad f'x = M_a/x, \quad f''x = -M_a/x^2, \quad \text{etc.},$$

$$f^n x = (-1)^{n-1}M_a\underline{|n-1}/x^n. \qquad \text{Hence,}$$

$$\log_a(x+y) = \log_a x + M_a\left(\frac{y}{x} - \frac{y^2}{2x^2} + \frac{y^3}{3x^3} - \ldots + (-1)^{n-1}\frac{y^n}{nx^n}\right) + R.$$

In which $R = \pm M_a y^{n+1}/(n+1)(x+\theta_n y)^{n+1}$ is an infinitesimal, as $n \ggg\!\rightarrow \infty$ when $x =$ or $> y$ numerically.

If $x = 0$, the formula fails. If $x = 1$, we have

$$\log_a(1 + y) = M_a\left(y - y^2/2 + \ldots + (-1)^{n-1}y^n/n\right) + R,$$
$$\log(1 + y) = y - y^2/2 + y^3/3 - y^4/4 + R.$$

7. $a^{x+y} = a^x(1 + \log a\, y + \log^2 a\, y^2/2 + \ldots + \log^n a\, y^n/\underline{|n}) + R.$

$$R = a^{(x+\theta_n y)} \log^{n+1} a\, y^{n+1}/\underline{|n+1}$$

is an infinitesimal as $n \ggg\!\!\rightarrow \infty$, since $\log a$ is a constant, while $\left[y/(n+1)\right] \ggg\!\!\rightarrow 0$. Making $x = 0$, we have

$$a^y = 1 + \log a\, y + \log^2 a\, y^2/2 + \ldots + \log^n a\, y^n/\underline{|n} + R_0.$$

8. Develop $(x^{1/2} + y^2)^5$.

The variables are $\sqrt{x}$ and y^2.

When the variables considered are not represented by the first powers of letters, substitute the first powers of other letters for the variables, develop the result, and resubstitute the variables for the auxiliary letters.

Thus, placing $x^{1/2} = r$, and $y^2 = s$, we have

$$(x^{1/2} + y^2)^5 = (r + s)^5, \quad f(r) = r^5, \quad f'(r) = 5r^4, \quad f''(r) = 20r^3, \text{ etc.}$$

Hence,

$$(x^{1/2} + y^2)^5 = (r + s)^5 = r^5 + 5r^4s + 10r^3s^2 + 10r^2s^3 + 5rs^4 + s^5$$
$$= x^{5/2} + 5x^2y^2 + 10x^{3/2}y^4 + 10xy^6 + 5x^{1/2}y^8 + y^{10}.$$

9. $\sqrt{a + x + y} = \sqrt{a + x} + \dfrac{y}{2\sqrt{a+x}} - \dfrac{y^2}{8\sqrt{(a+x)^3}} + R,$

which fails when $x = -a$ or $< -a$.

10. $(x - a + y)^{5/2} = (x - a)^{5/2} + 5(x-a)^{3/2}y/2 + 15(x - a + \theta y)^{1/2}y^2/8$

$x = a$ gives $\quad y^{5/2} = 15\theta^{1/2}y^{5/2}/8, \quad \therefore\ \theta = 64/225;$

but the development would fail with more terms.

If $x < a$, the formula fails.

11. $\tan(x + y) = \tan x + \sec^2 x\, y + 2\sec^2 x \tan x\, y^2/2$
$$+ 2\sec^2 x(1 + 3\tan^2 x)y^3/\underline{|3} + R.$$

Inapplicable when $x = \pi/2$.

12. $\log_a(x^2+y^2) = \log_a x^2 + M_a\left(\frac{y^2}{x^2} - \frac{y^4}{2x^4} + \frac{y^6}{3x^6} - \frac{y^8}{4x^8} + R\right).$

13. $\frac{c}{x+y} = \frac{c}{x} - \frac{c}{x^2}y + \frac{c}{x^3}y^2 \ldots \pm \frac{c}{x^{n+1}}y^n + R.$

14. $\sec^{-1}(x+y) = \sec^{-1} x + \frac{y}{x\sqrt{x^2-1}} - \frac{2x^2-1}{x^2(x^2-1)^{3/2}}\frac{y^2}{2} + R.$

15. $\cos^2(x-y) = \cos^2 x + y \sin 2x - y^2 \cos 2x - 2y^3 \sin 2x/3$
$+ y^4 \cos 2x/3 + 2y^5 \sin 2x/15 + R.$

16. $2(x+y)^3 - 3(x+y)^2 + 1 = x^2(2x-3) + 1$
$+ 6(x^2-x)y + 3(2x-1)y^2 + 2y^3.$

17. $\frac{a}{\sqrt[3]{x-y}} = \frac{a}{\sqrt[3]{x}} + \frac{a}{3\sqrt[3]{x^4}}y + \frac{4a}{9\sqrt[3]{x^7}}\frac{y^2}{2} + \frac{28a}{27\sqrt[3]{x^{10}}}\frac{y^3}{|\underline{3}} + R.$

18. $(ax + a/y^2)^3 = a^3(x^3 + 3x^2/y^2 + 3x/y^4 + 1/y^6).$

19. $\log \sin (x+y) = \log \sin x + y \cot x - y^2 \operatorname{cosec}^2 x/2$
$+ y^3 \cos x/3 \sin^3 x + R.$

20. $(x-y)^{1/3} = x^{1/3} - \frac{x^{-2/3}y}{3} - \frac{x^{-5/3}y^2}{9} - \frac{5x^{-8/3}y^3}{81} + R.$

21. $\left(\frac{a}{2x^2} - \frac{3b}{\sqrt{y}}\right)^{-1/3} = \left(\frac{a}{2x^2}\right)^{-1/3} + \frac{1}{3}\left(\frac{a}{2x^2}\right)^{-4/3}\left(\frac{3b}{\sqrt{y}}\right)$
$+ \frac{4}{9}\left(\frac{a}{2x^2}\right)^{-7/3}\left(\frac{3b}{\sqrt{y}}\right)^2 \Big/ 2 + \frac{28}{27}\left(\frac{a}{2x^2}\right)^{-10/3}\left(\frac{3b}{\sqrt{y}}\right)^3 \Big/ |\underline{3} + R.$

22. $(-y+x)^{-2} = y^{-2} - 2xy^{-3} + 3x^2y^{-4} - 4x^3y^{-5} + R.$

23. $\sinh(x+y) = \sinh x(1 + y^2/|\underline{2} + y^4/|\underline{4} + R)$
$+ \cosh x(y + y^3/|\underline{3} + R').$

24. $\cosh(x+y) = \cosh x(1 + y^2/2 + y^4/|\underline{4} + R)$
$+ \sinh x(y + y^3/|\underline{3} + v^5/|\underline{5} + R').$

126. Stirling's Formula.—In Taylor's formula interchange the symbols x and h, and place $h = 0$, giving

$$\mathbf{f}x = \mathbf{f}0 + \mathbf{f}'0x + \mathbf{f}''0\frac{x^2}{2} + \ldots + \mathbf{f}^n0\frac{x^n}{|\underline{n}} + R, \quad . \quad (a)$$

which is Stirling's formula for developing a function of a single variable into a series arranged according to the ascending powers of the variable, with constant coefficients.*

$f0$, $f'0$, etc., represent what the given function and its successive derivatives respectively become when the variable vanishes.

Placing $fx = u$, we write

$$u = u\Big]_0 + \frac{du}{dx}\Big]_0 x + \frac{d^2u}{dx^2}\Big]_0 \frac{x^2}{2} + \dots + \frac{d^nu}{dx^n}\Big]_0 \frac{x^n}{\underline{|n}} + R.$$

To apply the formula, *differentiate the function and its derivatives in succession until one of the highest order desired is obtained. In the function and its derivatives make the variable equal to zero, and substitute the results, respectively, for their corresponding symbols in the formula.*

Thus, develop $(1 + x)^m$.

$$f(x) = (1 + x)^m, \quad f'(x) = m(1 + x)^{m-1}, \dots$$
$$f^n(x) = m(m - 1) \dots (m - n + 1)(1 + x)^{m-n}.$$

$$f(0) = 1, \quad f'(0) = m, \quad f''(0) = m(m - 1), \text{ etc.,}$$
$$f_n(0) = m(m - 1) \dots (m - n + 1).$$

Hence,

$$(1 + x)^m = 1 + mx + m(m - 1)x^2/2 + \dots$$
$$+ m(m - 1) \dots (m - n + 1)x^n/\underline{|n} + R.$$

127. From § 125 we have, by interchanging x and h and making $h = 0$,

---

* This formula is generally known as Maclaurin's, but it was published by Stirling in 1717; and not by Maclaurin till 1742. It is a particular case of Taylor's formula which was published in 1715.

$$R = f^{n+1}(\theta_n x)x^{n+1}/\underline{|n+1},$$

in which $x^{n+1}/\underline{|n+1} \ggg\!\!\rightarrow 0$ as $n \ggg\!\!\rightarrow \infty$ (§ 41).

R is therefore an infinitesimal as $n \ggg\!\!\rightarrow \infty$, provided $f^n x$ is continuous for all values of x from zero to the value assumed.

The limiting values of x in any case are the numerically least negative and positive values of x that cause fx or any of its derivatives to become discontinuous.

It is important to note that if $f^n 0$ is imaginary or infinite for any value of n, the formula is inapplicable for all values of x.

To illustrate the use of R, take the example in § 126, whence

$$R = \frac{m(m-1)\ldots(m-n)}{\underline{|n+1}}(1+\theta_n x)^{m-n-1}x^{n+1}.$$

When m is fractional or negative, and n is increased without limit, $m - n$ will become negative, giving

$$(1+x)^{m-n} = 1/(1+x)^{n-m}.$$

$f^n(x)$ will remain continuous for all positive values of x, but will become unlimited for $x = -1$. Therefore -1 is the limiting value of x, for which the complete formula gives exact results.

Writing

$$R = \frac{m(m-1)\ldots(m-n)}{\underline{|n+1}} \frac{x^{n+1}}{(1+\theta_u x)^{n-m+1}},$$

it may be shown, in a manner similar to that employed in § 125, that R is infinitesimal provided $x < 1$ numerically, and the series will be converging; otherwise not.

EXAMPLES.

1. Develop $\log_a (1 + x)$.

$$f(x) = \log_a (1 + x). \qquad \therefore \quad f(0) = 0.$$

$$f'(x) = \frac{M_a}{1 + x}. \qquad \therefore \quad f'(0) = M_a.$$

$$f''(x) = -\frac{M_a}{(1 + x)^2}. \qquad \therefore \quad f''(0) = -M_a.$$

$$f'''(x) = \frac{2M_a}{(1 + x)^3}. \qquad \therefore \quad f'''(0) = 2M_a.$$

$$f^{iv}(x) = -\frac{\underline{|3} M_a}{(1 + x)^4}. \qquad \therefore \quad f^{iv}(0) = -\underline{|3} M_a.$$

etc. etc.

$$f^n(x) = (-1)^{n-1}\frac{\underline{|n - 1} M_a}{(1 + x)^n}. \quad f^n(0) = (-1)^{n-1}\underline{|n - 1} M_a.$$

Hence,

$$\log_a (1 + x) = M_a\left(x - \frac{x^2}{2} + \frac{x^3}{3} - \dots \pm \frac{x^n}{n}\right) + R,$$

and

$$\log (1 + x) = x - \frac{x^2}{2} + \frac{x^3}{3} - \dots \pm \frac{x^n}{n} + R',$$

in which $R' = (-1)^n \frac{1}{n + 1} \frac{x^{n+1}}{(1 + \theta_n x)^{n+1}}$ is an infinitesimal under the law that n increases without limit, provided $x =$ or < 1 numerically.

2. $$(1 - x^2)^{1/2} = 1 - x^2/2 - x^4/8 - \text{etc.} + R.$$

3. $$\sin x = x - x^3/\underline{|3} + x^5/\underline{|5} - x^7/\underline{|7} + R,$$

in which $R = \sin \left[(n + 1)\pi/2 + \theta_n x\right] x^{n+1}/\underline{|n + 1}$ is an infinitesimal.

4. $$\cos x = 1 - x^2/\underline{|2} + x^4/\underline{|4} - x^6/\underline{|6} + R$$

in which $R = \cos \left((n + 1)\pi/2 + \theta_n x\right) x^{n+1}/\underline{|n + 1}$.

Since $\cos x = d \sin x / dx$, the development of $\cos x$ may be obtained by differentiating that of $\sin x$

The radian measure of 1′ is 0.000291 —, for which value the developments of sin 1′ and cos 1′ converge rapidly, giving their values with great accuracy.

$$5.\quad a^x = 1 + \log a \,.\, x + \log^2 a \frac{x^2}{\underline{2}} + \ldots + \log^n a \frac{x^n}{\underline{n}} + R,$$

in which $R = a^{\theta_n x} \log^{n+1} a \; x^{n+1} / \underline{|n+1}$ diminishes as n increases without limit.

Placing $a = e$, we have

$$e^x = 1 + x + \frac{x^2}{2} + \ldots + \frac{x^n}{\underline{|n}} + \frac{e^{\theta_n x} x^{n+1}}{\underline{|n+1}},$$

in which $x = 1$ gives

$$e = 1 + 1 + \frac{1}{2} + \frac{1}{\underline{|3}} + \frac{1}{\underline{|4}} + \ldots + \frac{e^{\theta_n}}{\underline{|n+1}},$$

and $x = x\sqrt{-1}$ gives

$$e^{x\sqrt{-1}} = 1 + x\sqrt{-1} - \frac{x^2}{2} - \frac{x^3\sqrt{-1}}{\underline{|3}} + \frac{x^4}{\underline{|4}} + \text{etc.}$$

$$= \left(1 - \frac{x^2}{2} + \frac{x^4}{\underline{|4}} - \text{etc.}\right) + \left(x - \frac{x^3}{\underline{|3}} + \text{etc.}\right)\sqrt{-1}$$

$$= \cos x + \sqrt{-1} \sin x. \quad \ldots\ldots\ldots \quad (a)$$

Substituting $-x$ for x, we have

$$e^{-x\sqrt{-1}} = \cos x - \sqrt{-1} \sin x. \quad \ldots\ldots \quad (b)$$

Hence,

$$\left.\begin{aligned} \cos x &= \left(e^{x\sqrt{-1}} + e^{-x\sqrt{-1}}\right)/2, \\ \sin x &= \left(e^{x\sqrt{-1}} - e^{-x\sqrt{-1}}\right)/2\sqrt{-1}, \end{aligned}\right\} \quad \ldots\ldots \quad (c)$$

which are known as Euler's expressions for the sine and cosine.

From (c) we obtain

$$\sqrt{-1} \tan x = \frac{e^{x\sqrt{-1}} - e^{-x\sqrt{-1}}}{e^{x\sqrt{-1}} + e^{-x\sqrt{-1}}} = \frac{e^{2x\sqrt{-1}} - 1}{e^{2x\sqrt{-1}} + 1}. \quad \ldots \quad (d)$$

In (*a*) and (*b*) put $x = mx$, then

$$e^{\pm mx\sqrt{-1}} = \cos mx \pm \sqrt{-1} \sin mx,$$

or, since $e^{\pm mx\sqrt{-1}} = \left(e^{\pm x\sqrt{-1}}\right)^m = (\cos x \pm \sqrt{-1} \sin x)^m$,

$$\cos mx \pm \sqrt{-1} \sin mx = (\cos x \pm \sqrt{-1} \sin x)^m,$$

which is De Moivre's formula.

Expanding the second member by the binomial formula, and equating separately the real and imaginary parts, we obtain, m being a positive integer, finite expansions for $\cos mx$ and $\sin mx$ in terms of $\cos x$ and $\sin x$.

Thus, $m = 3$ gives

$$\cos 3x \pm \sqrt{-1} \sin 3x = \cos^3 x + 3 \cos^2 x(\pm \sqrt{-1} \sin x) - 3 \cos x \sin^2 x \mp \sqrt{-1} \sin^3 x.$$

Hence,
$$\cos 3x = \cos^3 x - 3 \cos x \sin^2 x,$$
$$\sin 3x = 3 \cos^2 x \sin x - \sin^3 x.$$

In (*c*) place $\sqrt{-1} = i$, giving

$$\cos x = (e^{xi} + e^{-xi})/2, \quad \sin x = (e^{xi} - e^{-xi})/2i.$$

From which, putting xi for x, and multiplying both members of the second by i^2, we have

$$\cos xi = (e^x + e^{-x})/2, \quad \sin xi = i(e^x - e^{-x})/2.$$

If $f(-x) = -f(x)$, the development of $f(x)$ will contain powers of x of an odd degree only. Such functions are called odd functions.* Entire functions all terms of which are of an odd degree with respect to the variable are examples, also $\sin x$, $\tan x$, $\cot x$, $\operatorname{cosec} x$, $\sin^{-1}x$, and $\tan^{-1}x$.

If $f(-x) = f(x)$, the development of $f(x)$ will contain powers of x of an even degree only. Such functions are

---

* Rice and Johnson's Calculus, § 169.

called even functions, of which $\cos x$, $\sec x$, $\text{vers}\, x$, and all entire functions containing powers of the variable of an even degree only, are examples.

6. Develop $\sec x$.

$$f'(x) = \sec x \tan x, \qquad \therefore\ f'(0) = 0.$$
$$f''(x) = \sec x\,(1 + 2\tan^2 x), \qquad \therefore\ f''(0) = 1.$$
$$f'''(x) = \sec x \tan x\,(5 + 6\tan^2 x), \qquad \therefore\ f'''(0) = 0.$$
$$f^{\text{iv}}(x) = \sec x\,(5 + 28\tan^2 x + 24\tan^4 x), \qquad \therefore\ f^{\text{iv}}(0) = 5.$$

etc. etc.

Hence, $\sec x = 1 + x^2/2 + 5x^4/24 + 61x^6/720 + R.$

7. Develop $\cos^3 x$.

$$f'(x) = 3\sin x\,(\sin^2 x - 1), \qquad \therefore\ f'(0) = 0.$$
$$f''(x) = 3\cos x\,(3\sin^2 x - 1), \qquad \therefore\ f''(0) = -3.$$
$$f'''(x) = 3\sin x\,(1 - 3\sin^2 x + 6\cos^2 x), \qquad \therefore\ f'''(0) = 0.$$
$$f^{\text{iv}}(x) = 3\cos x\,(1 - 9\sin^2 x - 12\sin^2 x + 6\cos^3 x), \qquad \therefore\ f^{\text{iv}}(0) = 21.$$

etc. etc

Hence, $\cos^3 x = 1 - 3x^2/2 + 7x^4/8 - \text{etc.}$

8. Develop $(1 + e^x)^n$.

$$f'(x) = n(1 + e^x)^{n-1} e^x, \qquad \therefore\ f'(0) = n2^{n-1}.$$
$$f''(x) = n(1 + e^x)^{n-2} e^x \Big[(1 + e^x) + e^x(n - 1)\Big],$$
$$\therefore\ f''(0) = n(n + 1)2^{n-2}.$$
$$f'''(x) = n(1 + e^x)^{n-3} e^x \left[\begin{aligned}&(1 + e^x)^2 + (n - 1)(1 + e^x)e^x \\ &+ 2(n - 1)(1 + e^x)e^x + e^{2x}(n - 1)(n - 2)\end{aligned}\right],$$
$$\therefore\ f'''(0) = n^2(n + 3)2^{n-3}.$$

etc. etc.

Hence,

$$(1 + e^x)^n = 2^n\left[1 + \frac{nx}{2} + \frac{n(n + 1)x^2}{2^2.2} + \frac{n^2(n + 3)}{2^3}\frac{x^3}{\underline{|3}} + \text{etc.}\right].$$

9. Develop $e^{m \sin^{-1}x}$ with respect to $\sin^{-1}x$.

$$e^{m \sin^{-1}x} = 1 + m \sin^{-1}x + m^2(\sin^{-1}x)^2/2 + m^3(\sin^{-1}x)^3/\underline{|3} + R.$$

10. Develop $e^{m \sin^{-1}x}$ with reference to x.

The general rule may be applied, but the following method * is perhaps simpler in this case.

Place $e^{m \sin^{-1}x} = y$, then $dy/dx = me^{m \sin^{-1}x}/\sqrt{1-x^2}$,

and $d^2y/dx^2 = m^2 e^{m \sin^{-1}x}/(1-x^2) + mxe^{m \sin^{-1}x}/(1-x^2)^{3/2}$.

Hence $(1-x^2)d^2y/dx^2 - x\,dy/dx = m^2y$. . . . . . . (1)

Assume $y = A_0 + A_1x + A_2x^2 + A_3x^3 + \ldots + A_nx^n + R$.

Then $dy/dx = A_1 + 2A_2x + \ldots + nA_nx^{n-1} + R'$,
$d^2y/dx^2 = 2A_2 + \ldots + n(n-1)A_nx^{n-2} + R''$.

Substituting in (1) and equating the coefficients of x^n in the two members, we find

$$A_{n+2} = A_n(m^2+n^2)/(n+1)(n+2), \quad . \ . \ . \ . \ . \ (2)$$

from which A_2, A_3, A_4, etc., may be determined in order when A_0 and A_1 are known.

$$A_0 = e^{m \sin^{-1}x}\Big]_0 = 1. \quad A_1 = me^{m \sin^{-1}x}/\sqrt{1-x^2}\Big]_0 = m.$$

Hence (2), $A_2 = m^2/\underline{|2}$. $A_3 = m(m^2+1)/\underline{|3}$, etc., and

$$e^{m \sin^{-1}x} = 1 + mx + \frac{m^2x^2}{2} + \frac{m(m^2+1)x^3}{\underline{|3}} + \frac{m^2(m^2+4)x^4}{\underline{|4}} + R.$$

Comparing this result with that of example 9, we have, by equating the coefficients of m,

11. $\sin^{-1}x = x + \frac{x^3}{\underline{|3}} + \frac{3^2x^5}{\underline{|5}} + \frac{3^25^2x^7}{\underline{|7}} + R.$

Similarly, by equating the coefficients of m^2,

---

* Todhunter's Calculus.

12. $(\sin^{-1}x)^2 = x^2 + \frac{2^2x^4}{3 \cdot 4} + \frac{2^2 \cdot 4^2x^6}{3 \cdot 4 \cdot 5 \cdot 6} + \frac{2^2 \cdot 4^2 \cdot 6^2x^8}{3 \cdot 4 \cdot 5 \cdot 6 \cdot 7 \cdot 8} + R.$

Equating the coefficients of m^3, we have

13. $(\sin^{-1}x)^3 = x^3 + 3^2\left(1 + \frac{1}{3^2}\right)\frac{\lfloor 3}{\lfloor 5}x^5 + 3^2 \cdot 5^2\left(1 + \frac{1}{3^2} + \frac{1}{5^2}\right)\frac{\lfloor 3}{\lfloor 7}x^7 + R.$

Dividing both members of example 12 by 2 and differentiating, we have

14. $\frac{\sin^{-1}x}{\sqrt{1 - x^2}} = x + \frac{2x^3}{3} + \frac{2 \cdot 4x^5}{3 \cdot 5} + \frac{2 \cdot 4 \cdot 6x^7}{3 \cdot 5 \cdot 7} + R.$

From which, multiplying both members by $1 - x^2$, we obtain

15. $(1 - x^2)^{1/2} \sin^{-1}x = x - \frac{x^3}{3} - \frac{2}{3}\frac{x^5}{5} - \frac{2}{3}\frac{4}{5}\frac{x^7}{7} + R,$

in which, putting $x = \sin\theta$, we have

16. $\theta \cot\theta = 1 - \frac{\sin^2\theta}{3} - \frac{2}{3}\frac{\sin^4\theta}{5} - \frac{2}{3}\frac{4}{5}\frac{\sin^6\theta}{7} + R.$

When the determination of the successive derivatives of a higher order is laborious, a simpler method may be employed provided the development of $f'(x)$ is known.

Thus, since $\sin^{-1}x$ is an odd function, which vanishes with x, we assume

$$f(x) = \sin^{-1}x = Ax + Bx^3 + Cx^5 + Dx^7 + \text{etc.}$$

Differentiating, we have

$$f'(x) = \frac{1}{\sqrt{1 - x^2}} = A + 3Bx^2 + 5Cx^4 + 7Dx^6 + \text{etc.} \qquad (1)$$

Developing $1/\sqrt{1 - x^2}$, we have, provided $x < 1$,

$$(1 - x^2)^{-1/2} = 1 + \frac{x^2}{2} + \frac{3x^4}{2 \cdot 4} + \frac{3 \cdot 5x^6}{2 \cdot 4 \cdot 6} + \cdots$$

$$+ \frac{3 \cdot 5 \cdot 7 \ldots (2n - 1)x^{2n}}{2 \cdot 4 \cdot 6 \ldots 2n} + \text{etc.} \quad . \quad . \quad (2)$$

(1) and (2) are identical. Hence,

$$A = 1, \quad B = \frac{1}{2 \cdot 3}, \quad C = \frac{3}{2 \cdot 4 \cdot 5}, \quad D = \frac{3 \cdot 5}{2 \cdot 4 \cdot 6 \cdot 7},$$

and coefficient of $n + 1$ term $= \dfrac{3 \cdot 5 \cdot 7 \ldots (2n - 1)}{2 \cdot 4 \cdot 6 \ldots 2n(2n + 1)}$,

and $$\sin^{-1} x = x + \frac{x^3}{2 \cdot 3} + \frac{3x^5}{2 \cdot 4 \cdot 5} + \frac{3 \cdot 5x^7}{2 \cdot 4 \cdot 6 \cdot 7} + \ldots$$
$$+ \frac{3 \cdot 5 \cdot 7 \ldots (2n - 1)x^{2n+1}}{2 \cdot 4 \cdot 6 \ldots 2n(2n + 1)} + R,$$

which is convergent for $x < 1$, since each term is less than the corresponding term in the geometrical progression $x + x^3 + x^5 +$ etc.

$$\frac{\pi}{6} = \sin^{-1}\frac{1}{2} = \frac{1}{2} + \frac{1}{2 \cdot 3 \cdot 8} + \frac{3}{2 \cdot 4 \cdot 5 \cdot 32} + R.$$

From which $\pi = 3.14159 \ldots$

17. Develop $\tan^{-1} x$.

Since $\tan^{-1} x$ is an odd function which vanishes with x, we assume

$$f(x) = \tan^{-1} x = Ax + Bx^3 + Cx^5 + Dx^7 + \text{etc.}$$

Differentiating, we have

$$f'(x) = 1/(1 + x^2) = A + 3Bx^2 + 5Cx^4 + 7Dx^6 + \text{etc.} \quad . \quad (a)$$

Developing $1/(1 + x^2)$, we have, provided $x < 1$,

$$(1 + x^2)^{-1} = 1 - x^2 + x^4 - x^6 + \text{etc.} \quad . \quad . \quad . \quad . \quad . \quad (b)$$

(a) and (b) are identical. Hence,

$$A = 1, \quad B = -1/3, \quad C = 1/5, \quad D = -1/7, \quad \text{etc.}$$

and $$\tan^{-1} x = x - \frac{x^3}{3} + \frac{x^5}{5} - \frac{x^7}{7} + \text{etc.} + (-1)^n \frac{x^{2n+1}}{2n + 1} + R,$$

in which R is an infinitesimal as $n \ggg\!\!\to \infty$ provided $x =$ or < 1 numerically.

If $x = 1 = \tan(\pi/4)$, we have

$$\pi/4 = 1 - 1/3 + 1/5 - 1/7 + R.$$

To obtain a development which converges more rapidly, let $\phi = \tan^{-1}(1/5)$, and $\theta = 4\phi$; then $\theta = 4 \tan^{-1}(1/5)$. Hence,

$$\tan\theta = \frac{4\tan\phi - 4\tan^3\phi}{1 - 6\tan^2\phi + \tan^4\phi} = \frac{120}{119}.$$

We also have

$$\tan(\theta - 45^\circ) = (\tan\theta - 1)/(1 + \tan\theta) = 1/239.$$

Therefore

$$\theta - 45^\circ = \tan^{-1}(1/239), \quad \text{and} \quad 45^\circ = \theta - \tan^{-1}(1/239),$$

or $$\pi/4 = 4\tan^{-1}(1/5) - \tan^{-1}(1/239).^*$$

Developing $\tan^{-1}(1/5)$ and $\tan^{-1}(1/239)$, and substituting in above,

$$\frac{\pi}{4} = 4\left(\frac{1}{5} - \frac{1}{3(5)^3} + \frac{1}{5(5)^5} - \frac{1}{7(5)^7} + \text{etc.}\right) - \left(\frac{1}{239} - \frac{1}{3(239)^3} + \frac{1}{5(239)^5} - \text{etc.}\right),$$

which converges rapidly. Seven terms of the first set and three in the second give $\pi = 3.141592653589793\ldots$†

18. $\dfrac{x}{1+x} + \log(1+x) = 2x - \dfrac{3x^2}{2} + \dfrac{4x^3}{3} - \dfrac{5x^4}{4} + R.$

19. $e^x \cos x = 1 + x - 2x^3/\underline{|3} - 4x^5/\underline{|5} + R.$

20. $e^{\sin x} = 1 + x + x^2/2 - x^4/8 - x^5/15 - x^6/240 + R.$

21. $\log\left(x + \sqrt{1 + x^2}\right) = x - x^3/\underline{|3} + 3^2x^5/\underline{|5} + R.$

22. $e^x x^2 = x^2 + x^3 + x^4/\underline{|2} + x^5/\underline{|3} + R.$

23. $(1 + x)/(1 - x) = 1 + 2x + 2x^2 + 2x^3 + R.$

24. $\log\dfrac{x}{x-1} = \dfrac{1}{x-1} - \dfrac{1}{2(x-1)^2} + \dfrac{1}{3(x-1)^3} + R.$

---

* Known as Machin's Formula.

† Haddon's Differential Calculus.

[*Suggestion.*—Put $x/(x-1) = 1 - z$, and in development of $(1-z)$ put $z = 1/(1-x)$.]

25. $(1 + 2x + 3x^2)^{-1/2} = 1 - x + 2x^3 - 7x^4/4 + 3x^5/2 + R.$

26. $\cos^{-1} x = \pi/2 - x - x^3/2.3 - 3x^5/2.4.5 + R.$

27. $x^2/(e^x - x) = x^2 - x^4/2 - x^5/\underline{|3} + R.$

28. $(1 - x^2)^{-2} = 1 + 2x^2 + 3x^4 + 4x^6 + R.$

29. $e^x/\cos x = 1 + x + x^2 + 2x^3/3 + x^4/2 + 3x^5/10 + R.$

30. $e^{x \sin x} = 1 + x^2 + x^4/3 + R.$

31. $\sqrt{1 + 4x + 12x^2} = 1 + 2x + 4x^2 + R.$

32. $e^{-1/x^2} = 1 - 1/x^2 + 1/2x^4 - 1/6x^6 + R.$

33. $e^{-x^2} = 1 - x^2 + x^4/2 + R.$

34. $(1 + x^2)^{5/3} = 1 + 5x^2/3 + 5x^4/9 - 5x^6/81 + R.$

35. $e^{\tan^{-1} x} = 1 + x + x^2/2 - x^3/6 - 7x^4/24 + R.$

36. $\sin^2 x = x^2 - x^4/3 + 2x^6/3^2.5 + R.$

37. $\log \sec x = x^2/2 + x^4/12 + x^6/45 + R.$

38. $(a^2 + bx)^{1/2} = a + bx/2a - b^2x^2/8a^3 + 3b^3x^3/48a^5 + R.$

39. $e^x \log(1 + x) = x + x^2/2 + 2x^3/\underline{|3} + 9x^5/\underline{|5} + R.$

40. $\tan x = x + x^3/3 + 2x^5/15 + 17x^7/315 + R.$

41. $\log(1 + \sin x) = x - x^2/2 + x^3/6 - x^4/12 + R.$

42. $\log(1 + e^x) = \log 2 + x/2 + x^2/2^3 - x^4/2^3\underline{|4} + R.$

43. $(e^x + e^{-x})^n = 2^n(1 + nx^2/2 + n(3n - 2)x^4/\underline{|4}) + R.$

44. $\cot x = 1/x - x/3 - x^3/3^2.5 - 2x^5/3^3.5.7 + R.$ (By method of Indeterminate Coefficients.)

45. $\tan^4 x = x^4 + 4x^6/3 + 6x^8/5 + R.$

46. By means of Taylor's and Stirling's formulas deduce the following:

$$\sin(x \pm y) = \sin x \cos y \pm \cos x \sin y,$$
$$\cos(x \pm y) = \cos x \cos y \mp \sin x \sin y.$$

47. $\sinh x = (e^x - e^{-x})/2 = x + x^3/\underline{|3} + x^5/\underline{|5} + R.$

48. $\cosh x = (e^x + e^{-x})/2 = 1 + x^2/2 + x^4/\lfloor 4 + R.$

49. $\cosh^n x = 1 + nx^2/2 + n(3n - 2)x^4/\lfloor 4 + R.$

50. $\log(\sinh x/x) = x^2/6 - x^4/180 + R.$

51. $\tanh^{-1} x = x + x^3/3 + x^5/5 + R.$

52. $\sinh^{-1}(x/a) = x/a - x^3/2 \cdot 3a^3 + 3x^5/2 \cdot 4 \cdot 5a^5 + R.$

53. Given $y^3 - 3y + x = 0$; develop y in terms of the ascending powers of x.

$$y = fx, \qquad f0 = 0 \quad \text{or} \quad \pm\sqrt{3}.$$

$3y^2\, dy/dx - 3dy/dx + 1 = 0$ gives

$$dy/dx\Big]_0 = -1/(3y^2 - 3)\Big]_0 = 1/3, \qquad \text{for } y = 0.$$

$2y(dy/dx)^2 + y^2(d^2y/dx^2) - d^2y/dx^2 = 0$ gives

$$d^2y/dx^2\Big]_0 = -6y(dy/dx)^2/(3y^2 - 3)\Big]_0 = 0.$$

$$2\left(\frac{dy}{dx}\right)^3 + 4y\left(\frac{dy}{dx}\right)\frac{d^2y}{dx^2} + 2y\frac{dy}{dx}\frac{d^2y}{dx^2} + y^2\frac{d^3y}{dx^3} - \frac{d^3y}{dx^3} = 0 \text{ gives}$$

$$d^3y/dx^3\Big]_0 = 2/27.$$

Hence, $$y = x/3 + x^3/3^4 + R.$$

54. Given $2y^3 - yx - 2 = 0$; show that

$$y = 1 + x/2 \cdot 3 - x^3/2^3 3^4 + R.$$

55. Given $y^2 - 8/y = 6x$; show that

$$y = 2 + x - x^3/2^2 3 + x^4/2 \cdot 3 \cdot 4 + R.$$

56. Given $y^3x - 8y - 8x = 0$; show that

$$y = -x - x^4/8 - 3x^7/2^6 + R.$$

57. Given $4y^3x - y - 4 = 0$; show that

$$y = -4 - 4^4x - 3(4)^7x^2 + R.$$

58. Given $y^3 - a^2y + ayx - x^3 = 0$; show that

$$y = -x^3/a^2 - x^4/a^3 - x^5/a^4 + R.$$

59. Given $\sin\phi = x\sin(a+\phi)$; show that

$$\phi = n\pi + \sin a \,.\, x + \sin 2a \,.\, x^2/2 + 2\sin a \,.\, (3 - 4\sin^2 a)x^3/\underline{|3} + R.$$

60. Develop $\dfrac{x}{2}\dfrac{e^x+1}{e^x-1}$.*

By Stirling's formula we write

$$f(x) = \frac{x}{2}\frac{e^x+1}{e^x-1} = f(0) + f'(0)x + f''(0)\frac{x^2}{2} + R.$$

Since $fx = f(-x)$, the development contains no power of x of an odd degree.

Writing $(e^x+1)/(e^x-1) = 1 + 2/(e^x-1)$, we have

$$f(x) = \frac{x}{2}\frac{e^x+1}{e^x-1} = \frac{x}{2} + \frac{x}{e^x-1} = \frac{x}{2} + xe^{-x} + \frac{xe^{-x}}{e^x-1},$$

$$e^x f(x) = \frac{e^x x}{2} + x + \frac{x}{e^x-1} = f(x) + \frac{x}{2} + \frac{xe^x}{2}$$

Differentiating, we have

$$e^x\Big(f(x) + f'(x)\Big) = f'(x) + \frac{1}{2} + \frac{e^x(x+1)}{2}. \quad \ldots \ldots \quad (1)$$

$$e^x\Big(f(x) + 2f'(x) + f''(x)\Big) = f''(x) + \frac{e^x(x+2)}{2}. \quad \ldots \ldots \quad (2)$$

$$e^x\Big(f(x) + 3f'(x) + 3f''(x) + f'''(x)\Big) = f'''(x) + \frac{e^x(x+3)}{2}. \quad (3)$$

etc. etc.

Making $x = 0$ in (1), (3), etc., we find

$$f(0) = 1,\quad f''(0) = 1/6,\quad f^{\text{iv}}(0) = -1/30,\quad f^{\text{vi}}(0) = 1/42,$$
$$f^{\text{viii}}(0) = -1/30, \quad \text{etc.}$$

Hence,

$$\frac{x}{2}\frac{e^x+1}{e^x-1} = 1 + \frac{1}{6}\frac{x^2}{2} - \frac{1}{30}\frac{x^4}{\underline{|4}} + \frac{1}{42}\frac{x^6}{\underline{|6}} - \frac{1}{30}\frac{x^8}{\underline{|8}} + R;$$

* Edward's Differential Calculus.

or, placing

$$\frac{1}{6} = B_1, \quad \frac{1}{30} = B_3, \quad \frac{1}{42} = B_5, \quad \frac{1}{30} = B_7, \quad \text{etc.,}$$

$$\frac{x}{2} + \frac{x}{e^x - 1} = 1 + B_1\frac{x^2}{2} - B_3\frac{x^4}{\underline{|4}} + B_5\frac{x^6}{\underline{|6}} - B_7\frac{x^8}{\underline{|8}} + R.$$

The coefficients represented by B_1, B_3, B_5, etc., are used in the higher branches of analysis, and are called Bernoulli's numbers.

128. Extension of Taylor's Formula to functions of two or more sums of two variables each.

Let $u = f(x, y)$, x and y being independent variables; and let it be required to develop $f(x+h, y+k)$, in which h and k are variable increments of x and y respectively.

If, in $f(x, y)$, x be increased by h, and $f(x+h, y)$ be developed by Taylor's formula; then if, in each term of the result, y be increased by k and developed in a similar manner as a function of $y+k$, we shall have $f(x+h, y+k) =$ sum of the latter developments.

Otherwise, develop $\phi(t) = f(x+ht, y+kt)$ as a function of t, by Stirling's formula, and in the result make $t = 1$.

$$\phi(t) = f(x+ht, y+kt), \quad \therefore \phi(0) = f(x, y) = u.$$

In order to express conveniently the successive derivatives with respect to t, place $x + ht = w$, and $y + kt = s$, giving $\phi(t) = f(w, s)$.

Hence (§ 102),

$$\frac{d\phi(t)}{dt} = \frac{\partial\phi(t)}{\partial w}\frac{\partial w}{dt} + \frac{\partial\phi(t)}{\partial s}\frac{\partial s}{dt}.$$

But $\frac{\partial\phi(t)}{\partial w} = \frac{\partial\phi(t)}{dx}, \qquad \frac{\partial\phi(t)}{\partial s} = \frac{\partial\phi(t)}{dy}$. (Ex. 7, p. 71.)

Also $\partial w/dt = h, \qquad \partial s/dt = k.$

Hence, $\frac{d\phi(t)}{dt} = \phi'(t) = \frac{\partial\phi(t)}{dx}h + \frac{\partial\phi(t)}{dy}k.$

$$\phi''(t) = \frac{d}{dt}\left(\frac{\partial\phi(t)}{dx}h + \frac{\partial\phi(t)}{dy}k\right) = h\frac{d}{dt}\left(\frac{\partial\phi(t)}{dx}\right) + k\frac{d}{dt}\left(\frac{\partial\phi(t)}{dy}\right).$$

$$\frac{d}{dt}\left(\frac{\partial\phi(t)}{dx}\right) = \frac{\partial^2\phi(t)}{dx\,\partial w}\frac{\partial w}{dt} + \frac{\partial^2\phi(t)}{dx\,\partial s}\frac{\partial s}{dt} = \frac{\partial^2\phi(t)}{dx^2}h + \frac{\partial^2\phi(t)}{dx\,dy}k.$$

Similarly, $\frac{d}{dt}\left(\frac{\partial\phi(t)}{dy}\right) = \frac{\partial^2\phi(t)}{dy^2}k + \frac{\partial^2\phi(t)}{dy\,dx}h.$

Hence,

$$\phi''(t) = \frac{\partial^2\phi(t)}{dx^2}h^2 + 2\frac{\partial^2\phi(t)}{dx\,dy}hk + \frac{\partial^2\phi(t)}{dy^2}k^2.$$

etc. etc.

$$\phi^{n+1}(t) = \partial^{n+1}\phi(t)h^{n+1}/dx^{n+1} + (n+1)\partial^{n+1}\phi(t)h^n k/dx^n dy$$
$$+ \ldots + \partial^{n+1}\phi(t)k^{n+1}/dy^{n+1}.$$

Therefore

$$\phi'(0) = (\partial u/dx)h + (\partial u/dy)k,$$

$$\phi''(0) = \frac{\partial^2 u}{dx^2}h^2 + 2\frac{\partial^2 u}{dx\,dy}hk + \frac{\partial^2 u}{dy^2}k^2,$$

etc. etc.

$$\phi^{n+1}(\theta_n t) = \left[\frac{\partial^{n+1}\phi(t)}{dx^{n+1}}h^{n+1} + (n+1)\frac{\partial^{n+1}\phi(t)}{dx^n\,dy}h^n k + \text{etc.}\right.$$
$$\left. + \frac{\partial^{n+1}\phi(t)}{dy^{n+1}}k^{n+1}\right]_{t=\theta_n t}.$$

Substituting in Stirling's formula, and making $t = 1$, we have

$$\left.\begin{aligned} f(x+h, y+k) &= u + \frac{\partial u}{dx}h + \frac{\partial u}{dy}k \\ &+ \frac{1}{2}\left[\frac{\partial^2 u}{dx^2}h^2 + 2\frac{\partial^2 u}{dx\,dy}hk + \frac{\partial^2 u}{dy^2}k^2\right] \\ &+ \frac{1}{\underline{|3}}\left[\frac{\partial^3 u}{dx^3}h^3 + 3\frac{\partial^3 u}{dx^2\,dy}h^2k + 3\frac{\partial^3 u}{dx\,dy^2}hk^2 + \frac{\partial^3 u}{dy^3}k^3\right] \\ &+ \qquad \text{etc.} \qquad\qquad \text{etc.} \\ &+ \frac{1}{\underline{|n+1}}\left[\frac{\partial^{n+1}\phi(t)}{dx^{n+1}}h^{n+1} + (n+1)\frac{\partial^{n+1}\phi(t)}{dx^n\,dy}h^nk + \text{etc.} \right. \\ &\qquad\qquad \left. + \frac{\partial^{n+1}\phi(t)}{dy^{n+1}}k^{n+1}\right]_{t=\theta_n}. \end{aligned}\right\} \quad (a)$$

It should be noticed that
$\phi^{n+1}(t)$ is, in general, a function of $x + ht$ and $y + kt$;
$\phi^{n+1}(0)$ is the same function of x and y;
$\phi^{n+1}(\theta_n t)$ is the same function of $x + h\theta_n t$ and $y + k\theta_n t$ ·
$\phi^{n+1}(\theta_n)$ is the same function of $x + \theta_n h$ and $y + \theta_n k$.
Therefore the remainder term in (a), which is equivalent to $\frac{1}{\underline{|n+1}}\phi^{n+1}(\theta_n)$, is the same function of $x + \theta_n h$ and $y + \theta_n k$ that $\frac{1}{\underline{|n+1}}\phi^{n+1}(0)$ is of x and y. Hence, the remainder term, denoted by R, may be written

$$R = \frac{1}{\underline{|n+1}}\left[\frac{\partial^{n+1}u}{dx^{n+1}}h^{n+1} + (n+1)\frac{\partial^{n+1}u}{dx^n\,dy}h^nk + \text{etc.} + \frac{\partial^{n+1}u}{dy^{n+1}}k^{n+1}\right]_{\substack{x=x+\theta_n h \\ y=y+\theta_n k}}.$$

From § 125 and § 127 we see that formula (a) develops the given function provided that, as n increases without limit, $\phi^n(0)$ is real and finite and R is an infinitesimal or, what is equivalent, provided that u and all of its successive partial derivatives of the nth order are continuous between all states, corresponding to values of x and y from any assumed values to $x + h$ and $y + k$, under the same law.

Having $u = f(x, y, z)$, we may, in a manner analogous to above, deduce

$$\left.\begin{aligned} f(x+h, y+k, z+l) &= u + \frac{\partial u}{dx}h + \frac{\partial u}{dy}k + \frac{\partial u}{dz}l \\ &+ \frac{1}{2}\left[\frac{\partial^2 u}{dx^2}h^2 + \frac{\partial^2 u}{dy^2}k^2 + \frac{\partial^2 u}{dz^2}l^2\right. \\ &\left. + 2\left(\frac{\partial^2 u}{dx\,dy}hk + \frac{\partial^2 u}{dx\,dz}hl + \frac{\partial^2 u}{dy\,dz}kl\right)\right] \\ &+ \quad \text{etc.} \qquad \text{etc.} \\ &+ \frac{1}{\lfloor n+1}\left[\frac{\partial}{dx}h + \frac{\partial}{dy}k + \frac{\partial}{dz}l\right]^{n+1} u_{\substack{x=x+\theta_n h \\ y=y+\theta_n k \\ z=z+\theta_n l}}, \end{aligned}\right\} \quad (b)$$

the remainder term being indicated by the symbolic form described in § 109.

In a similar manner, a formula for the development of a function of any number of sums of two variables each may be deduced.

129. Extension of Stirling's Formula.—In (*a*), § 128, put $x = 0$ and $y = 0$; then write x and y for h and k respectively, giving

$$\begin{aligned} u = f(x, y) = f(0, 0) &+ \left[\frac{\partial u}{dx}x + \frac{\partial u}{dy}y\right]_{(0,0)} \\ &+ \frac{1}{2}\left[\frac{\partial^2 u}{dx^2}x^2 + 2\frac{\partial^2 u}{dx\,dy}xy + \frac{\partial^2 u}{dy^2}y^2\right]_{(0,0)} \\ &+ \text{etc.} \\ &+ \frac{1}{\lfloor n+1}\left[\frac{\partial}{dx}x + \frac{\partial}{dy}y\right]^{n+1} u_{(\theta_n x,\, \theta_n y)} \end{aligned}$$

which is a formula for the development of any function of two variables in which $f(0, 0)$, $(\partial u/dx)_0$, $(\partial^2 u/dy^2)_0$, etc., denote constants resulting from making $x = 0$ and $y = 0$ in u, $\partial u/dx$, $\partial^2 u/dy^2$, etc., respectively.

The conditions of applicability, for any assumed values of x and y, are that u and all of its successive partial derivatives shall be continuous for all values of x and y from 0 to those assumed.

In a similar manner we may deduce from (b), § 128, and its extension, corresponding formulas for the development of any function of three or more variables.

EXAMPLES.

1. Develop $(x+h)^m(y+k)^n$.

$$u = f(x, y) = x^m y^n, \quad \frac{\partial u}{dx} = mx^{m-1}y^n, \quad \frac{\partial u}{dy} = nx^m y^{n-1},$$

$$\frac{\partial^2 u}{dx^2} = m(m-1)x^{m-2}y^n, \quad \frac{\partial^2 u}{dx\,dy} = mnx^{m-1}y^{n-1}, \quad \frac{\partial^2 u}{dy^2} = n(n-1)x^m y^{n-2},$$

etc. etc. etc.

Substituting in formula (a), § 128,

$$(x+h)^m(y+k)^n = x^m y^n + mx^{m-1}y^n h + nx^m y^{n-1}k$$
$$+ m(m-1)x^{m-2}y^n h^2/2 + mnx^{m-1}y^{n-1}hk$$
$$+ n(n-1)x^m y^{n-2}k^2/2 + R.$$

2. Develop $(x+h)^2[(a+y)+k]^3$.

$$u = f(x, y) = x^2(a+y)^3, \quad \frac{\partial u}{dx} = 2x(a+y)^3, \quad \frac{\partial u}{dy} = 3x^2(a+y)^2.$$

$$\frac{\partial^2 u}{dx^2} = 2(a+y)^3, \quad \frac{\partial^2 u}{dx\,dy} = 6x(a+y)^2, \quad \frac{\partial^2 u}{dy^2} = 6x^2(a+y),$$

$$\frac{\partial^3 u}{dx^3} = 0, \quad \frac{\partial^3 u}{dx^2\,dy} = 6(a+y)^2, \quad \frac{\partial^3 u}{dx\,dy^2} = 12x(a+y), \quad \frac{\partial^3 u}{dy^3} = 6x^2,$$

$$\frac{\partial^4 u}{dx^2\,dy^2} = 12(a+y), \quad \frac{\partial^4 u}{dx\,dy^3} = 12x, \quad \frac{\partial^5 u}{dx^2\,dy^3} = 12.$$

Hence, substituting in formula (a), § 128,

$$(x+h)^2(a+y+k)^3 = x^2(a+y)^3 + 2x(a+y)^3h + 3x^2(a+y)^2k$$
$$+ (a+y)^3h^2 + 6x(a+y)^2hk + 3x^2(a+y)^2k^2$$
$$+ 3(a+y)^2h^2k + 6x(a+y)hk^2 + x^2k^3$$
$$+ 3(a+y)h^2k^2 + 2xhk^3 + h^2k^3.$$

3. Develop $u = e^x \sin y$.

$$f(0, 0) = 0, \quad \left(\frac{\partial u}{dx}\right)_0 = \left(\frac{\partial^2 u}{dx^2}\right)_0 = \text{etc.} = 0, \quad \left(\frac{\partial u}{dy}\right)_0 = 1,$$

$$\left(\frac{\partial^2 u}{dx\,dy}\right)_0 = 1, \quad \left(\frac{\partial^2 u}{dy^2}\right)_0 = 0,$$

$$\left(\frac{\partial^3 u}{dx^2 dy}\right)_0 = 1, \quad \left(\frac{\partial^3 u}{dx\,dy^2}\right) = 0, \quad \left(\frac{\partial^3 u}{dy^3}\right)_0 = -1.$$

etc. etc.

Hence, $e^x \sin y = y + xy + \frac{x^2y}{2} - \frac{y^3}{6} + \frac{x^3y}{6} - \frac{xy^3}{6} + R.$

130. Theorems of Lagrange and Laplace.

Suppose
$$y = z + x\phi(y), \quad . \quad . \quad . \quad . \quad . \quad . \quad . \quad (1)$$
in which x and z are independent, and let it be required to develop $f(y)$ according to the ascending powers of x.

Placing $u = f(y)$, in which case u will also be a function of x and z, then developing by Stirling's formula, we have

$$u = u_{x=0} + \left(\frac{\partial u}{dx}\right)_{x=0}\frac{x}{1} + \left(\frac{\partial^2 u}{dx^2}\right)_{x=0}\frac{x^2}{1\,.\,2} + \ldots$$
$$+ \left(\frac{\partial^n u}{dx^n}\right)_{x=0}\frac{x^n}{\lfloor n} + R, \quad . \quad . \quad . \quad . \quad (2)$$

in which the coefficients of the different powers of x are functions of z. In order to determine their values, we transform $\frac{\partial u}{dx}$, $\frac{\partial^2 u}{dx^2}$, etc., before making $x = 0$.

Differentiating (1) with respect to x, we obtain

$$\partial y/dx = \phi(y) + x[\partial\phi(y)/\partial y](\partial y/dx),$$

whence $$\partial y/dx = \phi(y)/[1 - x\partial\phi(y)/\partial y].$$

Differentiating (1) with respect to z, we have

$$\partial y/dz = 1 + x[\partial\phi(y)/\partial y](\partial y/dz),$$

giving $$\partial y/dz = 1/[1 - x\partial\phi(y)/\partial y].$$

Hence $$\partial y/dx = \phi(y)(\partial y/dz).$$

Since $$\frac{\partial u}{dx} = \frac{\partial u}{\partial y}\frac{\partial y}{dx}, \quad \frac{\partial u}{dz} = \frac{\partial u}{\partial y}\frac{\partial y}{dz},$$

we have

$$\partial u/dx = \phi(y)(\partial u/dz). \quad . \quad . \quad . \quad . \quad (3)$$

Observing that

$$\frac{\partial}{dx}\left(\phi(y)\frac{\partial y}{dz}\right) = \frac{\partial}{dz}\left(\phi(y)\frac{\partial y}{dx}\right)$$
$$= \frac{\partial\phi(y)}{\partial y}\frac{\partial y}{dx}\frac{\partial y}{dz} + \phi(y)\frac{\partial^2 y}{dz\,dx}, \quad . \quad . \quad (4)$$

and that $u = f(y)$ gives

$$\frac{\partial u}{dz} = \frac{\partial f(y)}{\partial y}\frac{\partial y}{dz}, \quad \frac{\partial u}{dx} = \frac{\partial f(y)}{\partial y}\frac{\partial y}{dx},$$

we have, differentiating (3),

$$\frac{\partial^2 u}{dx^2} = \frac{\partial}{dx}\left(\phi(y)\frac{\partial u}{dz}\right) = \frac{\partial}{dx}\left(\phi(y)\frac{\partial f(y)}{\partial y}\frac{\partial y}{dz}\right)$$
$$= \frac{\partial}{dz}\left(\phi(y)\frac{\partial f(y)}{\partial y}\frac{\partial y}{dx}\right) = \frac{\partial}{dz}\left(\phi(y)\frac{\partial u}{dx}\right).$$

Hence, by (3), $\dfrac{\partial^2 u}{dx^2} = \dfrac{\partial}{dz}\left(\overline{\phi(y)}^2 \dfrac{\partial u}{dz}\right).$

Again,

$$\frac{\partial^3 u}{dx^3} = \frac{\partial^2}{dx\,dz}\left(\overline{\phi(y)}^2 \frac{\partial u}{dz}\right) = \frac{\partial^2}{dz^2}\left(\overline{\phi(y)}^2 \frac{\partial u}{dx}\right), \text{ by (4).}$$

Hence, by (3), $\dfrac{\partial^3 u}{dx^3} = \dfrac{\partial^2}{dz^2}\left(\overline{\phi(y)}^3 \dfrac{\partial u}{dz}\right).$

Similarly, from $\dfrac{\partial^n u}{dx^n} = \dfrac{\partial^{n-1}}{dz^{n-1}}\left(\overline{\phi(y)}^n \dfrac{\partial u}{dz}\right),$

we deduce $\dfrac{\partial^{n+1} u}{dx^{n+1}} = \dfrac{\partial^n}{dz^n}\left(\overline{\phi(y)}^{n+1} \dfrac{\partial u}{dz}\right),$

which shows that the law of formation is general.

$x = 0$ gives $y = z$, $u = f(z)$, and $\dfrac{\partial u}{dz} = \dfrac{\partial f(z)}{dz}.$

Hence,

$$\left.\frac{\partial u}{dx}\right]_{x=0} = \phi(z)\frac{\partial f(z)}{dz},$$

$$\left.\frac{\partial^2 u}{dx^2}\right]_{x=0} = \frac{\partial}{dz}\left(\overline{\phi(z)}^2 \frac{\partial f(z)}{dz}\right),$$

etc. etc.

$$\left.\frac{\partial^n u}{dx^n}\right]_{x=0} = \frac{\partial^{n-1}}{dz^{n-1}}\left(\overline{\phi(z)}^n \frac{\partial f(z)}{dz}\right).$$

Substituting these expressions in (2), we have

$$f(y) = f(z) + x\phi(z)\frac{\partial f(z)}{dz} + \frac{x^2}{2}\frac{\partial}{dz}\left(\overline{\phi(z)}^2 \frac{\partial f(z)}{dz}\right) + \frac{x^3}{\underline{|3}}\frac{\partial^2}{dz^2}\left(\overline{\phi(z)}^3 \frac{\partial f(z)}{dz}\right) + \dots + \frac{x^n}{\underline{|n}}\frac{\partial^{n-1}}{dz^{n-1}}\left(\overline{\phi(z)}^n \frac{\partial f(z)}{dz}\right) + R,$$

which is called *Lagrange's Theorem.*

Suppose $t = F\big(z + x\phi(t)\big)$.

Placing $y = z + x\phi(t)$, we have

$$t = F(y), \quad u = f(t) = f\big(F(y)\big), \quad \text{and} \quad y = z + x\phi\big(F(y)\big),$$

to which Lagrange's theorem is applicable provided we write $f(F)$ for f, and $\phi(F)$ in place of ϕ; therefore we have

$$f(y) = f\big(F(z)\big) + x\phi\big(F(z)\big)\frac{\partial f\big(F(z)\big)}{dz}$$
$$+ \frac{x^2}{2}\frac{\partial}{dz}\left[\overline{\phi\big(F(z)\big)}^2 \frac{\partial f\big(F(z)\big)}{dz}\right]$$
$$+ \ldots + \frac{x^n}{\underline{|n}}\frac{\partial^{n-1}}{dz^{n-1}}\left[\overline{\phi\big(F(z)\big)}^n \frac{\partial f\big(F(z)\big)}{dz}\right]$$
$$+ \ldots,$$

which is called *Laplace's Theorem.*

Since the theorems of Lagrange and Laplace depend upon that of Stirling's, they hold only when x is small enough to make the developments convergent.

EXAMPLES.

1. Develop $y = z + xe^y$.

In this case $\qquad f(y) = y, \qquad f(z) = z,$

$$\phi(y) = e^y, \quad \text{and} \quad \phi(z) = e^z.$$

$$\frac{\partial^{n-1}}{dz^{n-1}}\left(\overline{\phi(z)}^n \frac{\partial f(z)}{dz}\right) = \frac{d^{n-1}}{dz^{n-1}}(e^{nz}) = n^{n-1}e^{nz}.$$

Hence, from Lagrange's theorem, we obtain

$$z + xe^y = z + e^z x + 2e^{2z}\frac{x^2}{2} + 3^2e^{3z}\frac{x^3}{\underline{|3}} + \ldots + n^{n-1}e^{nz}\frac{x^n}{\underline{|n}} + R.$$

2. Given $\log y = xy$, develop y.

We may write $y = e^{xy}$, and putting $xy = y'$; we have $y' = xe^{y'}$; which may be developed by making $z = 0$ and $y = y'$ in example 1, giving

$$y' = x + x^2 + \frac{3^2 x^3}{\underline{|3}} + \ldots + \frac{n^{n-1}x^n}{\underline{|n}} + R'.$$

Replacing y' by xy and dividing by x, we have

$$y = 1 + x + 3\frac{x^2}{2} + \ldots + n^{n-2}\frac{x^{n-1}}{\underline{|n-1}} + R.$$

3. Develop $y = z + xy^n$.

Here $\qquad \phi(y) = y^n, \quad \phi(z) = z^n.$

Hence,

$$z + xy^n = z + z^n x + 2nz^{2n-1}\frac{x^2}{2} + 3n(3n-1)z^{3n-2}\frac{x^3}{\underline{|3}} + R.$$

4. Develop $y = z + e \sin y$.

Here $\qquad x = e, \quad \phi(y) = \sin y, \quad \phi(z) = \sin z,$

$$\frac{\partial}{dz}\left(\overline{\phi(z)}^2\right) = 2 \sin z \cos z = \sin 2z.$$

$$\frac{\partial^2}{dz^2}\left(\overline{\phi(z)}^3\right) = 6 \sin z \cos^2 z - 3 \sin^3 z = (3/4)(3 \sin 3z - \sin z).$$

etc. $\qquad$ etc.

Hence,

$$y = z + e \sin z + \frac{e^2}{2} \sin 2z + \frac{e^3}{8}(3 \sin 3z - \sin z) + R.$$

5. Having $y = z + e \sin y$, develop $\sin y$.

Here $\qquad f(y) = \sin y, \quad f(z) = \phi(z) = \sin z,$

$$\partial f(z)/dz = \cos z, \qquad x = e,$$

$$\phi(z)\, \partial f(z)/dz = \sin z \cos z = \sin 2z/2.$$

$$\frac{\partial}{dz}\left(\overline{\phi(z)}^2 \frac{\partial f(z)}{dz}\right) = \frac{d}{dz}(\sin^2 z \cos z) = (3 \sin 3z - \sin z)/4.$$

Hence,

$$\sin y = \sin z + \frac{e}{2}\sin 2z + \frac{e^2}{8}(3\sin 3z - \sin z) + R.$$

6. Having $y = z + e\sin y$, develop $\sin 2y$.

Here $\quad f(y) = \sin 2y, \quad f(z) = \sin 2z, \quad \phi(z) = \sin z,$

$$\partial f(z)/dz = 2\cos 2z, \quad x = e.$$

Hence,

$$\phi(z)\partial f(z)/dz = \sin z\ 2\cos 2z = \sin 3z - \sin z.$$

$$\sin 2y = \sin 2z + (\sin 3z - \sin z)e + \frac{d}{dz}(\sin^2 z\ 2\cos 2z)\frac{e^2}{2} + R.$$

7. Similarly, develop $\sin 3y$.

In this case $\quad f(y) = \sin 3y, \quad f(z) = \sin 3z, \quad \phi(z) = \sin z,$

$$\partial f(z)/dz = 3\cos 3z, \quad x = e.$$

Hence,

$$\sin 3y = \sin 3z + e\sin z\ 3\cos 3z + \frac{d}{dz}(\sin^2 z\ 3\cos 3z)\frac{e^2}{2} + R.$$

8. Similarly, develop $\cos y$.

Here $\quad f(y) = \cos y, \quad f(z) = \cos z, \quad \phi(z) = \sin z,$

$$\partial f(z)/dz = -\sin z, \quad x = e.$$

Hence,

$$\cos y = \cos z - e\sin^2 z - 3\sin^2 z\cos z\frac{e^2}{2} + R.$$

9. Having $u = nt + e\sin u$, develop u, $\sin u$, $\sin 2u$, $\sin 3u$, and $\cos u$ in terms of t and e. By comparison with examples 4, 5, 6, 7, and 8 we have

$$u = nt + (\sin nt)e + \sin 2nt\frac{e^2}{2} + (3\sin 3nt - \sin nt)\frac{e^2}{8} + R.$$

$$\sin u = \sin nt + \sin 2nt\frac{e}{2} + (3\sin 3nt - \sin nt)\frac{e^2}{8} + R.$$

$$\sin 2u = \sin 2nt + (\sin 3nt - \sin nt)e + R.$$

$$\sin 3u = \sin 3nt + R.$$

$$\cos u = \cos nt - (1 - \cos 2nt)\frac{e}{2} + (3\cos 3nt - 3\cos nt)\frac{e^2}{8} + R.$$

10. Kepler's Problem.*

Having
$$nt = u - \epsilon \sin u, \quad . \quad . \quad . \quad . \quad . \quad . \quad . \quad . \quad . \qquad (1)$$
$$\tan \frac{\theta}{2} = \left(\frac{1+\epsilon}{1-\epsilon}\right)^{1/2} \tan \frac{u}{2}, \quad . \quad . \quad . \quad . \quad . \quad . \quad . \qquad (2)$$
$$r = a(1 - \epsilon \cos u), \quad . \quad . \quad . \quad . \quad . \qquad (3)$$

find θ and r in terms of t.

First develop θ in terms of u. In (2) put $\left(\frac{1+\epsilon}{1-\epsilon}\right)^{1/2} = m$, and since from ($d$), Ex. 5, § 127,
$$\sqrt{-1} \tan x = \frac{e^{2x\sqrt{-1}} - 1}{e^{2x\sqrt{-1}} + 1},$$

we have
$$\frac{e^{\theta\sqrt{-1}} - 1}{e^{\theta\sqrt{-1}} + 1} = \frac{m(e^{u\sqrt{-1}} - 1)}{e^{u\sqrt{-1}} + 1}.$$

Hence,
$$e^{\theta\sqrt{-1}} = \frac{(m+1)e^{u\sqrt{-1}} - (m-1)}{(m+1) - (m-1)e^{u\sqrt{-1}}}.$$

Placing $\frac{m-1}{m+1} = \lambda$, and $\sqrt{-1} = i$, we have
$$e^{\theta i} = e^{ui}\,\frac{1 - \lambda e^{-ui}}{1 - \lambda e^{ui}}.$$

Taking the logarithms of both members,
$$\theta i = ui + \log(1 - \lambda e^{-ui}) - \log(1 - \lambda e^{ui})$$
$$= ui - \left(\lambda e^{-ui} + \frac{\lambda^2}{2} e^{-2ui} + \frac{\lambda^3}{3} e^{-3ui} + \ldots\right)$$
$$+ \left(\lambda e^{ui} + \frac{\lambda^2}{2} e^{2ui} + \frac{\lambda^3}{3} e^{3ui} + \ldots\right)$$
$$= ui + \lambda(e^{ui} - e^{-ui}) + \frac{\lambda^2}{2}(e^{2ui} - e^{-2ui}) + \frac{\lambda^3}{3}(e^{3ui} - e^{-3ui}) + \ldots$$
$$= ui + 2\lambda i \sin u + 2\frac{\lambda^2}{2} i \sin 2u + 2\frac{\lambda^3}{3} i \sin 3u + \ldots$$

---

* Price's Calculus, Vol. III. pp. 561–564.

Hence,

$$\theta = u + 2\left(\lambda \sin u + \frac{\lambda}{2} \sin 2u + \frac{\lambda^3}{3} \sin 3u + \ldots\right), \quad . \quad (4)$$

in which
$$\lambda = \frac{m-1}{m+1} = \frac{(1+\epsilon)^{1/2} - (1-\epsilon)^{1/2}}{(1+\epsilon)^{1/2} + (1-\epsilon)^{1/2}}$$

$$= \frac{1-(1-\epsilon^2)^{1/2}}{\epsilon} = \frac{\epsilon}{2} + \frac{\epsilon^3}{8} + \ldots$$

In (4) and (3) substitute for u, $\sin u$, $\sin 2u$, $\sin 3u$, and $\cos u$ their respective values in terms of t from example 9, replace λ by its value in terms of ϵ, omit terms involving powers of ϵ higher than the third, and we have

$$\theta = nt + 2\epsilon \sin nt + \frac{5\epsilon^2}{4} \sin 2nt + \frac{\epsilon^3}{12}(13 \sin 3nt - 3 \sin nt) + \ldots,$$

$$r = a\left(1 - \epsilon \cos nt + \frac{\epsilon^2}{2}(1 - \cos 2nt) - \frac{3\epsilon^3}{8}(\cos 3nt - \cos nt) + \ldots\right),$$

which are important equations in Astronomy.

CHAPTER XI.

MAXIMUM AND MINIMUM STATES.

FUNCTIONS OF A SINGLE VARIABLE.

131. A Maximum state of a continuous function of a single variable is a state greater than adjacent states which precede or follow it. Thus, fx has a maximum state corresponding to $x = a$, provided that as h vanishes we have ultimately and continuously $fa > f(a \pm h)$.

A maximum state is, therefore, a state through which, *as the variable increases continuously*, the function changes from an *increasing* to a *decreasing* function, and *its first differential coefficient changes its sign from plus to minus* (§ 63).

A Minimum state is one less than adjacent states which precede or follow it. Thus, fa is a minimum provided that, as h vanishes, we have ultimately and continuously $fa < f(a \pm h)$.

A minimum state is, therefore, a state through which, *as the variable increases continuously*, the function changes from a *decreasing* to an *increasing* function, and *its first differential coefficient changes its sign from minus to plus.*

A maximum is not necessarily the greatest, nor a minimum the least, state of a function.

A function may have several maximum and minimum states.

To illustrate, let *ABCD-IJ* be the graph of a function (§ 20).

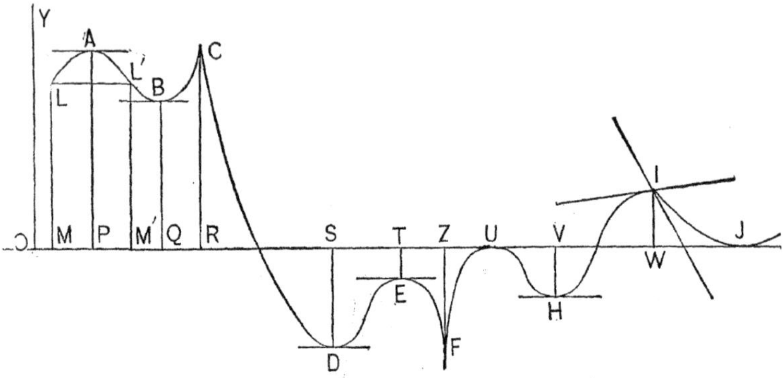

The ordinates *PA*, *RC*, $-TE$, and *WI* represent maxima states of the function, and *QB*, $-SD$, $-ZF$, and $-VH$ represent minima states.

$QB > IW$ illustrates the fact that a minimum state may be greater than a maximum.

The ordinate at *U* represents a zero maximum, and the ordinate at *J* a zero minimum.

The point *I* at which two branches of a curve terminate with separate tangents is called a *salient point*, and the points *C* and *F* at which two branches of a curve terminate with a common tangent are called *cusps*.

132. A continuous function must have at least one maximum or minimum state between any two equal states; for if, in passing through *any state*, the function is increasing it must change to decreasing, and if decreasing it must change to increasing, at least once before it can again arrive at *that state*. The maximum ordinate *PA*, between the equal ordinates *ML* and *M'L'*, illustrates the principle.

Similarly, it may be shown that a continuous function has at least one minimum state between any two maxima, and one or more maxima between any two minima. That is, as the variable increases, maxima and minima of a continuous function occur alternately.

133. The general definition given for maxima and minima assumes that the function is continuous, and that as the variable increases adjacent states precede and follow those considered. Some exceptional cases arise which are illustrated in the following figure.

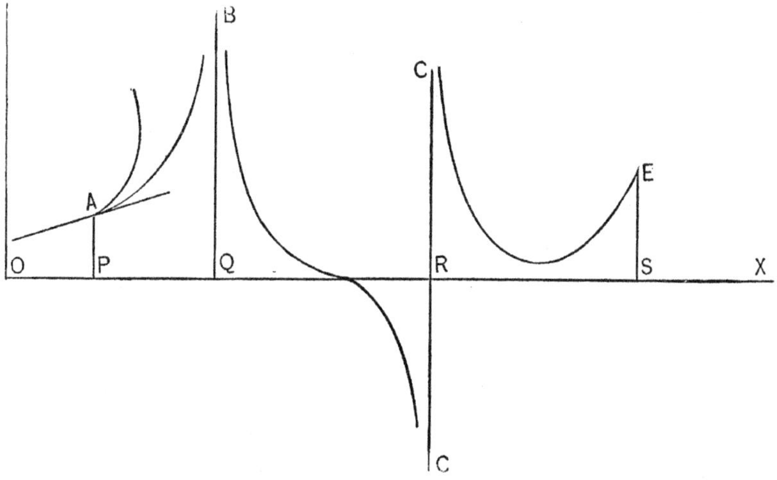

As x *increases*, the function represented by the ordinate of the curve ABC in passing through PA has no adjacent *preceding* states. $f'x$ does not change its sign, and is neither zero nor infinite; yet as PA is smaller than adjacent states it is generally considered a minimum.

At Q the positive ordinate is an asymptote to both branches of the curve; and although the unlimited value of the ordinate does not represent a possible value of the function, yet $f'x$ changes its sign; therefore the ordinate QB is said to be an *infinite* maximum.

At R the positive ordinate is an asymptote to one branch

of the curve, and the negative ordinate to the other. $f'x$ does not change its sign, therefore neither of the ordinates $\pm RC$, respectively, is considered as a maximum or a minimum.

E is called a *terminating point*, and the corresponding ordinate SE is generally considered as a maximum although $f'x$ does not change its sign.

METHODS OF DETERMINING MAXIMA AND MINIMA.

134. Any particular state of fx, as fa, may be examined directly by determining whether, as h vanishes from any definite value, we have ultimately and continuously

$$fa > f(a \pm h) \quad \text{or} \quad fa < f(a \pm h).$$

Thus, let $$fx = c + (x - a)^2.$$

$$x = a \text{ gives } fa = c, \text{ and } f(a \pm h) = c + h^2.$$

Hence, $fa < f(a \pm h)$ as h vanishes from any value, and $fa = c$ is a minimum.

Again, let $$fx = (x - 1)(x - 2)^2.$$

$$x = 2 \text{ gives } f2 = 0, \text{ and } f(2 \pm h) = (1 \pm h)h^2.$$

Hence, $f2 < f(2 \pm h)$ as h vanishes, and $f2 = 0$ is a zero minimum.

$x = 4/3$ gives $f(4/3) = 4/27$, and $f(4/3 \pm h) = \pm h^3 - h^2 + 4/27$.

Hence, $$f(4/3) > f(4/3 \pm h) \text{ as } h \text{ vanishes,}$$

and $$f(4/3) = 4/27 \text{ is a maximum.}$$

Let $fx = \sin x$.

$x = \pi/2$ gives $f(\pi/2) = 1$, and $f(\pi/2 \pm h) = \sin(\pi/2 \pm h)$.

Hence, $f(\pi/2) > f(\pi/2 \pm h)$ as h vanishes,

and $f(\pi/2) = 1$ is maximum.

$x = 0$ gives $f0 = 0$, and $f(0 \pm h) = \sin(\pm h)$.

Hence, $f0 < f(0 + h)$, and $f0 > f(0 - h)$, as h vanishes; therefore $f0 = 0$ is neither a maximum nor a minimum.

135. In general, maxima and minima are determined by finding those values of x corresponding to which, as the variable increases, $f'x$ changes its sign.

Assuming that *x increases continuously*, that fx is continuous, and that every state considered has adjacent preceding and following states, *a maximum state is characterized by a change of sign from plus to minus in the first differential coefficient, and a minimum state by a corresponding change from minus to plus.*

Conversely, if, in passing through $f'a$, $f'x$ changes from plus to minus, fa is a maximum, and if $f'x$ changes from minus to plus, fa is a minimum.

$f'x$, *if discontinuous*, may change its sign by passing through a double value, one positive and the other negative, as illustrated by a salient point or by passing through infinity as illustrated by a cusp with common tangent perpendicular to the axis of X.

$f'x$, *if continuous*, can change its sign only by passing through zero.

A maximum or minimum of fx corresponding to a salient point is an exceptional case distinguished by a double value for $f'x$.

Hence, in general, values of x corresponding to which $f'x$ changes its sign as *x increases*, are real roots of one or the other of the two equations

$f'x = 0$, . . . (1) and $f'x = \infty$. . . . (2)

The figure, p. 207, illustrates the fact indicated by (1) and (2), that, in general, the tangent corresponding to a maximum or minimum ordinate of a curve is parallel or perpendicular to the axis of abscissas.

$f'x$ does not necessarily change its sign as x passes through roots of (1) and (2), as may be seen in the cases represented by particular ordinates of the following curves.

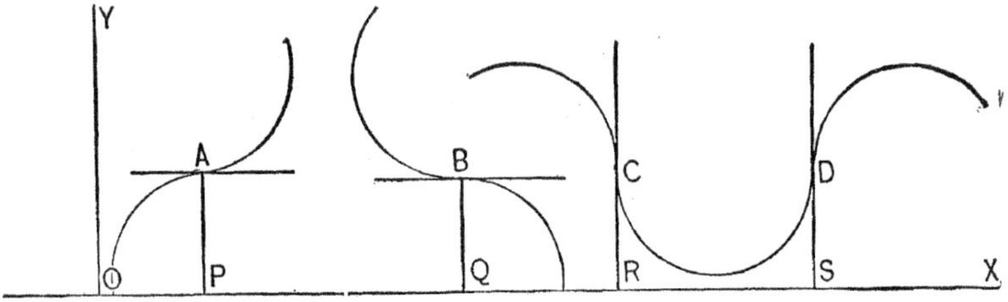

At the points A and B, where the tangents are parallel to X, $f'x = 0$; and at C and D, where the tangents are perpendicular to X, $f'x = \infty$, but $f'x$ does not change its sign as $f(x)$ passes through the corresponding states.

The points A, B, C, and D are called *points of inflexion.*

The real roots of (1) and (2) are, therefore, called *critical values*, and the next step is to determine which of them correspond to states at which *$f'x$ changes its sign as x increases*, and, therefore, correspond to maxima or minima of fx.

The general method, for any critical value, as a, is to determine whether, as h vanishes, $f'(a-h)$ and $f'(a+h)$ ultimately have and retain different signs If so, a corresponds to a maximum or a minimum, according as the sign of $f'(a-h)$ is plus or minus.

EXAMPLES.

1. Let $fx = b + (x - a)^{2/3}$.

Then $f'x = 2/3(x - a)^{1/3} = \infty$ gives the critical value a.

As h vanishes, $f'(a - h)$ is negative and $f'(a + h)$ is positive.

Hence, $fa = b$ is a minimum.

2. $fx = 6x + 3x^2 - 4x^3$.

$f'x = 6(1 + x - 2x^2) = 0$ gives the critical values 1 and $-1/2$.

$f'(1 - h) = 6h(3 - 2h)$, which is positive when $h < 3/2$.

$f'(1 + h) = 6h(-3 - 2h)$, which is negative when $h > 0$.

Hence, $f1 = 5$ is a maximum.

$f'(-1/2 - h)$ is negative when $h > 0$.

$f'(-1/2 + h)$ is positive when $h < 2/3$.

Hence, $f(-1/2) = -7/4$ is a minimum.

3. $fx = a + (x - b)^{1/3}$.

$f'x = 1/3(x - b)^{2/3} = \infty$ gives $x = b$.

$f'(b \mp h)$ are both positive for all values of h.

Hence, $f(b) = a$ is neither a maximum nor a minimum.

4. $fx = (x - 1)^4(x + 2)^3$.

$f'x = (x - 1)^3(x + 2)^2(7x + 5) = 0$ gives

$$x = 1, \quad x = -2, \quad x = -5/7.$$

$f1 = 0$ is a minimum.

$f(-2) = 0$ is neither a maximum nor a minimum.

$f(-5/7) = 124.93/77$ is a maximum.

5. $fx = (x + 2)^3/(x - 3)^2$.

$f'x = (x + 2)^2(x - 13)/(x - 3)^3 = 0$ and ∞ gives

$$x = -2, \quad x = 13, \quad x = 3.$$

$f13 = 135/4$ is a minimum.

$f3 = \infty$ is a maximum.

6. $fx = (a - x)^3/(a - 2x)$. $\quad f(a/4)$, min.

7. $fx = \dfrac{\sin x}{1 + \tan x}$. $\quad f\left(\dfrac{\pi}{4}\right)$, max.

8. $fx = x^{2/3}(2a - x)^{1/3}$. $\quad f0$, min. $\quad f(4a/3)$, max.

136. The preceding method, or that indicated in § 134, must be employed for testing critical values from the equation $f'x = \infty$; but when $f'x, f''x, f'''x$, etc., are continuous for values of x adjacent to critical values, those derived from the equation $f'x = 0$ may be examined by another method.

In Taylor's formula (§ 124) put $x = a$, and write $\pm h$ for h; then, since $f'a = 0$, we have

$$f(a \pm h) - fa = h^2 f''a/2 \pm h^3 f'''a/\underline{|3} + \dots$$
$$+ h^{n+1} f^{n+1}(a \pm \theta_n h)/\underline{|n+1}, \quad (1)$$

which form is exact for continuous values of h from zero to certain limits, provided $fa, f'a$, etc., to include $f^{n+1}a$, are real and finite.

In order that fa may be a maximum, we must have ultimately, as h vanishes,

$$fa > f(a \pm h);$$

and fa a minimum requires, under the same law,

$$fa < f(a \pm h).$$

That is, fa a maximum requires that, as h vanishes,

$$f(a \pm h) - fa,$$

and therefore that the second member of (1) shall ultimately become and remain *negative*; and fa a minimum requires that the second member of (1) shall, under the same law, ultimately become and remain *positive*.

As h vanishes, the sign of the second member of (1) will ultimately become and remain the same as that of $h^2 f''a/2$: and since $h^2/2$ is always positive,

$$fa \text{ a maximum requires } f''a < 0, \text{ and}$$

$$fa \text{ a minimum requires } f''a > 0.$$

In the exceptional case when $f''a = 0$, the sign of the second member of (1) will, under the law, ultimately depend upon that of $\pm h^3 f'''a/\underline{|3}$, which changes with that of h. fa cannot, therefore, be a maximum or a minimum unless $\pm h^3 f'''a/\underline{|3} = 0$, which requires $f'''a = 0$.

If also $f'''a = 0$, the sign of the second member of (1) will, under the law, ultimately depend upon that of $h^4 f^{\text{iv}}a/\underline{|4}$; and since $h^4/\underline{|4}$ is always positive,

$$fa \text{ a maximum requires } f^{\text{iv}}a < 0, \text{ and}$$

$$fa \text{ a minimum requires } f^{\text{iv}}a > 0.$$

By continuing the same method of reasoning it may be shown that, if $f^n a$ is the first derivative in order which does not reduce to 0, fa is neither a maximum nor a minimum if n is odd, and that it is a maximum or a minimum if n is even, according as $f^n a$ is negative or positive. Hence, we have the following rule:

Having $f'a = 0$, substitute a for x in the successive derivatives of fx in order, until a result other than 0 is obtained. If the corresponding derivative is of an odd order, fa is neither a maximum nor a minimum; but if it is of an even order, fa is a maximum or a minimum according as the result is negative or positive. If a result ∞ is obtained, the method of § 135 *should be employed.*

The relations between the corresponding states of $fx, f'x$, and $f''x$, in a case where $ABCDE$ is the graph of fx, are shown graphically in the following figure.*

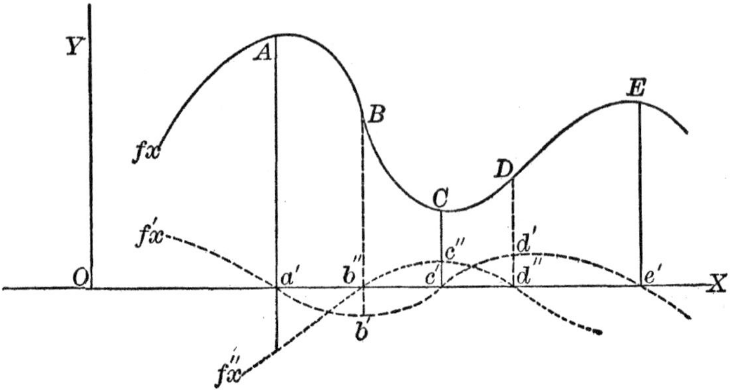

EXAMPLES.

Find the values of the variable which correspond to maxima or minima of the following functions:

1. $fx \quad = x^5 - 5x^4 + 5x^3 + 1.$

$f'x \quad = 5x^4 - 20x^3 + 15x^2 = 0$ gives the critical values 0, 1, 3.

$f''x \quad = 20x^3 - 60x^2 + 30x.$

$f''0 \quad = 0. \quad f''1 = -10. \quad f''3 = 90.$

Hence,

$f1 \quad = 2$, maximum. $f3 = -26$, minimum.

$f'''x = 60x^2 - 120x + 30, \quad \therefore \quad f'''0 = 30$, and

$f0 \quad = 1$ is neither a maximum nor a minimum.

2. $fx \quad = x^3 - 9x^2 + 24x - 7.$

$f'x \quad = 3(x^2 - 6x + 8) = 0$ gives $x = 2$ or $x = 4$.

$f''x = 3(2x - 6), \quad \therefore \quad f''2 < 0, \quad f''4 > 0.$

Therefore $f2$, maximum. $f4$, minimum.

---

* Calcul, par P. Haag, page 71.

3. $fx \quad = \sin^3 x \cos x.$

$f'x \quad = 3 \sin^2 x \cos^2 x - \sin^4 x = 0, \quad \text{gives} \quad x = 60°, \text{ etc.}$

$f''x = -10 \sin^3 x \cos x + 6 \sin x \cos^3 x.$

Since $\sin 60° = \sqrt{3}/2$, and $\cos 60° = 1/2$,

$$f'' 60° = -3\sqrt{3}/2, \quad \therefore \quad f 60° = 3\sqrt{3}/16, \text{ maximum.}$$

4. $x^4 - 8x^3 + 22x^2 - 24x + 12.$ $\quad x = 3$, min.
$x = 2$, max.
$x = 1$, min.

5. $x^2 - 4x + 9.$ $\quad x = 2$, min.

6. $x^3/3 + ax^2 - 3a^2x.$ $\quad x = a$, min.
$x = -3a$, max.

7. $x^3 - 6x^2 + 9x + 10.$ $\quad x = 3$, min.
$x = 1$, max.

8. $3a^2x^3 - b^4x + c^5.$ $\quad x = b^2/3a$, min.
$x = -b^2/3a$, max.

9. $\dfrac{x^2 - x + 1}{x^2 + x + 1}.$ $\quad x = -1$, max.
$x = 1$, min.

10. $\dfrac{x}{1 + x^2}.$ $\quad x = 1$, max.
$x = -1$, min.

11. $x^{1/x}.$ $\quad x = e = 2.71828 \ldots$ max.

12. $\sec x + \operatorname{cosec} x$ $\quad x = \pi/4$, min.
$x = 5\pi/4$, max.

13. $fx \quad = e^x + e^{-x} + 2 \cos x,$

$f'x \quad = e^x - e^{-x} - 2 \sin x, \quad f''x = e^x + e^{-x} - 2 \cos x,$

$f'''x = e^x - e^{-x} + 2 \sin x, \quad f^{iv} x = e^x + e^{-x} + 2 \cos x,$

$f'0 \quad = f''0 = f'''0 = 0, \qquad f^{iv} 0 = 4.$

Hence, $f0 = 4$ is a minimum.

14. $\dfrac{x}{1 + x \tan x}.$ $\quad x = \cos x$, max.

137. The following principles frequently facilitate the determination of maxima and minima:

If $F[f(x)]$ is an increasing function of $f(x)$, $F[f(a)]$ is a maximum or a minimum of $F[f(x)]$ according as $f(a)$ is a maximum or a minimum of $f(x)$.

If $F[f(x)]$ is a decreasing function of $f(x)$, $F[f(a)]$ is a maximum or a minimum according as $f(a)$ is a minimum or a maximum. Hence—

1°. $Cf(a)$, in which C is a positive constant, is a maximum or a minimum of $Cf(x)$ according as $f(a)$ is a maximum or a minimum of $f(x)$.

2°. $C+f(a)$ is a maximum or a minimum according as $f(a)$ is a maximum or a minimum.

3°. $C-f(a)$ is a minimum or a maximum according as $f(a)$ is a maximum or a minimum.

4°. The base being greater than unity, $\log f(a)$ is a maximum or a minimum according as $f(a)$ is a maximum or a minimum; also any value of x that makes $a^{f(x)}$ a maximum or a minimum makes $f(x)$ a maximum or a minimum.

5°. $1/f(a)$ is a maximum or a minimum according as $f(a)$ is a minimum or a maximum.

6°. n being any positive odd integer, $[f(a)]^n$ is a maximum or a minimum according as $f(a)$ is a maximum or a minimum.

n being any positive even integer, $[f(a)]^n$ is a maximum or a minimum according as $f(a)$ positive is a maximum or a minimum or $f(a)$ negative is a minimum or a maximum.

$[f(a)]^n$ may, however, be a maximum or a minimum when $f(a)$ is neither. Thus, $(a^3 - x^3)_a^2$ is a minimum, whereas $(a^3 - x^3)_a$ is neither a minimum nor a maximum.

A radical sign which affects the entire function may therefore be omitted, provided critical values which correspond to maximum and minimum states of the *power only* are rejected.

Thus, having $f(x) = \pm \sqrt{2ax^2 - x^3}$, we write

$$\phi(x) = [f(x)]^2 = 2ax^2 - x^3.$$

Then $\phi'(x) = 4ax - 3x^2 = 0$ gives $x = 0$ and $4a/3$.

$\phi(0)$ is a minimum and $\phi(4a/3)$ is a maximum of $\phi(x)$.

But $f'(x) = \pm (4a - 3x)/2\sqrt{2a - x} = 0$ gives $x = 4a/3$ only, and $f(0)$ is neither a maximum nor a minimum of $f(x)$.

Similarly, $(x - 2a)^2/(x^2 - a^2)$ is a minimum when $x = 2a$, but $(x - 2a)/\sqrt{x^2 - a^2}$ is not.

To illustrate the use of the foregoing principles:

Let $$f(x) = 1/(5 + \log \sqrt{4b^2x^2 - 2bx^3}).$$

By 5° we take $5 + \log \sqrt{4b^2x^2 - 2bx^3}$.

By 2° and 4° we take $\sqrt{4b^2x^2 - 2bx^3}$.

By 6° and 1° we take $2bx^2 - x^3 = \phi(x)$.

$$\phi'(x) = 4bx - 3x^2 = 0 \quad \text{gives} \quad x = 0,\ x = 4b/3.$$

$$\phi''(0) = 4b, \qquad \phi''(4b/3) = -4b.$$

Hence,

$f(0)$ is a maximum and $f(4b/3)$ is a minimum.

138. In certain cases it is not necessary to determine the second derivative.

1°. In case of one critical value only, and it is known that the function has a maximum state or that it has a minimum state.

2°. If a is the only critical value, $f(a)$ is a maximum or a minimum, provided $f(a)$ is greater than or less than both $f(a \pm h)$ for any assumed value of h.

3°. When $f'(x)$ is composed of two or more factors, one of which reduces to 0, for $x = a$, $f''(a)$ may be determined without using $f''(x)$.

Thus, let $\quad f'(x) = \psi(x)\phi(x)$.

Then $\quad f''(x) = \psi(x)\phi'(x) + \phi(x)\psi'(x)$.

Supposing that $\psi(a) = 0$, we have

$$f''(a) = \phi(a)\psi'(a).$$

Hence, to obtain $f''(a)$, *multiply the differential coefficient of that factor of $f'(x)$ which reduces to 0 by the other factors, and substitute a.*

To illustrate, let

$$f(x) = (x - a)^2 x^2.$$

$f'(x) = 2x(x - a)(2x - a) = 0$ gives

$$x = 0, \quad x = a, \quad x = a/2.$$

$f''(0) = 2(x - a)(2x - a)_0 = 2a^2$, indicating a minimum.

$f''(a) = 2x(2x - a)_a = 2a^2$, indicating a minimum.

$f''(a/2) = 4x(x - a)_{a/2} = -a^2$, indicating a maximum.

EXAMPLES.

Find the values of the variable which correspond to maxima or minima of the following functions:

1. $\dfrac{(x + 3)^3}{(x + 2)^2}$. $\quad x = 0$, min. $\quad x = -2$, max.

2. $\dfrac{(y - 1)^2}{(y + 1)^3}$. $\quad y = 1$, min. $\quad y = 5$, max.

3 $\sqrt{a^2x^2 - x^4}$. $x = 0$, min. of power.
$x = \pm a/\sqrt{2}$, max., min.

4. $\dfrac{2r\pi x^2 - \pi x^3}{3}$. $x = 0$, min.
$x = 4r/3$, max.

5. $a^2x - x^3$. $x = a/\sqrt{3}$, max.

6. $ax^3 - x^4$. $x = 3a/4$, max.

7. $x^2 - x^{5/2}$. $x = 0$, min.
$x = 16/25$, max.

8. $\dfrac{x^3 - 2x^2}{2x^2 + 8}$. $x = 0$, max.
$x = 1.19$, min.

9. x^x. $x = 1/e$, min.

10. $(2bx^4 + a^3bx)/a^2$. $x = -a/2$, min.

11. $\dfrac{a}{x^2} + \dfrac{b}{(c - x)^2}$. $x = \dfrac{c\sqrt[3]{a}}{\sqrt[3]{a} + \sqrt[3]{b}}$, min.

12. $\dfrac{b}{\sin\theta} + \dfrac{a}{\cos\theta}$. $\theta = \tan^{-1}\sqrt[3]{b/a}$, min.

13. $fx = x/2 - x^2 \sin(1/x)/2$.

$f'x = 1/2 + \cos(1/x)/2 - x\sin(1/x)$.

$$\frac{1}{2} + \frac{1}{2}\cos\frac{1}{x} = \cos^2\frac{1}{2x} \qquad x\sin\frac{1}{x} = 2x\sin\frac{1}{2x}\cdot\cos\frac{1}{2x}.$$

Hence, $f'x = \cos\dfrac{1}{2x}\cdot\left(\cos\dfrac{1}{2x} - 2x\sin\dfrac{1}{2x}\right) = 0$ gives

$$x = 1/\pi, \qquad x = \infty.$$

$f''(1/\pi) = -4/\pi^3$, indicating a maximum.

14. $\dfrac{a^2x}{(a - x)^2}$. $x = -a$, min.
$x = a$, max.

15. $\dfrac{ab}{x\sqrt{a^2 + b^2 - x^2}}$. $x = \sqrt{\dfrac{a^2 + b^2}{2}}$, min.

PROBLEMS.

1. Divide a number a into two such parts that the product of the mth power of one and the nth power of the other shall be a maximum.

$$fx = x^m(a - x)^n, \quad f'x = x^{m-1}(a - x)^{n-1}[ma - (m + n)x] = 0$$

gives $\quad x = 0, \quad x = a, \quad x = ma/(m + n),$

$$f''ma/(m + n) = -(m + n)c, \text{ indicating a maximum.}$$

2. Divide a number a into two such factors that the sum of their squares shall be a minimum.

$$fx = x^2 + a^2/x^2, \qquad x = \pm\sqrt{a}, \text{ minimum.}$$

3. Into how many equal parts must a number a be divided that their continued product may be a maximum?

Let x = the number of equal parts, then

$$fx = (a/x)^x, \qquad \therefore \log fx = x(\log a - \log x).$$

$$f'x = fx(-1 + \log a - \log x) = 0 \text{ gives } x = a/e.$$

$$f''x(a/e) = e^{a/e}(-e/a), \text{ indicating a maximum.}$$

4. Let h be the hypothenuse of a right triangle; find the lengths of the other sides when the area is a maximum.

Let x = one side, then $\sqrt{h^2 - x^2}$ = the other.

$$fx = \text{area} = x\sqrt{h^2 - x^2}/2. \quad f'x = 0 \text{ gives } h^2 - 2x^2 = 0,$$

whence $\quad x = h/\sqrt{2}, \quad f''(h/\sqrt{2}) = -4h^2.$

5. What fraction exceeds its nth power by the greatest number possible?

Let x = fraction, then $fx = x - x^n$.

$$f'x = 1 - nx^{n-1} = 0 \text{ gives } x = 1/\sqrt[n-1]{n}.$$

$$f''\left(1/\sqrt[n-1]{n}\right) = -n(n-1)\Big/ n^{\frac{n-2}{n-1}}, \text{ indicating a maximum.}$$

6. Of all isoperimetrical rectangles which has the greatest area?

7. On the right line $A \underline{\qquad\qquad c \qquad\qquad} B$ joining the two lights A and B, find the point between the lights of least illumination.

Let c = number of miles from A to B.

Let x = number of miles from A to required point.

Let a = intensity of the light A at 1 mile from A.

Let b = intensity of the light B at 1 mile from B.

Then

$$fx = \frac{a}{x^2} + \frac{b}{(c-x)^2} = \text{intensity of light at point required.}$$

$$f'x = \frac{2b}{(c-x)^3} - \frac{2a}{x^3} = 0 \quad \text{gives} \quad x = \frac{c\sqrt[3]{a}}{\sqrt[3]{a} + \sqrt[3]{b}}.$$

8. Find the relation between the radius of the base and the altitude of a cylinder, open at the top, which shall just hold a given quantity of water and have its surface a minimum.

Let x = radius of base, and y = altitude. Then

$$\pi y x^2 = \text{volume} = v. \quad \therefore\ y = v/\pi x^2.$$

$$\pi x^2 + 2\pi xy = \text{surface} = f(x).$$

Hence,

$$fx = \pi x^2 + 2\pi x v/\pi x^2 = \pi x^2 + 2v/x.$$

$$f'x = 2\pi x - 2v/x^2 = 0 \text{ gives } x = \sqrt[3]{v/\pi} \text{ and } y = \sqrt[3]{v/\pi}.$$

$$f''(\sqrt[3]{v/\pi}) = 6\pi, \text{ indicating a minimum.}$$

9. Find the maximum rectangle that can be inscribed in a given circle.

Take the origin at the centre, and let $2x$ and $2y$ respectively equal the sides of the required rectangle. Then $xy = 1/4$ area of rectangle.

$x^2 + y^2 = R^2$, in which R represents the radius of the circle, gives

$$y = \pm\sqrt{R^2 - x^2}.$$

Hence, $$fx = \pm x\sqrt{R^2 - x^2}.$$

$$f'x = \pm(2R^2x - 4x^3) = 0 \quad \text{gives} \quad x = 0 \quad \text{and} \quad x = R/\sqrt{2}.$$

$$f''(R/\sqrt{2}) = -4R^2, \text{ indicating a maximum.}$$

10. Determine the maximum rectangle which can be inscribed in a triangle of given base and altitude.

11. Show that the difference between the sine and the versed sine is a maximum when the angle is $45°$.

12. The base and the vertical angle of a triangle being given, show that its area is a maximum when it is isosceles.

13. Of all isoperimetrical plane triangles which has the maximum area?

14. Divide a triangle whose sides are a, b, and c, respectively, into two equal parts by the minimum right line.

Ans. $\sqrt{(c-a+b)(c+a-b)/2}$.

15. Through a given point (a, b) draw the shortest straight line terminating in X and Y.

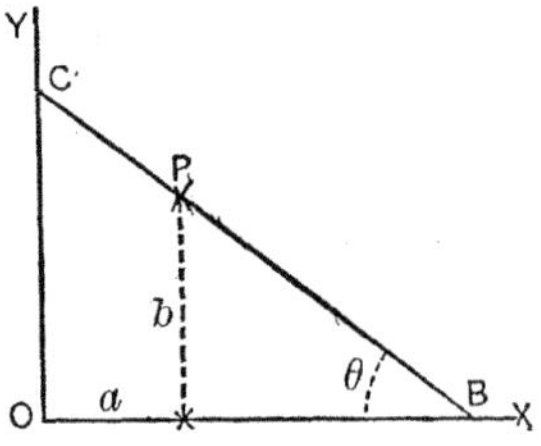

Let $\theta =$ angle required line CB makes with X.

Then

$$CB = CP + PB = a/\cos\theta + b/\sin\theta,$$

which is a minimum when $\theta \tan^{-1}\sqrt[3]{b/a}$, giving $CB = (a^{2/3} + b^{2/3})^{3/2}$

Similarly, show that

$OB + OC$ is a minimum when $\theta = \tan^{-1}\sqrt{b/a}$;

$OB \times OC$ is a minimum when $\theta = \tan^{-1}(b/a)$;

$OB + OC + CB$ is a minimum when $\theta = \tan^{-1}\dfrac{b+\sqrt{2ab}}{a+\sqrt{2ab}}$;

$OB \times OC \times CB$ is a minimum when

$$2a\tan^3\theta - b\tan^2\theta + a\tan\theta - 2b = 0.$$

16. Determine the maximum right cone which can be inscribed in a sphere whose radius is R.

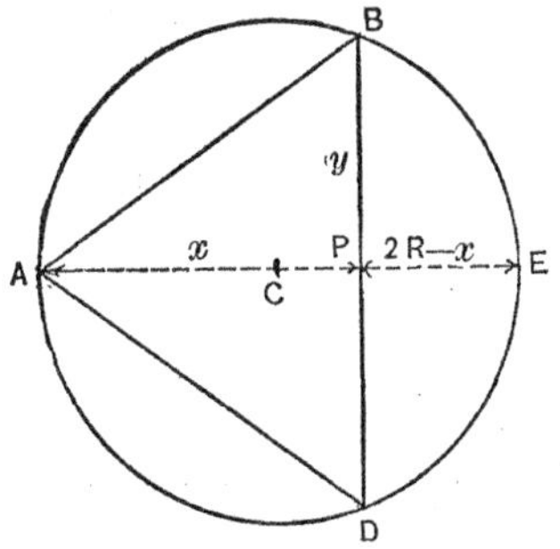

Let $x = AP$, and $y = PB$.

Then

vol. $= v = \dfrac{\pi y^2 x}{3}$, but $y^2 = 2Rx - x^2$.

Therefore $v = (2R\pi x^2 - \pi x^3)/3$.

$dv/dx = \pi x(4R - 3x)/3 = 0$

gives $x = 0$, $x = 4R/3$.

$$d^2v/dx^2 = \pi(4R - 6x)/3 = -4R\pi/3\Big]_{x=4R/3}.$$

17. Find the radius of a circle such that the segment corresponding to an arc of a given length shall be a maximum.

Let $2a$ = length of arc, and r = radius.

Draw CD bisecting the arc, then

$\angle DCA = \theta/2 = a/r$, and $\theta = 2a/r$.

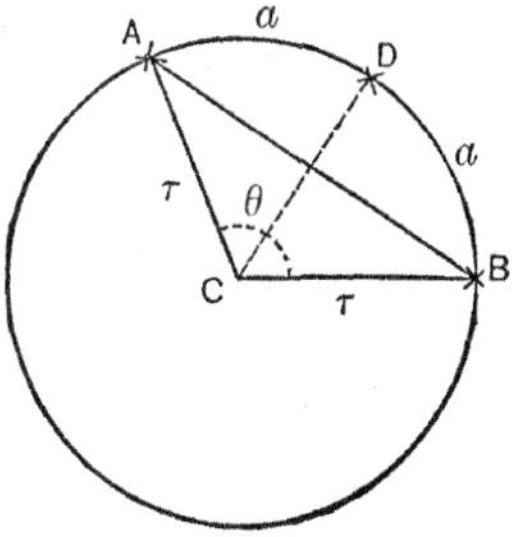

Segment $=$ sector $BCAD - \Delta BCA$

$= r^2\theta/2 - r^2 \sin \theta/2$

$= ra - r^2 \sin (2a/r)/2,$

which is a maximum when $r = 2a/\pi$, and the segment is a semi-circle.

18. With a given perimeter find the radius which makes the corresponding circular sector a maximum. *Ans.* radius $= 1/4$ perimeter.

19. Find the maximum right cylinder which can be inscribed in a given right cone.

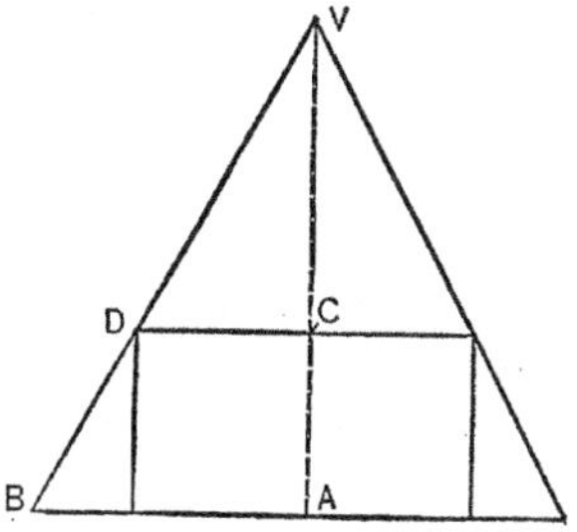

Let $VA = a$, $BA = b$, $AC = x$, $CD = y$, $CV = a - x$.

Hence, vol. cylinder $= v = \pi y^2 x$.

$VA : AB :: VC : DC$, $\therefore y = b(a - x)/a$.

Therefore $v = \pi b^2(a - x)^2 x/a^2$.

Omitting $\pi b^2/a^2$, we have

$$f(x) = a^2 x - 2ax^2 + x^3.$$

$$f'(x) = a^2 - 4ax + 3x^2 = 0 \quad \text{gives} \quad x = a \quad \text{or} \quad a/3.$$

$$f''(a/3) = -2a, \quad \text{therefore} \quad v = 4\pi ab^2/27 \text{ is a maximum.}$$

20. Circumscribe the minimum isosceles triangle about the parabola $y^2 = 4ax$.

Let $x = OP = AO$, $y = PM$, $h = OD$. Then

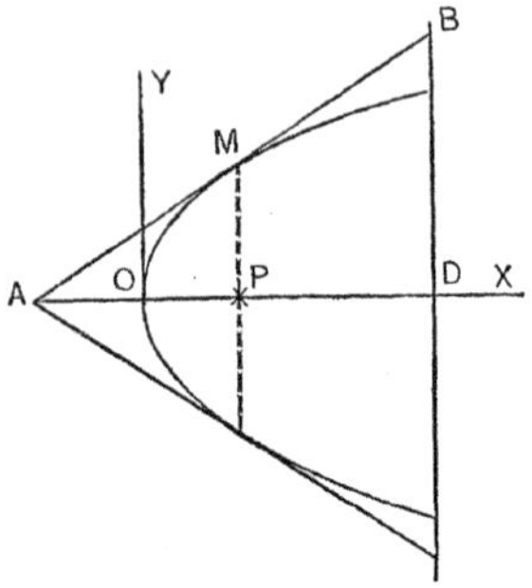

$$BD = (h + x)y/2x = (h + x)\sqrt{ax}/x,$$

and area Δ

$$= (h + x)^2 \sqrt{ax}/x. \quad x = h/3, \text{ min.}$$

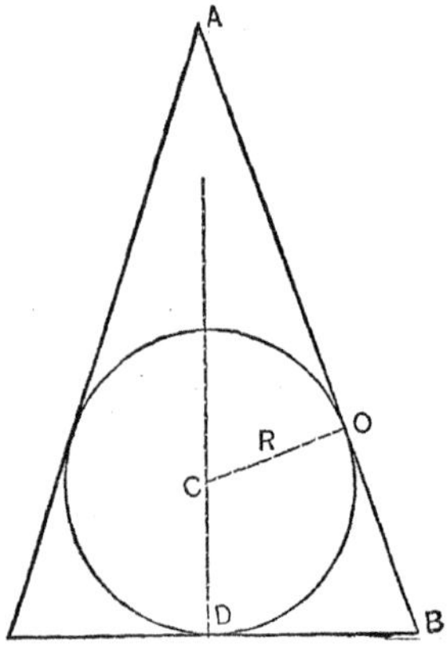

21. Determine the minimum right cone circumscribing a given sphere.

Let $x =$ alt. $= AD$, $y =$ radius of base, $R =$ radius of sphere.

Then $V =$ vol. of cone $= \pi y^2 x/3$.

$$y : R :: \sqrt{x^2 + y^2} : x - R.$$

From which $y^2 = R^2x/(x - 2R)$,

and $V = \pi R^2x^2/3(x - 2R)$.

$$x = 4R, \text{ min.}$$

22. Find the maximum parabola that can be cut from a given right cone.

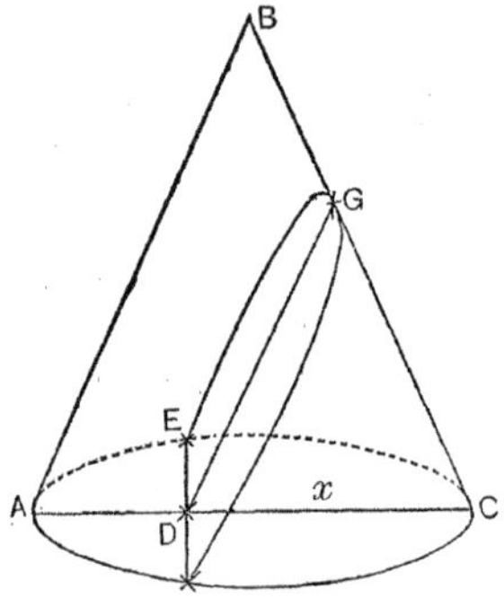

Let $AC = a$, $AB = b$, $DC = x$.

Then $AD = a - x$, $DE = \sqrt{(a - x)x}$.

Also, $a : x :: b : DG$. $\therefore$ $DG = bx/a$.

$$\text{Parabola} = 4b\sqrt{ax^3 - x^4}/3a.$$

$$x = 3a/4, \text{ max.}$$

23. Find the maximum isosceles triangle inscribed in a given circle.

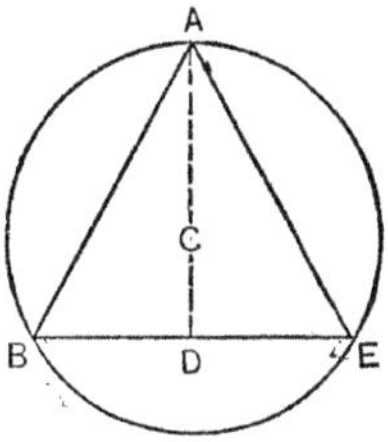

Let $r =$ radius CA, $AB = AE = x$, $BE = 2y$.

Then area $\triangle = u = y\sqrt{x^2 - y^2}$.

Also, $u =$

$$\frac{AB \times AE \times BE}{4r} = \frac{x^2 y}{2r}. \quad \therefore \quad y = \frac{x\sqrt{4r^2 - x^2}}{2r}.$$

And $u = \dfrac{x^2}{2r} \dfrac{x\sqrt{4r^2 - x^2}}{2r}$. $\quad x = r\sqrt{3}$, max.

24. Find the maximum cylinder that can be inscribed in a given prolate spheroid.

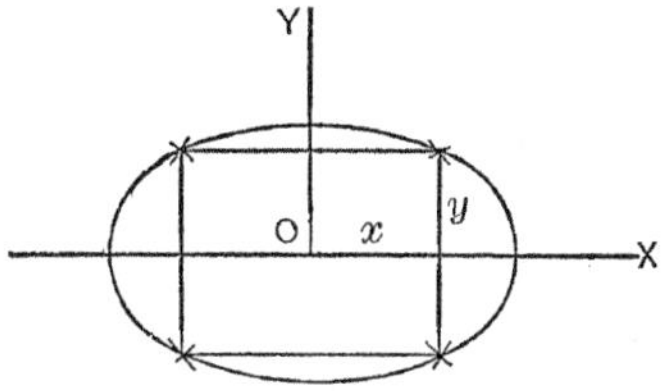

Let $2x =$ axis, and $y =$ radius of base, of required cylinder.

Then, vol. of cylinder

$$= v = 2\pi y^2 x = 2\pi x b^2(a^2 - x^2)/a^2,$$

which is a maximum for $x = a/\sqrt{3}$.

25. Find the minimum isosceles triangle circumscribing a given circle.

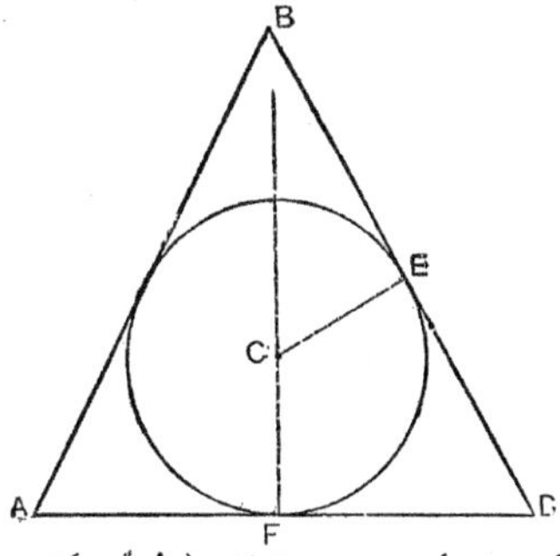

Let $r =$ radius CE, $\quad x = BF$, $2y = AD$.

Then area $\triangle = xy$.

Similar triangles, BCE, BFD, give

$$y : r : \sqrt{x^2 + y^2} : x - r.$$

Hence, $y = r\sqrt{x/(x - 2r)}$, and

$$\text{area } \triangle = xr\sqrt{x/(x - 2r)},$$

which is a minimum for $x = 3r$.

26. * A boatman 3 mi from shore goes to a point 5 mi. down the shore in the shortest time. He rows 4 mi. and walks 5 mi an hour. Where did he land?

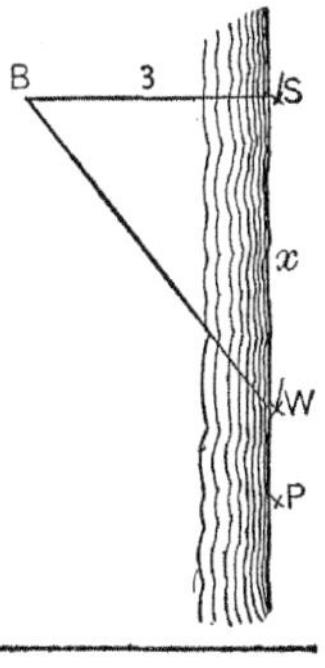

Let B be the boat 3 mi. from S, which is 5 mi. from the point P. Let W be the landing-place, and $x = SW$.

Then, number of hours

$$= t = \sqrt{9 + x^2}/4 + (5 - x)/5,$$

which is a minimum for $x = 4$.

---

* Todhunter's Diff. Calc., p. 213.

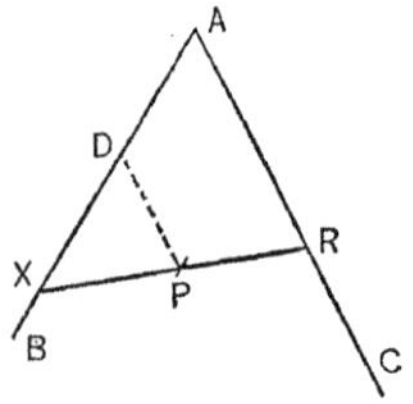

27. Through a given point P within an angle BAC draw a right line so that the triangle formed shall be a minimum.

Draw PD parallel to AC, and let $AD = a$.

$$PD = b,\ AX = x. \quad \therefore \quad DX = x - a.$$

Then $\qquad x - a : b :: x : AR.$

$AXR = x AR \sin A/2$, which is a minimum for $x = 2a$. $\therefore$ $XP = PR$.

28. The volume of a cylinder being constant, find its form when the surface is a minimum. *Ans.* Altitude = diameter of base.

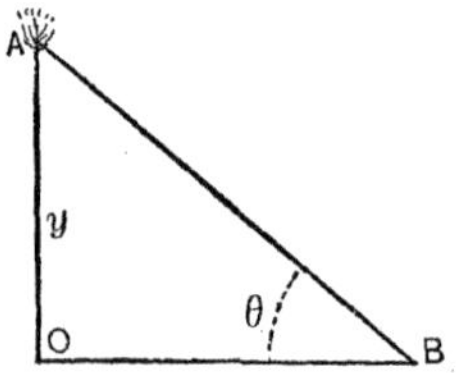

29. Find the height of a light A above the straight line OB when its intensity at B is a maximum.

Let a = intensity of light at 1 foot from the light.

$$OA = y, \quad OB = b, \quad \angle OBA = \theta.$$

The intensity varies directly as $\sin\theta$, and inversely as $\overline{BA}^2$.

Intensity at $B = ay/(b^2 - y^2)^{3/2}$, which is a maximum when

$$y = b\sqrt{2}/2.$$

30. Having $y = x\tan\alpha - x^2/4h\cos^2\alpha$; 1°. Find the maximum value of y. 2°. Considering $y = 0$ and α as varying, find maximum value of x.

Ans. $\begin{cases} \text{1st. } y = h\sin^2\alpha, \text{ a maximum, } x = h\sin 2\alpha. \\ \text{2d. } x = 2h, \text{ a maximum, } \alpha = 45^\circ. \end{cases}$

31. On the right line $CC_1 = a$ joining the centres of two spheres (radii R, r) find point from which the maximum spherical surface is visible.

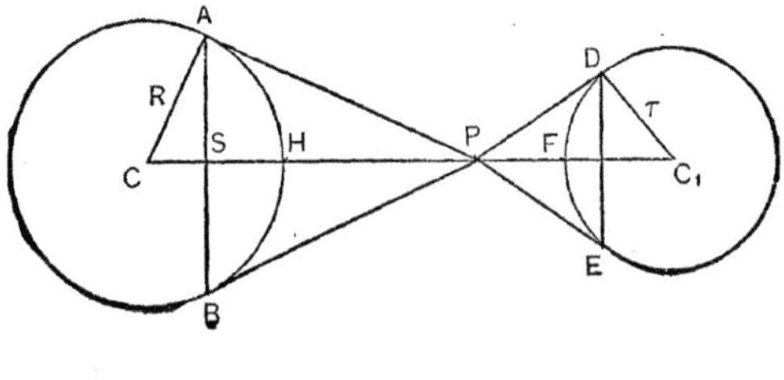

Let $CP = x. \quad \therefore \quad PC_1 = a - x.$

Area zone $ASHB = 2\pi R \times HS.$

$x : R :: R : CS. \quad \therefore \quad CS = R^2/x$ and $HS = R - R^2/x.$

Hence, zone $ASBH = 2\pi R(R - R^2/x).$

Similarly, zone $DEF = 2\pi r\left(r - r^2/(a - x)\right).$

Visible surface $= 2\pi\left[R^2 + r^2 - \left(R^3/x + r^3/(a - x)\right)\right]$, which is a maximum for $x = aR^{3/2}/(R^{3/2} + r^{3/2})$.

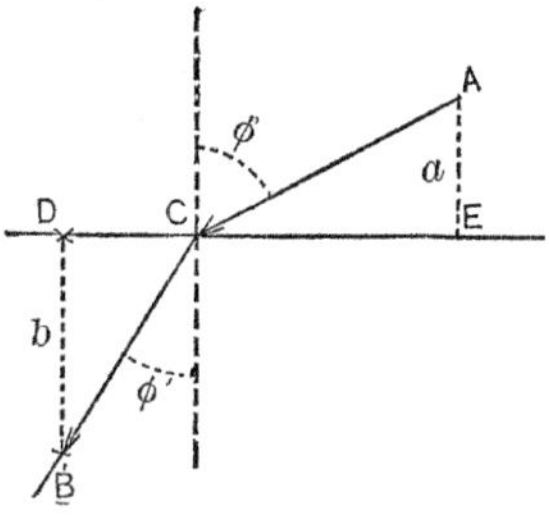

32. Find the path of a ray of light from a point A in one medium to a point B in another medium, such that a minimum time will be required for light to pass from A to B; the velocity of light in the first medium being V, and in the second V'. [Fermat's Problem.]

It is assumed that the required path is in a plane through A and B perpendicular to the plane separating the media.

Let ACB be the required path. Through A, C, and B draw perpendiculars to DE.

Let $a = AE, \quad b = DB, \quad d = DE.$

Then $AC = a/\cos\phi, \quad BC = b/\cos\phi',$

$CE = a\tan\phi, \quad CD = b\tan\phi'.$

$a\tan\phi + b\tan\phi' = d. \quad \therefore \quad d\phi'/d\phi = -a\cos^2\phi'/b\cos^2\phi.$

Time $= t$, from A to $B = \dfrac{a}{V\cos\phi} + \dfrac{b}{V'\cos\phi'}$

$$\frac{dt}{d\phi} = \frac{d\left(\dfrac{a}{V\cos\phi}\right)}{d\phi} + \frac{d\left(\dfrac{b}{V'\cos\phi'}\right)}{d\phi'}\left(\frac{d\phi'}{d\phi}\right). \quad (\S\ 77).$$

$$\frac{dt}{d\phi} = \frac{\sin\phi}{V} - \frac{\sin\phi'}{V'} = 0 \text{ gives}$$

$$\sin\phi = \frac{V}{V'}\sin\phi', \quad \text{or} \quad \frac{\sin\phi}{\sin\phi'} = \frac{V}{V'}, \quad \ldots\ldots\ldots \quad (a)$$

for t a minimum.

Equation (a) expresses an important law, which is known as Snell's law of refraction.

33. To inscribe in a given sphere a right cone with a maximum convex surface.

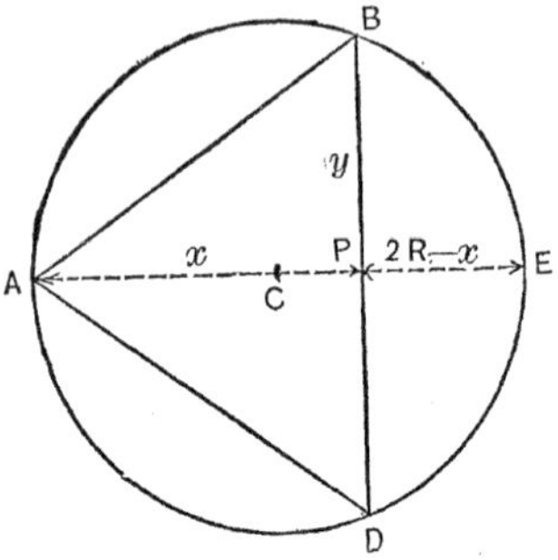

Let $R =$ radius AC, $x = AP$,

$$y = \sqrt{2Rx - x^2} = PB,$$

$$s = 2\pi PB\, AB/2$$

$= $ convex surface.

Then $AB = \sqrt{2Rx}$, and

$$s = \pi\sqrt{4R^2x^2 - 2Rx^3},$$

which is a maximum for $x = 4R/3$.

34. From two points A and B draw two right lines to a point P in a given right line OX, so that $AP + BP$ shall be a minimum.

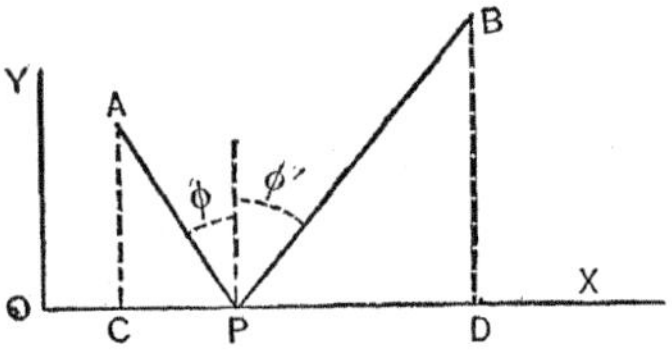

Let $x = OP$, $a = OC$, $b = CA$, $a' = OD$, $b' = DB$.

Ans. $\angle APC = DPB$, or $\phi = \phi'$.

In some cases the general method (§135) apparently fails when it is obvious that maxima and minima states exist.

35. Find the maximum and minimum distances from a given external point to a given circumference.

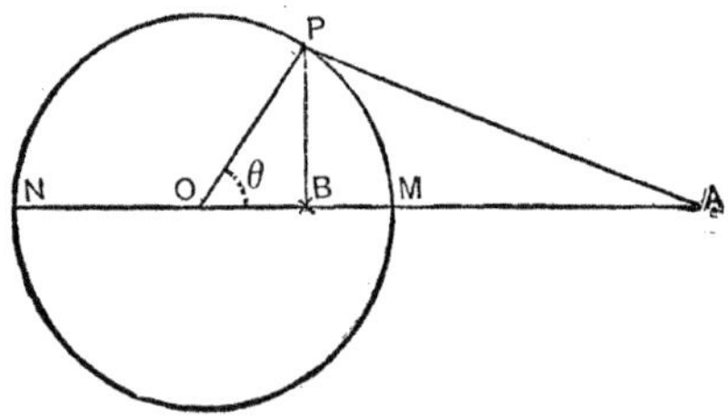

Let A be the point, $r =$ radius, $a = AO$, $OB = x$.

Then $\quad PA = \sqrt{r^2 + a^2 - 2ax}$, and $dPA/dx = -2a$.

It is obvious, however, that AM is a minimum and AN a maximum. The method of § 135 depends upon the assumption that the variable *increases continuously*; whereas x in the above expression changes from increasing to decreasing, or *vice versa*, as PA passes through AM and AN.

Let $\theta =$ angle AOP. Then $x = r\cos\theta$, and

$$PA = \sqrt{r^2 + a^2 - 2ar\cos\theta}.$$

$$dPA/d\theta = 2ar\sin\theta = 0. \qquad \begin{matrix} \theta = 0, & \text{min.} \\ \theta = \pi, & \text{max.} \end{matrix}$$

Otherwise PA may be expressed in terms of y and the problem solved.

36. Find those conjugate diameters in an ellipse whose sum is a maximum or a minimum.

Let x' and y' be any two semi-conjugate diameters, and let $s = x' + y'$ and $a^2 + b^2 = c^2$. Then [Anal. Geom.] $x'^2 + y'^2 = c^2$.

$$\therefore\ y' = \sqrt{c^2 - x'^2} \quad \text{and} \quad s = x' + \sqrt{c^2 - x'^2}.$$

$$\therefore\ ds/dx' = 1 - x'/\sqrt{c^2 - x'^2} = 0 \quad \text{gives} \quad x'^2 = c^2/2 = y'^2.$$

That is, equal conjugate diameters are those whose sum is a maximum.

Expressing x' and y' in terms of the inclination of x' to the transverse axis, denoted by θ, we have $ds/d\theta = (ds/dx')(dx'/d\theta)$.

a and b are, respectively, maximum and minimum states of x', giving $dx'/d\theta = 0$, and therefore $ds/d\theta = 0$. Hence the sum of the axes is a minimum.

37. "A rectangular hall 80 feet long, 40 feet wide, and 12 feet high has a spider in one corner of the ceiling. How long will it take the spider to crawl to the opposite corner on the floor if he crawls a foot in one second on the wall and two feet in a second on the floor?"*

Ans. 55.4754 seconds, minimum.

139. *To find the maximum and minimum distances from a given plane curve to a given point in its plane.*

Let $y = fx$ be the equation of any plane curve, (a, b) the coordinates of any point in its plane, and R the distance from (a, b) to the point (x', y') on the curve.

---

* Problem proposed by Professor H. C. Whitaker in *American Mathematical Monthly*, Vol. I. No. 8.

If (x', y') move along the curve, R in general becomes a varying distance measured on the radius vector joining (a, b) with the moving point (x', y') and $R^2 = (x' - a)^2 + (y' - b)^2$. We wish to find the maximum and minimum values of R.

Placing the first derivative of $(x' - a)^2 + (y' - b)^2$ equal to zero, we have

$$x' - a + (y' - b)(dy'/dx') = 0. \quad . \quad . \quad . \quad (1)$$

The equation of the normal to $y = fx$ at (x', y') is $y - y' = -(dx'/dy')(x - x')$. Hence (1) expresses the condition that (x', y') is on the normal through (a, b).

The required value of R is therefore estimated along the normal through (a, b), is a maximum or a minimum according as the second derivative,

$$1 + (dy/dx)^2 + (y - b)(d^2y/dx^2),$$

is negative or positive, and, in general, is neither a maximum nor a minimum when the second derivative reduces to zero (§ 136).

As (a, b) may be any point upon any normal, we conclude that the radial distance of each point of a normal from the curve is, in general, a maximum or a minimum when measured upon the normal. (See figure, page 232.)

Thus, let BAM be a normal to the curve NMO at M. With A and B as centres, and with the radii AM and BM respectively, describe the circumferences rMr and RMR. The figure shows that the radial distance of A from NMO is a minimum when measured upon the normal AM, and that the corresponding distance of the point B is a maximum. This is evident from the fact that the circumference rMr, in the vicinity of and on both sides of M, lies within the curve NMO, while the corresponding part of the circumference RMR lies without.

Consider the point (a, b) to move upon the normal, and let $(\bar{x}, \bar{y})$ be its variable coordinates. When the normal

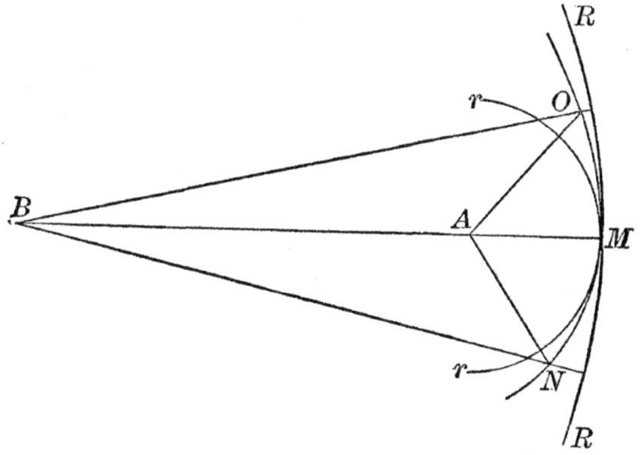

distance of $(\bar{x}, \bar{y})$ from the curve is neither a maximum nor a minimum, we have

$$1 + (dy/dx)^2 + (y - \bar{y})(d^2y/dx^2) = 0,$$

whence by combination with (1) we obtain

$$\left.\begin{aligned} \bar{x} &= x - \left(1 + \frac{dy^2}{dx^2}\right)\frac{dy}{dx} \Big/ \frac{d^2y}{dx^2}, \\ \bar{y} &= y + \left(1 + \frac{dy^2}{dx^2}\right) \Big/ \frac{d^2y}{dx^2}, \end{aligned}\right\} \quad . \quad . \quad . \quad . \quad (2)$$

for the coordinates of a point on the normal whose distance from the curve measured along the normal is, in general, neither a maximum nor a minimum. Representing this distance by ρ, we have

$$\rho^2 = (x - \bar{x})^2 + (y - \bar{y})^2,$$

which combined with (2) gives

$$\rho = \left(1 + \frac{dy^2}{dx^2}\right)^{3/2} \Big/ \frac{d^2y}{dx^2} \quad . \quad . \quad . \quad . \quad . \quad (3)$$

In the figure, page 232, $(\bar{x}, \bar{y})$ lies somewhere between A and B. It separates those points of the normal each of which has a minimum radial distance from the curve lying on the normal, from those points of the normal each of which has a corresponding maximum distance on the same line.

It is important to observe that a circumference described with $(\bar{x}, \bar{y})$ as a centre and with a radius equal to ρ will, in general, intersect the curve NMO at M.

This circle is important in the discussion of curves, and equations (2) and (3) will be referred to hereafter.

IMPLICIT FUNCTIONS.

140. Having y given as an implicit function of x, by an equation $f(x, y) = 0$ not readily solved with respect to y, we may differentiate as indicated in § 110 and obtain an expression for dy/dx. Placing it equal to 0, we may combine the resulting equation with the given, and find critical values of x.

Otherwise, let $$u = f(x, y) = 0. \quad . \quad . \quad . \quad . \quad . \quad . \quad (1)$$

Then (1), (§ 111),

$$du/dx = \partial u/dx + (\partial u/\partial y)(dy/dx) = 0. \quad . \quad (2)$$

Maxima and minima values of y in general require $dy/dx = 0$. Hence,

$$u = f(x, y) = 0, \text{ combined with } \partial u/dx = 0,$$

gives critical values of x.

Eq. (6) (§ 111) gives

$$d^2y/dx^2 = -(\partial^2 u/dx^2)/(\partial u/\partial y),$$

which, if not zero or infinity, is positive for a minimum and negative for a maximum of y.

Having $y = fz, \quad z = \phi x,$

$$dy/dx = (dy/dz) \times (dz/dx) = 0,$$

will give critical values of x.

EXAMPLES.

1. $u = x^3 + y^3 - 3a^2x = 0.$

$$\partial u/dx = 3x^2 - 3a^2 = 0. \qquad \therefore \quad x = \pm a.$$

Substituting in given equation, we have $y = \pm a \sqrt[3]{2}.$

$$\partial^2 u/dx^2 = 6x, \qquad \partial u/\partial y = 3y^2.$$

$$\left.\frac{d^2y}{dx^2}\right]_{\substack{x=a \\ y=a\sqrt[3]{2}}} = \frac{-6a}{3a^2\sqrt[3]{4}}, \qquad \therefore \quad y = a\sqrt[3]{2} \text{ is a maximum.}$$

$$\left.\frac{d^2y}{dx^2}\right]_{\substack{x=-a \\ y=-a\sqrt[3]{2}}} = \frac{6a}{3a^2\sqrt[3]{4}}, \qquad \therefore \quad y = -a\sqrt[3]{2} \text{ is a minimum.}$$

2. $x^3 - 3axy + y^3 = 0.$ $\quad x = 0, \; y = 0,$ is a minimum.

$x = a\sqrt[3]{2}, \quad y = a\sqrt[3]{4},$ is a maximum.

3. $x^2 + y^2 - 2bxy - a^2 = 0.$

$$x = \frac{ab}{\sqrt{1-b^2}}, \quad y = \frac{a}{\sqrt{1-b^2}}, \text{ is a maximum.}$$

4. $4xy - y^4 - x^4 = 2.$ $\quad x = \pm 1, \; y = \pm 1,$ no max. or min.

5. $y^2 - 3 = -2x(yx + 2).$ $\quad x = -1/2, \; y = 2,$ a maximum.

6. $y = \pi\sqrt{z/a}, \quad z = (k^2 + x^2)/x, \quad x = k$ makes y a minimum.

141. Having v given as an implicit function of x, by two equations $v = \phi(x, y)$ and $u = f(x, y) = 0$, from which y is not readily eliminated, we may proceed as follows:

$$\frac{dv}{dx} = \frac{\partial v}{dx} + \frac{\partial v}{\partial y}\frac{dy}{dx}.$$

$$\frac{dy}{dx} = -\frac{\partial u}{dx}\bigg/\frac{\partial u}{\partial y}. \qquad (2) \quad (\S\ 111)$$

Hence, $$\frac{dv}{dx} = \frac{\partial v}{dx} - \frac{\partial v}{\partial y}\frac{\partial u}{dx} \Big/ \frac{\partial u}{\partial y},$$

and $$\frac{dv}{dx} = 0 \text{ gives } \frac{\partial v}{dx}\frac{\partial u}{\partial y} - \frac{\partial v}{\partial y}\frac{\partial u}{dx} = 0, \quad . \quad . \quad (1)$$

which combined with $u = f(x, y) = 0$ gives critical values of x. The sign of the corresponding value of d^2v/dx^2 will, in general, determine whether v is a maximum or a minimum.

EXAMPLES.

1. $v = x^2 + y^2, \quad (x-a)^2 + (y-b)^2 - c^2 = 0.$

$$\partial v/dx = 2x, \qquad \partial v/\partial y = 2y,$$
$$\partial u/dx = 2(x-a), \qquad \partial u/\partial y = 2(y-b).$$

Substituting in (1), we have $ay = bx$; which combined with $u = 0$ gives

$$x = a \pm ac/\sqrt{a^2+b^2}.$$

The positive sign gives a maximum, and the negative a minimum, for v.

2. Find the points in the circumference of a given circle which are at a maximum or minimum distance from a given point.

3 Given the four sides of a quadrilateral, to find when its area is a maximum.

Let a, b, c, d be the lengths of the sides, ϕ the angle between a and b, ψ that between c and d.

Then $$\text{area} = v = ab \sin \phi/2 + cd \sin \psi/2,$$

and $$a^2 + b^2 - 2ab \cos \phi = c^2 + d^2 - 2cd \cos \psi,$$

each member being the square of the same diagonal.

$$\frac{\partial v}{d\phi} = \frac{ab}{2} \cos \phi, \qquad \frac{\partial v}{\partial \psi} = \frac{cd}{2} \cos \psi,$$

$$\frac{\partial u}{d\phi} = 2ab \sin \phi, \qquad \frac{\partial u}{\partial \psi} = -2cd \sin \psi.$$

Substituting in (1), we have

$$\tan \phi = -\tan \psi. \qquad \therefore \quad \phi = 180° - \psi.$$

That is, the quadrilateral is inscribable in a circle.

$$\frac{dv}{d\phi} = \frac{ab}{2}\cos \phi + \frac{cd}{2}\cos \psi \frac{d\psi}{d\phi} = 0,$$

$$ab \sin \phi = cd \sin \psi (d\psi/d\phi);$$

from which $$d\psi/d\phi = ab \sin \phi / cd \sin \psi.$$

Substituting in above, we have

$$dv/d\phi = ab \sin(\phi + \psi)/2 \sin \psi.$$

$$\frac{d^2v}{d\phi^2} = \frac{ab \cos(\phi + \psi)}{2 \sin \psi}\left(1 + \frac{d\psi}{d\phi}\right) - F[\sin(\phi + \psi)].$$

$$\left.\frac{d^2v}{d\phi^2}\right]_{\phi + \psi = \pi} = \frac{-ab}{2 \sin \psi}\left(1 + \frac{ab}{cd}\right), \text{ indicating a maximum.}$$

142. Having w given as an implicit function of x, by three equations

$$w = F(x, y, z), \quad v = \phi(x, y, z) = 0, \quad \text{and} \quad u = f(x, y, z) = 0,$$

and placing $dw/dx = 0$ for a maximum or a minimum, we write

$$\left.\begin{aligned} \frac{dw}{dx} &= \frac{\partial w}{dx} + \frac{\partial w}{\partial y}\frac{dy}{dx} + \frac{\partial w}{\partial z}\frac{dz}{dx} = 0, \\ \frac{dv}{dx} &= \frac{\partial v}{dx} + \frac{\partial v}{\partial y}\frac{dy}{dx} + \frac{\partial v}{\partial z}\frac{dz}{dx} = 0, \\ \frac{du}{dx} &= \frac{\partial u}{dx} + \frac{\partial u}{\partial y}\frac{dy}{dx} + \frac{\partial u}{\partial z}\frac{dz}{dx} = 0. \end{aligned}\right\} \quad . \; . \quad (1)$$

Eliminating dy/dx and dz/dx, we have a single equation which, combined with $w = F(x, y, z)$, $v = 0$, and $u = 0$, gives critical values of x and the corresponding values of y, z, and w.

By differentiating equations (1), and eliminating

$$dy/dx, \quad dz/dx, \quad d^2y/dx^2, \quad d^2z/dx^2,$$

an expression for d^2w/dx^2 may be determined.

Example. * A Norman window consists of a rectangle surmounted by a semicircle. With a given perimeter, find the height and width of the window when its area is a maximum.

Let $y =$ height, $2x =$ width, $w =$ area, $P =$ perimeter.

Then $w = \frac{\pi x^2}{2} + 2xy, \quad v = 2(x + y) + \pi x - P = 0.$

$$\left.\begin{aligned} dw/dx = \pi x + 2y + 2x dy/dx = 0, \\ 2 + \pi + 2dy/dx = 0. \end{aligned}\right\} \quad . \quad . \quad . \quad (1)$$

Eliminating dy/dx, and combining result with $v = 0$,

$$x = y = P/(4 + \pi).$$

Differentiating equation (1), we have

$$d^2w/dx^2 = \pi + 4dy/dx + 2xd^2y/dx^2,$$

$$2d^2y/dx^2 = 0.$$

Hence, $d^2w/dx^2\big]_{x=y} = -\pi - 4$, indicating a maximum.

FUNCTIONS OF TWO OR MORE VARIABLES.

143. A Maximum state of a continuous function of two independent variables is one greater than any adjacent state. Thus, $z = f(x, y)$ is a maximum corresponding to $x = a, y = b$, provided as h and k vanish from any values, we have ultimately and continuously $f(a, b) > f(a \pm h, b \pm k)$.

A maximum state is, therefore, one through which, as *either or both variables increase continuously*, the function changes from an *increasing* to a *decreasing* function, and *its partial differential coefficients of the first order change their signs from plus to minus.* (§ 71.)

---

* Todhunter's Diff. Calc., p. 214.

A Minimum state is one less than adjacent states. Thus, $f(a, b)$ is a minimum, provided as h and k vanish from any values, we have ultimately and continuously

$$f(a, b) < f(a \pm h, b \pm k).$$

A minimum state is, therefore, one through which, as *either or both variables increase continuously*, the function changes from a *decreasing* to an *increasing* function, and *its partial differential coefficients of the first order change their signs from minus to plus.* (§ 71.)

Any particular state of $z = f(x, y)$, as $f(a, b)$, may be examined directly by determining whether as h and k vanish from any assumed values, we have ultimately

$$f(a, b) > f(a \pm h, b \pm k), \quad \text{or} \quad f(a, b) < f(a \pm h, b \pm k).$$

144. In general, however, maxima and minima are determined by testing those values of the variables corresponding to which, as the variables increase, both partial derivatives of the first order change their signs from plus to minus, or minus to plus.

Sets of roots of the equations

$$\partial z/dx = 0, \quad \partial z/dy = 0, \quad . \quad . \quad . \quad . \quad (1)$$

$$\left(\begin{matrix}\text{Exceptional} \\ \text{cases.}\end{matrix}\right) \begin{cases} \partial z/dx = \infty, \ \partial z/dy = \infty, \quad . \quad . \quad . \quad . \quad (2) \\ \partial z/dx = 0, \quad \partial z/dy = \infty, \quad . \quad . \quad . \quad . \quad (3) \\ \partial z/dy = \infty, \ \partial z/dx = 0, \quad . \quad . \quad . \quad . \quad (4) \end{cases}$$

are, therefore, critical, and may be tested as indicated in § 143.

A maximum or minimum state of a function of two variables is illustrated by an ordinate of a surface which is either greater or less than all adjacent ordinates. The conditions $\partial z/dx = 0$ and $\partial z/dy = 0$ indicate, in general, that the corresponding tangent plane is parallel to XY.

145. Lagrange's Condition.—When the successive partial differential coefficients to include those of the $n + 1$ order are real and finite for a set of critical values, as (a, b), derived from equations (1), § 144, a condition for a maximum or a minimum may be deduced from (a), § 128.

Placing

$$\left(\frac{\partial^2 z}{dx^2}\right)_{(a,\, b)} = A, \quad \left(\frac{\partial^2 z}{dx\, dy}\right)_{(a,\, b)} = B, \quad \left(\frac{\partial^2 z}{dy^2}\right)_{(a,\, b)} = C,$$

we have

$$f(a \pm h,\, b \pm k) - f(a,\, b) = (Ah^2 \pm 2Bhk + Ck^2)/2 + R. \quad (1)$$

In general, as h and k vanish, the sign of the second member will ultimately depend upon that of $Ah^2 \pm 2Bhk + Ck^2$, and when $A \neq 0$ it may be written

$$[(Ah \pm Bk)^2 + (AC - B^2)k^2]/A.$$

A maximum or a minimum state will then require that the sign of this expression shall ultimately become fixed and remain so, while h and k change their signs by passing through zero.

If $(AC - B^2) < 0$, the numerator of the above expression will be positive when $k = 0$, and it will be negative for values of h and k other than zero which make $Ah \pm Bk = 0$. Hence, a condition for a maximum or a minimum state is

$$(AC - B^2) > 0, \quad \text{or} \quad AC > B^2;$$

that is,
$$\left.\frac{\partial^2 z}{dx^2}\right]_{(a,\, b)} \times \left.\frac{\partial^2 z}{dy^2}\right]_{(a,\, b)} > \left.\left(\frac{\partial^2 z}{dx\, dy}\right)^2\right]_{(a,\, b)} . \quad . \quad . \quad (2)$$

This condition being satisfied, $f(a, b)$ is a maximum or a minimum according as A and C are both negative or both positive.

If $AC < B^2$, there is neither a maximum nor a minimum.

If $A = 0$ and $B \neq 0$, we have the same result, for the sign of $2Bhk + Ck^2$, in the second member of eq (1), varies for a fixed value of k as h passes through the value $-Ck/2B$.

If $A = B = 0$, then

$$AC - B^2 = 0, \text{ and } Ah^2 \pm 2Bhk + Ck^2 = Ck^2,$$

which vanishes when $k = 0$, for all values of h leaving the question in doubt. The same result follows when $A = B = C = 0$.

EXAMPLES.

Find sets of values of the variables which correspond to maxima or minima of the following functions:

1. $z = x^2 + xy + y^2 + a^3/x + a^3/y$.

$$\partial z/dx = 2x + y - a^3/x^2 = 0, \quad \partial z/dy = 2y + x - a^3/y^2 = 0.$$

From which we obtain $x = y = a/\sqrt[3]{3}$,

$$\partial^2 z/dx^2 = 2 + 2a^3/x^3, \quad \partial^2 z/dy^2 = 2 + 2a^3/y^3, \quad \partial^2 z/dx\,dy = 1.$$

In which substituting values of x and y, condition (2) is satisfied and z is a minimum.

2. $z = \cos x \cos \alpha + \sin x \sin \alpha \cos (y - \beta)$.

$$\partial z/dx = -\sin x \cos \alpha + \cos x \sin \alpha \cos (y - \beta) = 0,$$

$$\partial z/dy = -\sin \alpha \sin x \sin (y - \beta) = 0,$$

give $x = \alpha, \quad y = \beta$.

$$\left(\frac{\partial^2 z}{dx^2}\right)_{(\alpha, \beta)} = -1, \quad \left(\frac{\partial^2 z}{dy^2}\right)_{(\alpha, \beta)} = -\sin^2\alpha, \quad \left(\frac{\partial^2 z}{dx\,dy}\right)_{(\alpha, \beta)} = 0.$$

Hence z is a maximum.

3. $x^3y^2(a-x-y)$. $\begin{cases} x = a/2, \\ y = a/3, \end{cases}$ max.

4. x^3+y^3-3axy. $x = y = a$, min.

5. $x^4+y^4-x^2+xy-y^2$. $x = y = 0$, max.

$x = y = \pm\frac{1}{2}$, min.

$x = \pm\frac{1}{2}\sqrt{3}$, $y = \mp\frac{1}{2}\sqrt{3}$, min.

6. $x^3y^2(6-x-y)$. $\begin{cases} x = 3, \\ y = 2, \end{cases}$ max.

7. $x^4+y^4-2x^2+4xy-2y^2$. $\begin{cases} x = \pm\sqrt{2}, \\ y = \mp\sqrt{2}, \end{cases}$ min.; $x = 0 = y$, max.

8. $(2ax-x^2)(2by-y^2)$. $\begin{cases} x = a, \\ y = b, \end{cases}$ max.

9. $e^{-x^2-y^2}(ax^2+by^2)$. $x = y = 0$, min.

$x = 0$, $y = \pm 1$, $a < b$, max.

$x = \pm 1$, $y = 0$, $a > b$, max.

10. $\sin x + \sin y + \cos(x+y)$. $x = y = 3\pi/2$, min.

$x = y = \pi/6$, max.

11. Divide a number a into three parts, such that the mth power of the first, by the nth power of the second, by the rth power of the third shall be a maximum.

Ans. $\dfrac{ma}{m+n+r}$, $\dfrac{na}{m+n+r}$, $\dfrac{ra}{m+n+r}$.

12. Find the minimum distance from a given point to a given plane.

13. The volume of a rectangular parallelopipedon being given, find its edges when the surface is a minimum. Each edge $= \sqrt[3]{\text{vol.}}$

14. An open tank to contain a given volume of water is to be constructed in the form of a rectangular parallelopipedon. Determine its edges so that the surface to be lined shall be a minimum.

Each edge of base $= \sqrt[3]{2\text{ vol.}}$; altitude $= \sqrt[3]{2\text{ vol.}}/2$.

15. Determine the maximum rectangular parallelopipedon which can be inscribed in a given sphere. Each edge $= 2R/\sqrt{3}$.

146. Functions of Three Variables.—Let $u = f(x, y, z)$. Reasoning as in the preceding cases, it may be shown that sets of roots of the equations

$$\partial u/dx = 0, \quad \partial u/dy = 0, \quad \partial u/dz = 0, \quad . \quad . \quad (1)$$

and $$\partial u/dx = \infty, \quad \partial u/dy = \infty, \quad \partial u/dz = \infty, \quad . \quad . \quad (2)$$

are critical.

Denoting a set of critical values from (1) by a, b, and c, we have, (b), § 128,

$$f(a \pm h, b \pm k, c \pm l) - f(a, b, c) = [Ah^2 + Bk^2 + Cl^2]/2$$
$$\pm Dhk \pm Ehl \pm Fkl + R,$$

in which A, B, C, D, E, F, represent the values of

$$\frac{\partial^2 u}{dx^2}, \quad \frac{\partial^2 u}{dy^2}, \quad \frac{\partial^2 u}{dz^2}, \quad \frac{\partial^2 u}{dx\,dy}, \quad \frac{\partial^2 u}{dx\,dz}, \quad \frac{\partial^2 u}{dy\,dz},$$

respectively, when $x = a$, $y = b$, $z = c$.

In order that $f(a, b, c)$ may be a maximum or a minimum,

$$\tfrac{1}{2}(Ah^2 + Bk^2 + Cl^2) \pm Dhk \pm Ehl \pm Fhl,$$

if not zero, should be either always negative or always positive, as h, k, and l vary through zero between certain positive and negative limits.

147. Functions of n Variables.—By extending the above method of reasoning, it may be shown that the sets of roots of the equations formed by placing the partial derivatives of the first order separately equal to zero are critical. Each set of critical values when substituted in the corresponding expansion should render the quadratic function of h, k, l, etc., always negative for a maximum, or always positive for a minimum, as h, k, l, etc., vary through zero between certain limits.

PART III.

GEOMETRIC APPLICATIONS.

CHAPTER XII.

TANGENTS AND NORMALS.

RECTANGULAR COÖRDINATES.

148. Equations of a Tangent and Normal.—The equation of a straight line passing through (x', y') on the curve $y = fx$ is (Anal. Geom.) $y - y' = m(x - x')$.

Placing $m = f'x' = dy'/dx'$, we have for the tangent line at (x', y') (§ 71)

$$y - y' = (dy'/dx')(x - x'), \quad . \quad . \quad . \quad . \quad (1)$$

and for the corresponding normal

$$y - y' = -(dx'/dy')(x - x'). \quad . \quad . \quad . \quad (2)$$

Thus, having $y^2 = 9x$, then $f'x = 9/2y$, $f'4 = 3/4$. Hence, $y - 6 = (3/4)(x - 4)$ is the tangent, and
$y - 6 = -(4/3)(x - 4)$ is the normal at $(4, 6)$.

If the equation of a line is in the form $u = \phi(x, y) = 0$, we have (§ 111)

$$f'x' = dy'/dx' = -(\partial u/dx')/(\partial u/dy').$$

Substituting in above, we have

$$(\partial u/dy')(y - y') = -(\partial u/dx')(x - x') \text{ for the tangent,}$$
$$(\partial u/dx')(y - y') = (\partial u/dy')(x - x') \text{ for the normal.}$$

Thus, having $u = y^2 - 9x = 0$,

$$(\partial u/dy) = 2y, \qquad (\partial u/dx) = -9,$$

and at the point (4, 6)

$$(\partial u/dy') = 12, \qquad (\partial u/dx') = -9.$$

Therefore, we have

$$12(y - 6) = 9(x - 4) \text{ for the tangent,}$$

and $$-9(y - 6) = 12(x - 4) \text{ for the normal.}$$

EXAMPLES.

Find the equations of the tangent and normal at the point (x', y') on each of the following curves:

1. $a^2y^2 + b^2x^2 = a^2b^2$. $\begin{cases} y - y' = -(b^2x'/a^2y')(x - x'), \\ y - y' = (a^2y'/b^2x')(x - x'). \end{cases}$

2. $y^2 = 2px$. $\begin{cases} yy' = p(x + x'), \\ y - y' = -(y'/p)(x - x'). \end{cases}$

3. $x^2 + y^2 = R^2$. $\begin{cases} y'y + x'x = R^2, \\ y = (y'/x')x. \end{cases}$

4. $y^2 = 2Rx - x^2$. $y - y' = (R - x')(x - x')/y'$.

5. $y = a \log \sec (x/a)$. $\begin{cases} y - y' = -\cot (x'/a)(x - x'), \\ y - y' = \tan (x'/a)(x - x'). \end{cases}$

6. $x^{2/3} + y^{2/3} = a^{2/3}$. $y = -x \sqrt[3]{y'/x'} + \sqrt[3]{a^2y'}$.

7. $y^2 = 9x^3$. $(1, 3) \begin{cases} y = 9x/2 - 3/2, \\ y = -2x/9 + 29/9. \end{cases}$

8. $x = r \operatorname{vers}^{-1}\frac{y}{r} - \sqrt{2ry - y^2}$. $y - y' = \dfrac{2r - y'}{\sqrt{2ry' - y'^2}}(x - x')$.

9. $y^2 = \frac{x^3}{2a - x}$. $\quad y - y' = \pm \frac{\sqrt{x'}(3a - x')}{(2a - x')^{3/2}}(x - x')$.

10. $\tan^{-1}\frac{y}{x} = k \log \sqrt{x^2 + y^2}$. $\quad y - y' = \frac{kx' + y'}{x' - ky'}(x - x')$.

11. $y = ae^{x/c}$. $\quad y - y' = (y'/c)(x - x')$.

12. $ax^2 + 2bxy + cy^2 + 2dx + 2ey + f = 0$.

$$y - y' = -\frac{ax' + by' + d}{bx' + cy' + e}(x - x').$$

13. $e^y = x$. $\quad y - y' = (x - x')/x'$.

14. $e^y = \sin x$. $\quad y - y' = \cot x(x - x')$.

15. $xy = a$. $\quad y - y' = -(y'/x')(x - x')$.

16. $y^2 = 2px + r^2x^2$. $\quad y - y' = \frac{p + r^2x'}{\sqrt{2px' + r^2x'^2}}(x - x')$.

17. $y = \frac{8a^3}{4a^2 + x^2}$. $(x' = 2a)$ $\quad \begin{cases} x + 2y = 4a. \\ y - 2x = -3a. \end{cases}$

18. $y = \frac{ax^2}{x^2 + a^2}$. $(y' = a/4)$ $\quad \begin{cases} y = \pm 3\sqrt{3}x/8 - a/8. \\ y = \mp 8x/3\sqrt{3} + 41a/36. \end{cases}$

19. $y(x - 1)(x - 2) = x - 3$. $\quad (x' = 3 \pm \sqrt{2})$

$$y = \sqrt{2}/(4 + 3\sqrt{2}),\ x = 0/0.$$

20. $3y^2 + x^2 = 5$. $(x' = 1)$ $\quad y = \mp .29x \pm 1.44$.

21. $y^2 = 6x - 5$. $\quad \begin{cases} y = [3(x + x') - 5]/y'. \\ y = -y'(x - x')/3 + y'. \end{cases}$

22. $y^2 = 2x^2 - x^3$. $(x' = 1)$ $\quad \begin{cases} y = \pm (x + 1)/2. \\ y = \mp 2x \pm 3. \end{cases}$

23. $y = (e^{x/c} + e^{-x/c})c/2$. $\quad y - y' = (e^{x'/c} - e^{-x/c})(x - x')/2$.

24. $x = a \log [(a + \sqrt{a^2 - y^2})/y] - \sqrt{a^2 - y^2}$.

$$y - y' = -[y'/(a^2 - y'^2)^{1/2}](x - x').$$

25. Find the angle at which $x^2 = y^2 + 5$ intersects $8x^2 + 18y^2 = 144$.

Ans. $\pi/2$.

26. Find the equation of the tangent to the curve $x^2(x + y) = a^2(x - y)$ at the origin. Ans. $y = x$.

27. Find the angle of intersection made by the two curves $x^2+y^2=(2a)^2$, and $y^2-x^2=-a^2$. Ans. $\tan^{-1}\sqrt{15}$.

28. Find the angle at which the curve $(y^2+x^2)^2=2a^2(x^2-y^2)$ cuts X. Ans. $\pi/2$.

29. Find that point on an ellipse at which the angle between the normal and diameter is a maximum. Ans. $(a/\sqrt{2},\ b/\sqrt{2})$.

30. Find the point on a parabola where the angle between a straight line to the vertex and the curve is a maximum. Ans. $x'=p$.

To find the equation of the tangent to $y=f(x)$ which makes a given angle, ϕ, with X:

Let $a=\tan\phi$; then $dy'/dx'=a$, which combined with $y=f(x)$ will determine (x', y') for the point of tangency.

Thus, to find the equation of the tangent to $y^2=6x$ which makes an angle of 45° with X:

$$dy'/dx'=3/y'=1; \quad \therefore y'=3.$$

Substituting in $y^2=6x$, we have $x'=3/2$, and the required equation is $y-3=x-3/2$.

EXAMPLES.

1. Find the equation of the tangent to $y=a+(c-x)^2$ which is parallel to X. Ans. $y=a$, $x=0/0$.

2. Find the equation of the tangent to $x^2+y^2=r^2$ which is parallel to $y=2x+7$. Ans. $y\pm r/\sqrt{5}=2(x\mp 2r/\sqrt{5})$.

3. Find the equation of the normal to $y^2/4+x^2/9=1$ which is parallel to $2x-y=3$. Ans. $y=2x\mp 2$.

4. Find the equation of a tangent to $18y^2+8x^2=72$ which is perpendicular to the right line passing through the positive ends of the axes. Ans. $2y=3x\mp\sqrt{97}$.

5. Find the equations of the tangents to $9y^2-25x^2=-225$ which makes an angle of 60° with the transverse axis.

6. Find the points on $y=x^3-3x^2-24x+85$ where the tangents are parallel to X. Ans. $(4, 5)$ and $(-2, 113)$.

7. Find the equation of the perpendicular through the focus of a parabola to a tangent at (x', y'). Ans. $y=-y'(x-p/2)/p$.

8. $y^2 = 4ax + 4a^2$; find the locus of the points of intersection of tangents, and perpendiculars upon them from the focus.

Ans. $x = -a$; $y = 0/0$.

9. $y - 2 = (x - 1)\sqrt{x - 2}$; find the angle at which the curve cuts X, and the point where the tangent is perpendicular to X.

Ans. $\text{Tan}^{-1} 2$; $(2, 2)$.

10. Show that normal to $y^2 = 4ax$ is tangent to $y^2 = 4(x - 2a)^3/(27a)$.

With oblique axes, inclined at an angle β, dy'/dx' represents the ratio of the sines of the angles which the tangent and curve at (x', y') make with the axes. The equation of the tangent remains unchanged in form, while that of the normal becomes

$$y - y' = \left[-\left(1 + \cos\beta\frac{dy'}{dx'}\right)(x - x')\right] \Big/ \left[\cos\beta + \frac{dy'}{dx'}\right].*$$

149. Let TM, PM, and NM, respectively, be the tangent, ordinate, and normal at any point M whose coördinates are (x', y').

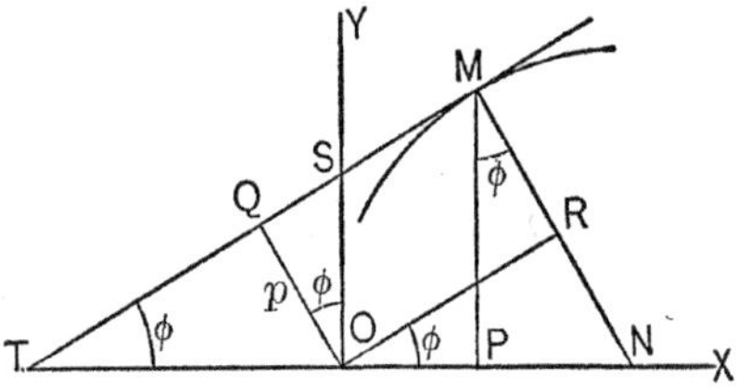

Let $OQ = p$ and $OR = q$ be perpendiculars from the origin to the tangent and normal respectively.

Let $XTM = \phi$. Then, since $dy'/dx' = \tan\phi$, we have

Subtangent $= PT = y' \cot\phi = y'dx'/dy'$.

Subnormal $= PN = y' \tan\phi = y'dy'/dx'$.

Normal $= NM = y' \sec\phi = y'\sqrt{1 + \tan^2\phi}$
$= y'\sqrt{1 + (dy'/dx')^2}$.

Tangent $= TM = y' \operatorname{cosec}\phi = y'\sqrt{1 + \cot^2\phi}$
$= y'\sqrt{1 + (dx'/dy')^2}$.

---

* Todhunter's Calculus, page 283.

Also, from figure, or equations (1) and (2) (§ 148),

OT = Intercept of tangent on $X = x' - y'dx'/dy'$.

OS = Intercept of tangent on $Y = y' - x'dy'/dx'$.

ON = Intercept of normal on $X = x' + y'dy'/dx'$.

Hence,

$$p = OQ = \text{Perpendicular to tangent} = OS \cos \phi$$

$$= OS \frac{1}{\sqrt{1+\tan^2 \phi}} = \frac{y' - x'dy'/dx'}{\sqrt{1+(dy'/dx')^2}} = \frac{y'dx' - x'dy'}{\sqrt{dx'^2 + dy'^2}},$$

$$q = OR = \text{Perpendicular to normal} = ON \cos \phi$$

$$= \frac{x' + y'dy'/dx'}{\sqrt{1+(dy'/dx')^2}} = \frac{x'dx' + y'dy'}{\sqrt{dx'^2 + dy'^2}}.$$

To apply these formulas, obtain general expression for dy'/dx' from the equation of the curve, and substitute values of x' and y'.

EXAMPLES.

1. $y^2/b^2 + x^2/a^2 = 1$. $\quad p = a^2b^2/(b^4x'^2 + a^4y'^2)^{1/2}$.

Subt $= -a^2y'^2/(b^2x') = (x'^2 - a^2)/x'$. $\quad$ Subn $= -b^2x'/a^2$.

If $a = b$, we have for the circle $r = a$, $p = a$, Subt $= -y'^2/x'$, Subn $= -x'$.

2. $y^2/b^2 - x^2/a^2 = -1$. $\quad p = -a^2b^2/(b^4x'^2 + a^4y'^2)^{1/2}$.

Subt $= a^2y'^2/(b^2x') = (x'^2 + a^2)/x'$. $\quad$ Subn $= b^2x'/a^2$.

If $a = b$, we have for an equilateral hyperbola $p = a^2/\sqrt{x'^2 + y'^2}$, Subt $= y'^2/x'$, Subn $= x'$.

3. $y^2 = 2px + r^2x^2$.

$$\text{Subt} = (2px' + r^2x'^2)/(p + r^2x'); \quad \text{Subn} = p + r^2x',$$

$$\text{Tan} = \sqrt{2px' + r^2x'^2 + \left(\frac{2px' + r^2x'^2}{p + r^2x'}\right)^2},$$

$$\text{Nor} = \sqrt{2px' + r^2x'^2 + (p + r^2x')^2}.$$

If $r^2 = 0$, we have, for the parabola referred to axis and tangent at vertex,

$$\text{Subt} = 2x', \quad \text{Subn} = p, \quad \text{Tan} = y'\sqrt{p^2 + y'^2}/p,$$

Nor $= \sqrt{y'^2 + p^2}$, Perp. to tan$= y'^2/2\sqrt{y'^2+p^2}$, $q = y'(x'+p)/\sqrt{y'^2+p}$.

4. $xy = m$.

Subt $= -x'$, Subn $= -y'^2/x' = -y'^3/m$,

Nor $= y'\sqrt{x'^2 + y'^2}/x'$, Tan $= \sqrt{y'^2 + x'^2}$.

5. $y = a^x$. Subt $= 1/\log a = M_a$.

6. $x = r \text{ vers}^{-1}(y/r) - \sqrt{2ry - y^2}$.

Subt $= y'^2/\sqrt{2ry' - y'^2}$, Subn $= \sqrt{2ry' - y'^2}$

Tan $= y'\sqrt{2ry'}/\sqrt{2ry' - y'^2}$, Nor $= \sqrt{2ry'}$.

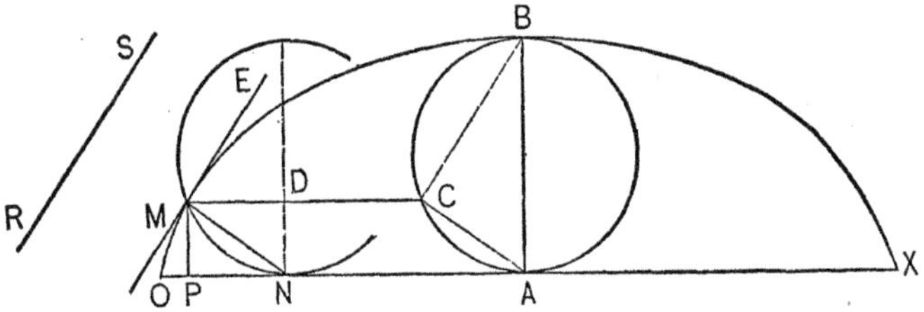

Since the subnormal $PN = MD$, the normal at any point passes through the foot of the vertical diameter of the corresponding position of the generating circle, and the tangent passes through the other extremity.

The tangent and normal at any point of a cycloid are therefore readily constructed when the corresponding position of the generating circle is drawn.

Otherwise, when the circle AB, upon the greatest ordinate as a diameter, is drawn, through the given point M draw MC parallel to the base OX; from C where it cuts the circle draw the chords CB and CA to the ends of its vertical diameter; through M draw the tangent ME parallel to CB, and the normal MN parallel to CA.

To construct a tangent parallel to any right line as RS, draw the chord BC parallel to it, through C draw CM parallel to OX, and through M draw ME parallel to RS.

7. $y^n = a^{n-1}x$. Subn $= y'^2/nx'$, Subt $= nx'$.

8. $y^3 = ax^2 + x^3$

Intercept of tan on $Y = [x'/(a+x')]^{2/3}(a/3)$.

9. $y^2 = 2x$. $x' = 8$, Tan $= 4\sqrt{17}$.

10. $x = \sec 2y$. $p = [(x^2-1)^{1/2}\sec^{-1}x - 1]/[1 + 4x^2(x^2-1)^{1/2}]$.

11. $y^2 = x^3/(2a - x)$.

Subt $= x'(2a - x')/(3a - x')$, Subn $= x'^2(3a - x')/(2a - x')^2$.

12. $y = c(e^{x/c} + e^{-x/c})/2$.

Nor $= y'^2/c$, Subn $= c(e^{2x'/c}) - e^{-2x'/c})/4$.

Tan $= y'^2/\sqrt{y'^2 - c^2}$, Subt $= cy'/\sqrt{y'^2 - c^2}$.

13. $x^4 + y^4 = c^4$. $p = c^4/\sqrt{x'^6 + y'^6}$.

14. $x^3 - 3axy + y^3 = 0$.

Eq. of tan $y = -x(x'^2 - ay')/(y'^2 - ax') + (ax'y')/(y'^2 - ax')$,

Subt $= (y'^3 - ax'y')/(ay' - x'^2)$.

15. $x^{2/3} + y^{2/3} = a^{2/3}$.

Eq. of tan $y = -x\sqrt[3]{(y'/x')} + \sqrt[3]{a^2y'}$.

Intercept on $X = \sqrt[3]{a^2x'}$, on $Y = \sqrt[3]{a^2y'}$.

Length of tan between axes $= a$.

Area between axes and tan $= \sqrt[3]{a^4x'y'}/2$.

$p = \sqrt[3]{ax'y'}$.

16. $y = ce^{x/a}$. Subt $= a$, Subn $= y'^2/a$.

17. $y^2 = 2a^2 \log x$. Subn $= a^2/x'$.

18. $xy^2 = a^2(a - x)$. Subt $= -2(ax' - x'^2)/a$.

19. $3ay^2 + a^3 = 2x^3$. Subn $= x'^2/a$.

20. $y^2 = 3x^2 - 12$. $x' = 4$, Subt $= 3$.

21. $y^2 = ax^3$. Subt $= 2x'/3$, Subn $= 3ax'^2/2$.

POLAR COÖRDINATES.

150. Polar Tangent, Subtangent, Normal, and Subnormal.

Let $PM = r$, corresponding to the point of tangency M of any tangent as TM. From the pole P draw PT per-

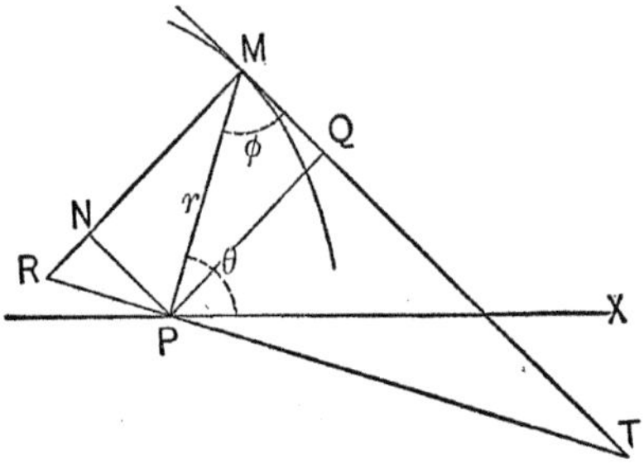

pendicular to PM. Draw PQ and MR perpendicular to TM.

PT, the part of the perpendicular to r, from P to its intersection with the corresponding tangent TM, is the **polar subtangent** corresponding to M.. PR is the **polar subnormal,** and RM is the **normal.**

Let $\phi = PMT$, the angle made by r with TM.

Then (§ 55, § 70)

$$\tan \phi = \lim_{\Delta\theta \to 0}[r\Delta\theta/(r' - r)] = rd\theta/dr.$$

From Trigonometry,

$$\sin \phi = \tan \phi/ \sqrt{1 + \tan^2 \phi}, \quad \cos \phi = \sqrt{1/(1 + \tan^2 \phi)}.$$

Therefore, since $\sqrt{dr^2 + r^2d\theta^2} = ds$, (§ 92,)

$$\sin \phi = rd\theta/ds, \quad \cos \phi = dr/ds, \quad \cot \phi = dr/rd\theta.$$

Hence,

$$PT = \textbf{Subt} = r\tan\phi = r^2 d\theta/dr,$$

$$TM = \textbf{Tan} = r\sqrt{1 + (rd\theta/dr)^2} = rds/dr,$$

$$PR = \textbf{Subn} = r\cot\phi = dr/d\theta,$$

$$RM = \textbf{Nor} = \sqrt{r^2 + (dr/d\theta)^2} = ds/d\theta,$$

$PQ = p =$ perpendicular to tangent

$$= r\sin\phi = r^2 d\theta/ds = \frac{r^2}{\sqrt{(dr/d\theta)^2 + r^2}}.$$

Since $ds/d\theta$ is assumed to be positive, positive values only of p are considered.

$$1/p^2 = 1/r^2 + (dr/d\theta)^2/r^4.$$

Putting $1/r = u$, from which $dr^2 = r^4 du^2$, we have

$$1/p^2 = u^2 + (du/d\theta)^2. \quad . \quad . \quad . \quad . \quad . \quad (1)$$

$PN = q =$ perpendicular to normal $= r\cos\phi$

$$= rdr/ds = \frac{r}{\sqrt{1 + (rd\theta/dr)^2}} = \sqrt{r^2 - p^2}.$$

Since $\quad ds = dr/(dr/ds) = rdr/r\cos\phi = rdr/q,$

$$ds/dr = r/q = r/\sqrt{r^2 - p^2}; \quad \therefore\ ds = rdr/\sqrt{r^2 - p^2}.$$

But $\quad p = r^2 d\theta/ds, \quad \therefore\ ds = r^2 d\theta/p.$

Therefore $\quad r^2 d\theta = prdr/\sqrt{r^2 - p^2}.$

EXAMPLES.

1. $r = a\theta$; find the angle ϕ between r and the tangent, the subtangent, the subnormal, the normal, and p.

$$\phi = \tan^{-1}\theta, \quad \text{Subt} = r^2/a, \quad \text{Subn} = a,$$

$$\text{Nor} = \sqrt{a^2 + r^2}, \quad p = r^2/\sqrt{a^2 + r^2}.$$

2. $r = a(1 + \cos\theta)$.

$$\text{Subt} = -r^2/(a\sin\theta), \quad \text{Subn} = -a\sin\theta, \quad p = \sqrt{r^3/2a}.$$

3. $r = a^\theta$. $\quad$ Subt $= M_a r$, $\quad$ Subn $= r/M_a$, $\quad$ Tan $\phi = M_a$.

$$\text{Tan} = r\sqrt{1 + M_a^2}, \quad \text{Nor} = r\sqrt{1 + 1/M_a^2}, \quad p = M_a r/\sqrt{M_a^2 + 1}.$$

4. $r^2 = a^2/\theta$. Subt $= 2a\sqrt{\theta}$, $p = 2a^2r/\sqrt{r^4 + 4a^4} = 2a\sqrt{\theta}/\sqrt{1 + 4\theta^2}$.

5. $r = a\theta^{1/2}$. $\quad$ Subt $= 2r^3/a^2$, $p = 2r^3/\sqrt{a^4 + 4r^4}$.

6. $r = 2a\cos\theta$.

Subt $= -2a\cot\theta\cos\theta$, $\quad$ Subn $= -2a\sin\theta$,

Tan $= 2a\cot\theta$, $\quad$ Nor $= 2a$.

7. $r = a\theta^{-1}$. $\quad$ Subt $= -a$, $\quad$ Tan $\phi = -\theta$, $\quad$ Subn $= -r^2/a$.

8. $r = a\sin\theta$. $\quad$ $\phi = \theta$.

9. $r^2 = a^2\cos 2\theta$.

$\phi = \pi/2 + 2\theta$, $\quad$ Subt $= -r^3/(a^2\sin 2\theta)$, $\quad$ Tan $= ra^2/\sqrt{a^4 - r^4}$.

Subn $= -a^2\sin 2\theta/r$, $\quad$ Nor $= a^2/r$,

$p = r^3/a^2$, $\quad$ Perp. to Nor $= r\sin 2\theta$.

10. $r = a\theta/\sin\theta$. $\quad$ Subt $= a\theta^2/(\sin\theta - \theta\cos\theta)$;

11. $r = ab/(ae^\theta + be^{-\theta})$. $\quad$ Subt $= -ab/(ae^\theta - be^{-\theta})$.

12. $r = a(1 - \cos\theta)$.

$\phi = \theta/2$. $\quad$ Subt $= 2a\sin^2(\theta/2)\tan(\theta/2)$, $\quad$ $p = 2a\sin^3(\theta/2)$;

13. $r = ae^{\theta\cot\alpha}$. $\quad$ $p = r\sin\alpha$.

14. $r = \dfrac{a(1 - e^2)}{1 - e\cos\theta}$. $\quad$ $p^2 = \dfrac{a^2(1 - e^2)}{2a/r - 1} = \dfrac{a^2(1 - e^2)^2}{1 - 2e\cos\theta + e^2}$.

15. $r = a - \theta^{-1/2}$. $\qquad p = 2a^2 r/\sqrt{r^4 + 4a^4}$.

16. $r = \dfrac{a(e^2 - 1)}{1 + e \cos \theta}$. $\qquad p = \dfrac{b\sqrt{r}}{\sqrt{2a + r}}$.

151. Polar Equation of a Tangent.—Let $PM = r'$, and $XPM = \theta'$, be the coördinates of M, the point of tangency of any tangent, as NM, and let $PN = r$, and $XPN = \theta$, be the coördinates of any other point of NM, as N. Let $PMN = \phi$; whence (§150) $\tan \phi = r'd\theta'/dr'$.

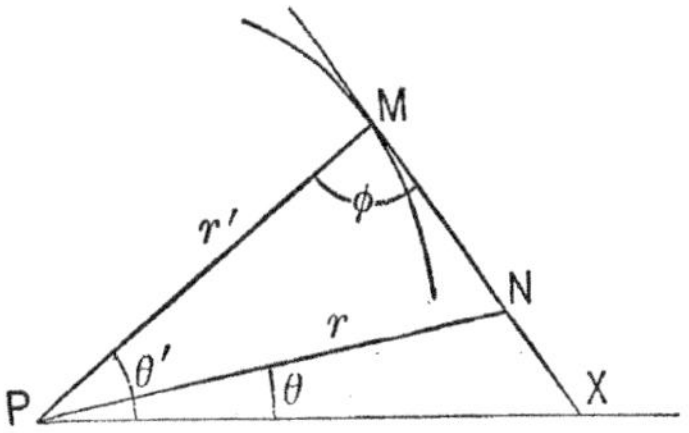

Triangle PMN gives

$$\frac{r'}{r} = \frac{\sin MNP}{\sin PMN} = \frac{\sin [(\theta' - \theta) + \phi]}{\sin \phi}$$

$$= \sin (\theta' - \theta) \cot \phi + \cos (\theta' - \theta)$$

$$= \sin (\theta' - \theta)\, dr'/r'd\theta' + \cos (\theta' - \theta). \quad . \quad . \quad (1)$$

Putting $1/r = u$ and $1/r' = u'$, whence

$$du'/d\theta' = - dr'/r'^2 d\theta',$$

we have, dividing both members of (1) by r',

$$u = u' \cos (\theta' - \theta) - \sin (\theta' - \theta)\, du'/d\theta' \qquad (2)$$

for the tangent.

Otherwise it may be obtained from eq. (1) (§ 148) by changing the reference to a system of polar coördinates.

Let r' and θ' be the polar coördinates of the point of tangency (x', y'), and r and θ those of all other points of the tangent. Taking the pole at the origin and the reference line to coincide with the axis of X, we have, as in ex. 11, § 115,

$$x = r\cos\theta, \qquad\qquad x' = r'\cos\theta'.$$

$$y = r\sin\theta, \qquad\qquad y' = r'\sin\theta'.$$

$$\frac{dy'}{dx'} = \frac{\sin\theta' dr' + r'\cos\theta' d\theta'}{\cos\theta' dr' - r'\sin\theta' d\theta'}.$$

Substituting in (1) (§ 148), we may write

$$r\sin\theta - r'\sin\theta' = \frac{\sin\theta' dr'/d\theta' + r'\cos\theta'}{\cos\theta' dr'/d\theta' - r'\sin\theta'}(r\cos\theta - r'\cos\theta'),$$

Whence

$$r\frac{dr'}{d\theta'}\sin\theta\cos\theta' - r'\frac{dr'}{d\theta'}\sin\theta'\cos\theta' - rr'\sin\theta\sin\theta' + r'^2\sin^2\theta'$$

$$= r\frac{dr'}{d\theta'}\sin\theta'\cos\theta + rr'\cos\theta\cos\theta' - r'\frac{dr'}{d\theta'}\sin\theta'\cos\theta' - r'^2\cos^2\theta',$$

or

$$\frac{rdr'}{d\theta}(\sin\theta\cos\theta' - \sin\theta'\cos\theta) - rr'(\sin\theta\sin\theta' + \cos\theta\cos\theta') = -r'^2,$$

or, changing signs of terms,

$$\frac{rdr'}{d\theta'}(\sin\theta'\cos\theta - \sin\theta\cos\theta') + rr'(\cos\theta'\cos\theta + \sin\theta'\sin\theta) = r'^2,$$

or $\quad \sin(\theta' - \theta)dr'/r'd\theta' + \cos(\theta' - \theta) = r'/r. \quad . \quad (1)$

Putting $1/r = u$, and $1/r' = u'$, whence,

$$du'/d\theta' = -dr'/(r'^2 d\theta'),$$

we have, dividing both members of (1) by r',

$$u = u'\cos(\theta' - \theta) - \sin(\theta' - \theta)du'/d\theta'.$$

Representing the polar coördinates of all points of a normal except the point of tangency (x', y') by r and θ, we deduce in a similar manner from (2) (§ 148)

$$u = u'\cos(\theta - \theta') - u'^2\sin(\theta - \theta')d\theta'/du'$$

for the **normal.**

CHAPTER XIII.

ASYMPTOTES.

RECTANGULAR COÖRDINATES.

152. An Asymptote to a curve is a *definite limiting* position of a tangent to the curve, under the law that the point of tangency recedes from the origin without limit.

Unless otherwise mentioned rectilinear asymptotes only will be considered.

Asymptotes Parallel to the Coördinate Axes.

153. If in the equation of a curve $y \ggg\!\rightarrow \infty$ as $x \ggg\!\rightarrow a$, then $(dy/dx)_a = \infty$ (§ 71), and $x = a$ is the equation of an asymptote parallel to Y. If, otherwise, $y \ggg\!\rightarrow a$ as $x \ggg\!\rightarrow \infty$, then $y = a$ is an asymptote parallel to X.

EXAMPLES.

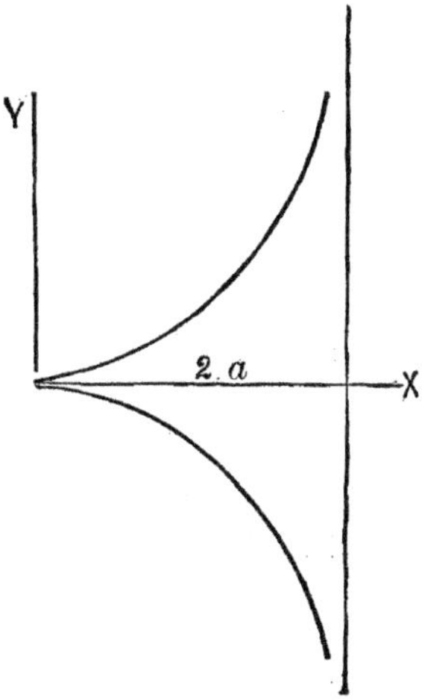

1. $y^2 = x^3/(2a - x)$.

$y \ggg\!\rightarrow \pm \infty$ as $x \ggg\!\rightarrow 2a$.

Hence $x = 2a$ is an asymptote.

2. $e^{1/x} = y^{-1}$. $y \rightarrow 1$ as $x \rightarrow \pm\infty$.

Hence $y = 1$ is an asymptote to both branches.

$$y \rightarrow \infty \text{ as } -x \rightarrow 0.$$

Hence Y is an asymptote to the left-hand branch.

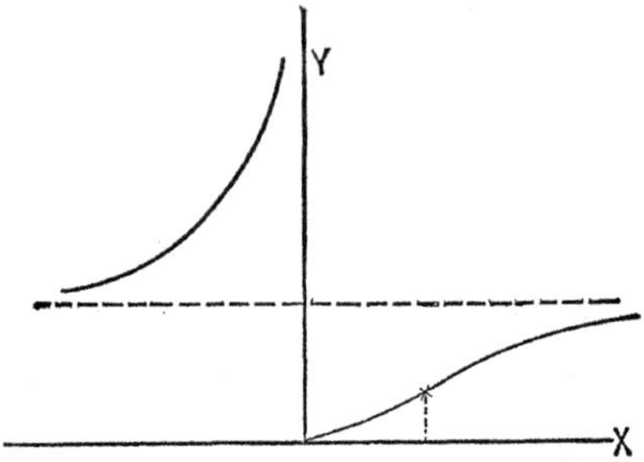

3. $y = a^x$. $y \rightarrow 0$ as $x \rightarrow -\infty$

Hence, X is an asymptote.

4. $y = \dfrac{2b^2x + b^2c}{a^2 - x^2}$. $\therefore\ y = \dfrac{2b^2/x + b^2c/x^2}{a^2/x^2 - 1}$.

$$y \rightarrow \infty \text{ as } x \rightarrow \pm a, \quad \text{and} \quad y \rightarrow 0 \text{ as } x \rightarrow \infty.$$

Hence, X and $x = \pm a$ are asymptotes.

CURVES.	ASYMPTOTES.
5. $y^2 = (x^3 + ax^2)/(x - a)$.	$x = a$.
6. $xy - ay - bx = 0$.	$x = a, \quad y = b$.
7. $a^2y - x^2y = a^3$.	$x = \pm a, \quad y = 0$.
8. $y = x/(1 + x^2)$.	$y = 0$.
9. $x = a \log\left[\left(a + \sqrt{a^2 - y^2}\right)/y\right] - \sqrt{a^2 - y^2}$.	$y = 0$.
10. $y = a^2x/(x - a)^2$.	$x = a, \quad y = 0$.
11. $y = \log x$.	$x = 0$.
12. $x^3 + aby = axy$.	$x = b$.
13. $y = a^2x/(a^2 + x)$.	$y = 0$.
14. $xy^2 = 4a^2(2a - x^2)$.	$x = 0$.

15. $x^2y + a^2y = a^3$.	$y = 0$.
16. $x = \pm (y + b) \sqrt{a^2 - y^2} / y$.	$y = 0$.
17. $y = a^3/(a^2 - x^2)$.	$x = \pm a, \quad y = 0$.
18. $x^2y^2 = a^2(x^2 - y^2)$.	$y = \pm a$.
19. $y(a^2 + x^2) = a^2(a - x)$.	$y = 0$.
20. $y = e^{-x}$.	$y = 0$.
21. $y = a + b^3/(x - c)^2$.	$y = a, \quad x = c$.
22. $a^2y + x^2y = a^2x$.	$y = 0$.
23. $x^3 - b^3 = bxy$.	$x = 0$.
24. $y(a^2 - x^2) = b(2x + c)$.	$y = 0, \quad x = \pm a$.
25. $y^2 = \dfrac{a^2(x - a)(x - 3a)}{x^2 - 2ax}$.	$x = 2a, \quad x = 0, \quad y = \pm a$.

ASYMPTOTES OBLIQUE TO THE COÖRDINATE AXES.

154. Let TM be a tangent and SP an asymptote to any plane curve, as NM.

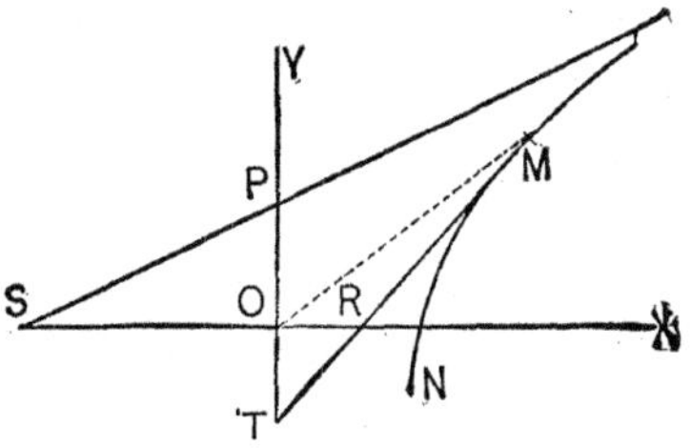

The equation of TM is (§ 148)

$$y - y' = (x - x')(dy'/dx');$$

and its intercepts are

$$OR = x' - y'(dx'/dy'), \quad OT = y' - x'(dy'/dx').$$

As the point of tangency M recedes from the origin without limit, the tangent TM approaches the asymptote SP, and the intercepts OR and OT approach OS and OP, respectively, as limits.

Assuming that in the equation of the curve $y' \ggg\!\!\rightarrow \infty$ as $x' \ggg\!\!\rightarrow \infty$, and placing

$$\tan OSP = K, \quad OP = Y_0, \quad \text{and} \quad OS = X_0,$$

we have, omitting the dashes,

$$K = \underset{\substack{x \ggg\rightarrow \infty \\ y \ggg\rightarrow \infty}}{\text{limit}} \left[\frac{dy}{dx}\right], \quad Y_0 = \underset{\substack{x \ggg\rightarrow \infty \\ y \ggg\rightarrow \infty}}{\text{limit}} \left[y - x\frac{dy}{dx}\right],$$

$$X_0 = \underset{\substack{x \ggg\rightarrow \infty \\ y \ggg\rightarrow \infty}}{\text{limit}} \left[x - y\frac{dx}{dy}\right],$$

any two of which will serve to determine in regard to an asymptote.

If $K = 0$ or ∞, or if $Y_0 = X_0 = \infty$, or if either Y_0 or X_0 is imaginary, there is no corresponding real asymptote within a finite distance from the origin; otherwise there is and its equation is

$$y = Kx + Y_0, \quad \text{or} \quad x = y/K + X_0,$$

or

$$x/X_0 + y/Y_0 = 1.$$

If either Y_0 or X_0 is zero, the corresponding asymptote passes through the origin and its equation is

$$y = Kx.$$

EXAMPLES.

1. $y^2 = 2px$.

$x = \infty = y, \quad dy/dx = p/y. \qquad K = 0.$

Hence, a parabola has no asymptote.

2. $y^3 = 2ax^2 - x^3$.

$x = \pm \infty, \quad y = \mp \infty. \qquad dy/dx = (4ax - 3x^2)/3y^2.$

$$X_0 = \left[x - \frac{3(2ax^2 - x^3)}{4ax - 3x^2}\right]_\infty = 2a/3 = Y_0 = \left[y - \frac{x(4ax - 3x^2)}{3y^2}\right]_\infty.$$

Hence, $y = -x + 2a/3$ is an asymptote.

3. $y^3 = 10 - x^3$.

$x = \pm \infty, y = \mp \infty. \qquad dy/dx = -x^2/y^2.$

$$X_0 = \left[x + y^3/x^2\right]_\infty = \left[10/x^2\right]_\infty = 0.$$

$$K = \left[-x^2/y^2\right]_\infty = -\left[(10 - y^3)/y^3\right]_\infty^{2/3} = -\left[10/y^3 - 1\right]_\infty^{2/3} = -1.$$

Hence, $y = -x$ is an asymptote.

4. $y^2 = 2px + r^2x^2$

$x = \infty = y. \qquad dy/dx = (p + r^2x)/\sqrt{2px + r^2x^2}.$

$$X_0 = \left[x - \frac{2px + r^2x^2}{p + r^2x}\right]_\infty = \left[\frac{-p}{p/x + r^2}\right]_\infty = \frac{-p}{r^2}.$$

$$Y_0 = \left[y - \frac{px + r^2x^2}{y}\right]_\infty = \left[\frac{p}{\pm\sqrt{2p/x + r^2}}\right]_\infty = \frac{\pm p}{\sqrt{r^2}}.$$

When $r^2 < 0$, Y_0 is imaginary. Hence, an ellipse has no asymptote.

When $r^2 = 0$, both results are unlimited, showing that a parabola has no asymptote.

When $r^2 > 0$, both results are finite, and since Y_0 has two values an hyperbola has two asymptotes.

Putting $p = b^2/a$ and $r^2 = b^2/a^2$, we have

$$X_0 = -a, \quad Y_0 = \pm b.$$

5. $y^3 - x^3 = ax^2$

$x = \infty = y. \qquad dy/dx = (2ax + 3x^2)/3y^2.$

$$X_0 = \left[x - \frac{3y^3}{2ax + 3x^2}\right]_\infty = \left[\frac{-ax}{2a + 3x}\right]_\infty = \left[\frac{-a}{(2a/x + 3)}\right]_\infty = -\frac{a}{3}.$$

$$Y_0 = \left[y - \frac{2ax^2 + 3x^3}{3y^2}\right]_\infty = \left[\frac{ax^2}{3(ax^2 + x^3)^{2/3}}\right]_\infty$$

$$= \left[\frac{a}{3(a/x + 1)^{2/3}}\right]_\infty = \frac{a}{3}.$$

Hence, $y = x + a/3$ is an asymptote

6. $y^3 - y = x^3 - 4x.$ $\qquad dy/dx = (3x^2 - 4)/(3y^2 - 1).$

Put $x = ty$, giving $y^3 - y = t^3y^3 - 4ty.$

Hence, $y = \pm\sqrt{(4t - 1)/(t^3 - 1)}$ and $t = 1$ give $y = \pm\infty = x.$

$$X_0 = \left[x - y\left(\frac{3y^2 - 1}{3x^2 - 4}\right)\right]_{\pm\infty} = \left[\frac{-3x}{3x^2 - 4}\right]_{\pm\infty}$$

$$= \left[-3/(3x - 4/x)\right]_{\pm\infty} = 0.$$

$$K = \left[\frac{3x^2 - 4}{3y^2 - 1}\right]_{\pm\infty} = \left[\frac{3 - 4/x^2}{3 - 1/x^2}\right]_{\pm\infty} = 1.$$

Hence, $y = x$ is an asymptote.

7. $x^3 - 3axy + y^3 = 0.$ $\qquad dy/dx = (ay - x^2)/(y^2 - ax).$

Put $x = ty$, divide by y^2, and we find $y = 3at/(1 + t^3)$, in which $= -1$ gives $y = \infty = -x.$

$$X_0 = \left[x - \frac{y^3 - axy}{ay - x^2}\right]_\infty = \left[\frac{-ax^2}{x^2 + ax}\right]_\infty = \left[\frac{-a}{1 + a/x}\right]_\infty$$

$$= -a = -\left[\frac{-a}{1 - a/x}\right]_\infty = \left[y - \frac{ayx - x^3}{y^2 - ax}\right]_\infty = Y_0.$$

Hence, $y = -x - a$ is an asymptote.

CURVES.	ASYMPTOTES.
8. $a^2y^2 - b^2x^2 = -a^2b^2.$	$y = \pm bx/a.$
9. $y^3 = x^2(a - x).$	$y = -x + a/3.$
10. $y^2 = ax + bx^2.$	$y = \sqrt{b}x + a/2\sqrt{b}.$
11. $y^2 = x^2/(x - 1).$	$x = 1.$
12. $y^3 + x^3 = 3x^2.$	$y = -x + 1.$
13. $y = x^3/(x^2 + 3a^2).$	$y = x.$
14. $y^3 = a^3 - x^3.$	$y = -x.$

15. $y^3 = a^2x - x^3$. $\qquad y = -x$.

16. $y^3 = 6x^2 + x^3$. $\qquad y = x + 2$.

17. $x^2y = x^3 + x + y$. $\qquad x = -1, \pm y = x$.

18. Find the perpendicular distance from the focus of an hyperbola to an asymptote. Ans. Semi-conjugate axis.

19. Find a tangent to a given curve which forms with the coördinate axes a maximum or a minimum triangle.

Let $u/2$ = area of triangle. Then

$$u = \left(x' - y'\frac{dx'}{dy'}\right)\left(y' - x'\frac{dy'}{dx'}\right) = -\left(y' - x'\frac{dy'}{dx'}\right)^2 \frac{dx'}{dy'}.$$

$$\frac{du}{dx'} = \left(y' - x'\frac{dy'}{dx'}\right)\left(x'\frac{dy'}{dx'} + y'\right)\frac{d^2y'}{dx'^2} \Big/ \left(\frac{dy'}{dx'}\right)^2$$

$\dfrac{d^2y'}{dx'^2} = 0$ and $y' - x'\dfrac{dy'}{dx'} = 0$ are exceptional cases. Hence, in general, $\dfrac{du}{dx} = 0$ requires $x'\dfrac{dy'}{dx'} + y' = 0$, whence $y' - x'\dfrac{dy'}{dx'} = 2y'$.

Therefore, in general, when the triangle is a maximum or a minimum, the portion of the tangent between the axes is bisected at the point of tangency.

Let the equation of the curve be $x^2 + y^2 = R^2$.

Then $\qquad \dfrac{dy'}{dx'} = \dfrac{-x'}{y'}$, and $\dfrac{x'dy'}{dx'} = \dfrac{-x'^2}{y'}$.

Hence, $\qquad \dfrac{-x'^2}{y'} + y' = 0$ gives $y' = x'$,

and $u/2 = R^2$ is a minimum.

155. The equation of any algebraic curve of the nth degree may be arranged in sets of homogeneous terms and written in the form

$$x^n f_0(y/x) + x^{n-1} f_1(y/x) + x^{n-2} f_2(y/x) + \ldots = 0. \quad (1)$$

Thus, having

$$y^3 - x^2y + 2y^2 + 4y + x = 0,$$

then $$(y^3 - x^2y) + 2y^2 + (4y + x) = 0,$$

and $$x^3(y^3/x^3 - y/x) + x^2(2y^2/x^2) + x(4y/x + 1) = 0.$$

Let $y = mx + c$ be the equation of a right line; combine it with (1) by substituting $mx + c$ for y, giving

$$x^n f_0(m + c/x) + x^{n-1} f_1(m + c/x) + x^{n-2} f_2(m + c/x) + \ldots = 0, \quad . \quad (2)$$

the n roots of which are the abscissas of the n points common to the two lines.

By causing m and c to vary, the right line may be made to have any position in the plane XY.

Developing each term of (2) by Taylor's formula, we have

$$x^n f_0(m) + x^{n-1}[cf_0'(m) + f_1(m)] + x^{n-2}\left[\frac{c^2}{\underline{|2}} f_0''(m) + cf_1'(m) + f_2(m)\right] + \text{etc.} = 0. \quad (3)$$

Any set of values of m and c which satisfy the two equations

$$f_0(m) = 0, \quad . \quad . \quad . \quad . \quad (4) \qquad cf_0'(m) + f_1(m) = 0, \quad (5)$$

cause the two terms in (3) containing the highest powers of x to disappear, and the resulting equation to have two infinite roots. That is, two of the points common to the right line and curve are thus made to coincide at an infinite distance from the origin, and the corresponding right line consequently is an asymptote.

Let $m_1, m_2, \ldots m_n$ be the n roots of (4); the corresponding values of c from (5) will be $-f_1(m_1)/f_0'(m_1)$, etc., and

$$y = m_1 x - f_1(m_1)/f_0'(m_1),$$

$$y = m_2 x - f_1(m_2)/f_0'(m_2),$$

etc. etc.

$$y = m_n x - f_1(m_n)/f_0'(m_n),$$

are the equations of the n asymptotes.

Hence we have the following rule:

In the equation of the curve substitute $mx + c$ for y, place the coefficients of the two highest powers of x equal to zero, and find corresponding sets of values for m and c. Each set will determine an asymptote real or imaginary.

Thus, having $x^3 - 2x^2y - 2x^2 - 8y = 0$, then

$$x^3 - 2x^2(mx + c) - 2x^2 - 8(mx + c) = 0,$$

and $$(1 - 2m)x^3 - (2c + 2)x^2 - 8mx - 8c = 0.$$

$$1 - 2m = 0, \quad -2c - 2 = 0, \text{ give } m = 1/2, \quad c = -1.$$

Hence, $y = x/2 - 1$ is an asymptote.

Having $xy^2 - x^3 - ay^2 - ax^2 = 0$, then

$$x(mx + c)^2 - x^3 - a(mx + c)^2 - ax^2 = 0,$$

and $$(m^2 - 1)x^3 + (2mc - am^2 - a)x^2 + \text{etc.} = 0.$$

$$m^2 - 1 = 0, \quad 2mc - am^2 - a = 0, \text{ give } m = \pm 1, \quad c = \pm a.$$

Hence, $y = \pm (x + a)$ are asymptotes.

EXAMPLES.

CURVES.	ASYMPTOTES.
1. $y^2(x - 2a) = x^3 - a^3$.	$x = 2a, \quad y = \pm (x + a)$.
2. $y = x(x + 1)^2/(x - 1)^2$.	$x = 1, \quad y = x + 4$.
3. $x^4 - y^4 = a^2xy$.	$y = x, \quad y = -x$.
4. $x^4 - x^2y^2 + a^2x^2 + b^4 = 0$.	$x = -y, \quad x = y, \quad x = 0$.
5. $x^3 + y^3 + 3a^2x + 3b^2y + c^3 = 0$.	$x = -y$.
6. $x^5 + y^5 = 5ax^2y^2$.	$x + y = a$.
7. $xy^4 - ay^4 + x^2y^3 = b^5$.	$x = a, \quad y = -x - a, \quad y = 0$.

8. $x^4y^2 - x^6 + x^4y + x^5 + x^3 = 0.$ $y = -x,\quad y + 1 = x,\quad x = 0.$

9. $(y^2 - 3xy + 2x^2)x = 4ay^2.$ $x = 4a,\ y = x - 4a,\ y = 2(x + 8a).$

10 $x(y + x)^2(y - 2x) = a^4.$ $x = 0,\quad y = 2x,\quad y = -x.$

11. $y^2 = x^2(x^2 - 1)/(x^2 + 1).$ $y = \pm x.$

12. $y^3 - x^3 = a^2x.$ $y = x.$

13. $y^3 = (x^4 - a^2x^2)/(2x - a).$ $x = a/2,\quad y = (x + a/6)/\sqrt[3]{2}.$

14. $x^3 - xy^2 + ay^2 = 0.$ $x = a,\quad y = \pm\ (x + a/2).$

15. $y^2 = x^2(x^2 - 4a^2)/(x^2 - a^2).$ $y = \pm x,\quad x = \pm a.$

16. $y = x(x - 2a)/(x - a).$ $y = x - a,\quad x = a.$

17. $x^3 + 2x^2y - xy^2 - 2y^3 + 4y^2 + 2xy + y = 1.$

$x + 2y = 0,\quad x + y = 1,\quad x - {}' = -1.$

18. $y^3 - 6xy^2 + 11x^2y - 6x^3 + x + y = 0.$ $y = x,\ y = 2x,\ y = 3x.$

19. $y^3 - x^2y + 2y^2 + 4y + x = 0.$ $y = 0,\ y = x - 1,\ y = -x - 1.$

20. $y^4 - x^4 + 2ax^2y - b^2x = 0.$ $y = x - a/2,\ y = -x - a/2.$

21 $x^4 - y^4 - a^2xy - b^2y^2 = 0,\ x = \mp y.$

22. $x^2y + y^2x = c^3,\quad x = 0,\quad y = 0,\quad x = -y.$

23. $2x^3 - x^2y - 2xy^2 + y^3 + 2x^2 + xy - y^2 + x + y + 1 = 0.$

$y = x + 1,\quad y = -x,\quad y = 2x.$

156. Let the equation of the curve (1) (§ 155) be arranged according to the descending powers of x; thus:

$$ax^n + (by + d)x^{n-1} + \ldots = 0. \quad \ldots \quad (6)$$

If $a = 0$ and y be assumed equal to $-d/b$, two of the roots of (6) are infinite and the right line $y = -d/b$ is an asymptote. Hence, when x^n is missing, the coefficient of the next highest power placed equal to zero is the equation of an asymptote parallel to X.

If both x^n and x^{n-1} are missing, the coefficient of x^{n-2}, which will be of the second degree with respect to y, placed equal to zero, determines two asymptotes real or imaginary parallel to X.

In a similar manner asymptotes parallel to Y may be determined. Thus, having $x^2y^2 - x^2y - xy^2 + x + y + 1 = 0$, in which x^4, x^3, y^4, y^3 are missing, $y^2 - y$ is the coefficient of x^2 and $x^2 - x$ that of y^2.

Hence, $y^2 - y = 0$, $x^2 - x = 0$, give the asymptotes

$$y = 0, \quad y = 1, \quad x = 0, \quad \text{and} \quad x = 1.$$

EXAMPLES.

	CURVES.	ASYMPTOTES.
1.	$y^2x - ay^2 - x^3 - ax^2 - b^3 = 0.$	$x = a, \quad y = x + a, \quad y = -x - a.$
2.	$xy^3 + x^3y - a^4 = 0.$	$y = 0, \quad x = 0.$
3.	$x^2y^2 - a^2(x^2 + y^2) = 0.$	$x = \pm a, \quad y = \pm a.$
4.	$x^2y^2 - a^2y^2 - x = 0.$	$x = \pm a, \quad y = 0.$
5.	$ay^2 - xy^2 - x^3 = 0.$	$x = a.$

Polar Coördinates.

157. If for any finite value of θ, designated by α, r is infinite, and the corresponding subtangent $PT' = (r^2 d\theta/dr)_{\theta=\alpha}$,

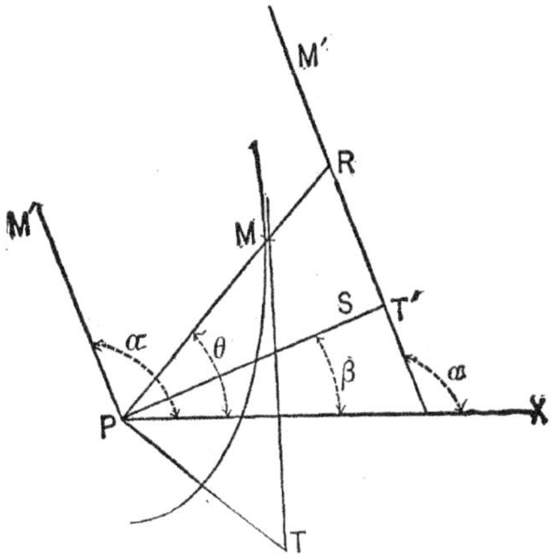

designated by S, is finite, the curve has an asymptote parallel to the radius vector corresponding to α.

Knowing α and S, the asymptote may be constructed as follows:

Through P draw PM', making the angle $XPM' = \alpha$; it will be the direction of the corresponding infinite radius vector. Draw PT' perpendicular to PM', and lay off

$$PT' = S = (r^2 d\theta/dr)_{\theta=\alpha}.$$

$T'M'$ drawn parallel to PM' is the corresponding asymptote.

Designate the angle $XPT' = \alpha - \pi/2$ by β, and let r and θ be the polar coördinates of all points of *the asymptote.* Then from the triangle $T'PR$ we have

$$r = S/\sin(\alpha - \theta). \quad . \quad . \quad . \quad . \quad . \quad (1)$$

Hence, knowing α and S, the polar equation of the asymptote is known.

In some cases α and S may be readily found. Thus, having the curve $r = a\theta/\sin\theta$, $\theta = \pi = \alpha$ gives $r = \infty$, and

$$S = \left(\frac{r^2 d\theta}{dr}\right)_{\theta=\alpha} = \left(\frac{a\theta^2}{\sin\theta - \theta\cos\theta}\right)_{\theta=\pi} = a\pi.$$

Hence, $r\sin\theta = a\pi$ is the polar equation of the *asymptote.*

Positive values of S are laid off in the direction

$$\beta = \alpha - \pi/2,$$

and negative values in the direction

$$\beta + 180^\circ = \alpha + \pi/2.$$

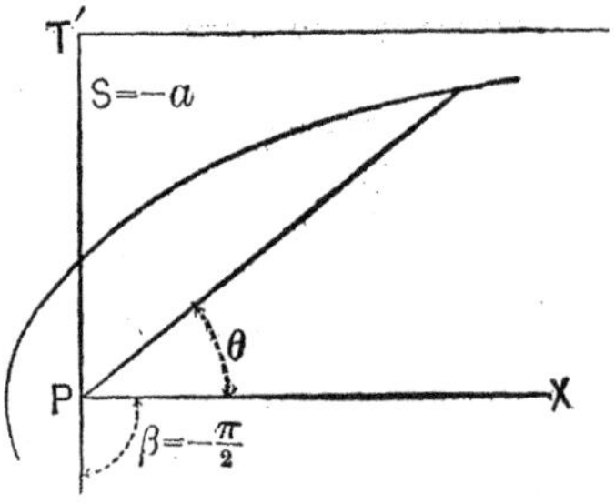

Thus, having $r = a/\theta$, $\theta = 0 = \alpha$ gives $r = \infty$, and $S = -a$. Hence, $\beta = -\pi/2$, and direction of S is $\pi/2$. $r\sin\theta = a$ is the equation of the asymptote.

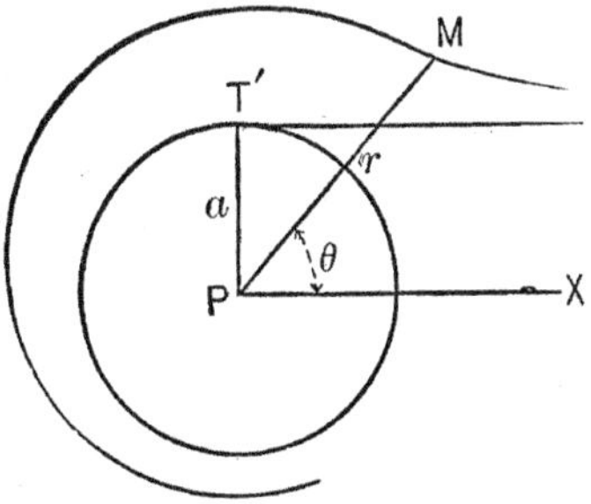

Having $r = a + a/\theta$, $\theta = 0 = \alpha$ gives $r = \infty$, and $S = (r^2 d\theta/dr)_{\theta=0} = -a$. Hence, $r = a/\sin\theta$ is an asymptote.

As $\theta \Rightarrow \infty$, $r \Rightarrow a$, and the circle $R = a$ is called *an asymptotic circle.*

158. In general, let the polar equation of a curve be

$$r^n f_n(\theta) + r^{n-1} f_{n-1}(\theta) + \ldots + r f_1(\theta) + f_0(\theta) = 0. \quad (1)$$

Placing $u = 1/r$, clearing of fractions, and arranging with respect to u, we have

$$u^n f_0(\theta) + u^{n-1} f_1(\theta) + \ldots + u f_{n-1}(\theta) + f_n(\theta) = 0. \quad (2)$$

For any point corresponding to $r = \infty$ u must be zero, and the roots of $f_n(\theta) = 0$ are the values of θ corresponding to $r = \infty$.

Differentiating (2) with respect to θ, making $u = 0$ and $\theta = \alpha$, we have

$$(du/d\theta)_{u=0} f_{n-1}(\alpha) + f_n'(\alpha) = 0. \quad . \quad . \quad . \quad (3)$$

Therefore (§ 151),

$$\left(\frac{r^2 d\theta}{dr}\right)_{\theta=\alpha} = \left(-\frac{d\theta}{du}\right)_{u=0} = \frac{f_{n-1}(\alpha)}{f_n'(\alpha)} = S.$$

Hence (1) (§ 157), the polar equation of the asymptote is

$$r \sin(\alpha - \theta) = f_{n-1}(\alpha)/f_n'(\alpha). \quad . \quad . \quad (4)$$

When $n = 1$,

$$r \sin(\alpha - \theta) = f_0(\alpha)/f_1'(\alpha). \quad . \quad . \quad . \quad . \quad (5)$$

To illustrate, having $r \cos \theta - b \sin \theta = 0$.

$$n = 1, \quad f_1(\theta) = \cos \theta, \quad f_0(\theta) = -b \sin \theta.$$

$$\cos \theta = 0 \text{ gives } \alpha = \pi/2, \ 3\pi/2, \text{ etc.}$$

$$f_0(\alpha) = -b \sin \alpha, \quad f_1'(\alpha) = -\sin \alpha.$$

Hence, $\quad r \sin (\pi/2 - \theta) = -(b \sin \alpha)/-\sin \alpha = b,$

or $r \cos \theta = b$ for the corresponding asymptote.

Let $r = a \sec \theta + b \tan \theta$.

Here $\quad 1/r = u = \cos \theta/(a + b \sin \theta) = 0$

gives $\quad \alpha = \pi/2, \quad (du/d\theta)_{\theta=\pi/2} = -1/(a+b),$

and $r \cos \theta = a + b$ for the corresponding asymptote.

(4) may be deduced from (2) (§ 151) by substituting in it $u' = 0$, $\theta' = \alpha$, and $-du'/d\theta' = f_n'(\alpha)/f_{n-1}(\alpha)$.

EXAMPLES.

Find the asymptotes of the following curves:

1. $r = a \tan \theta$. $\qquad r \cos \theta = \pm a.$
2. $r \cos \theta = a \cos 2\theta$. $\qquad r \cos \theta = -a.$
3. $(r - a) \sin \theta = b$. $\qquad r \sin \theta = b.$
4. $r \cos 2\theta = a$. $\qquad r = a/[2 \sin (45° - \theta)].$
5. $r \sin 4\theta = a \sin 3\theta$.
6. $r = a\theta^{-1/2}$. $\qquad r \sin \theta = 0.$
7. $r^2 \sin \theta = a^2 \cos 2\theta$. $\qquad r \sin \theta = 0.$
8. $r = a \operatorname{cosec} (\theta/2n)$ $\qquad r \sin \theta = \pm 2na.$
9. $r \sin n\theta = a$. $\qquad nr \sin \left(\theta - \frac{k\pi}{n}\right) = \frac{a}{\cos k\pi}.$

(In which k is any integer.)

10. $r = a(\sec\theta \pm \tan\theta)$. $r\cos\theta = 2a$.

11. $r = 2a \sin\theta \tan\theta$. $r\cos\theta = 2a$.

12. $r = a \sec m\theta + b \tan m\theta$. $r \sin[\pi/2m - \theta] = (a + b)/m$.

13. $r^2 \cos\theta = a^2 \sin 3\theta$. $r\cos\theta = 0$.

14. $r = a\theta^2/(\theta^2 - 1)$. $r = -a/[2\sin(180°/\pi - \theta)]$.

15. $r = a/\cos\theta + b$. $r = a/\sin(90° - \theta)$.

Find the asymptotic circles of the following curves :

16. $\theta = \dfrac{1}{\sqrt{2ar - r^2}}$. circle $r = 2a$.

17. $r = a\theta^2/(\theta^2 - 1)$. circle $r = a$.

CHAPTER XIV.

DIRECTION OF CURVATURE. SINGULAR POINTS.

DIRECTION OF CURVATURE.

159. Let $y = f(x)$ be the equation of any plane curve, and let θ represent the angle made by any tangent to the curve with X.

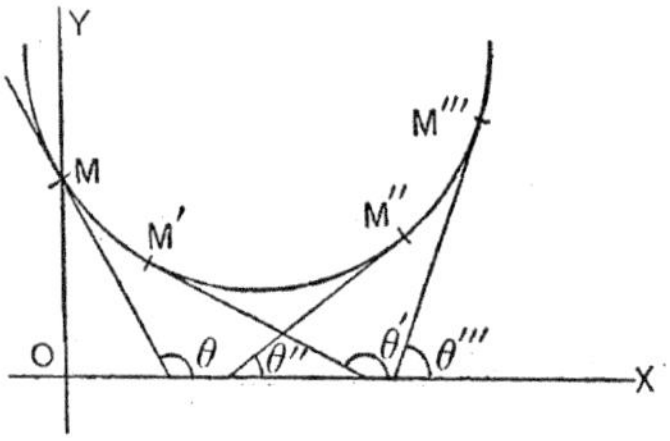

Then $dy/dx = \tan\theta$,

and $d^2y/dx^2 = (d\tan\theta)/dx$.

When d^2y/dx^2 *is positive*, $\tan\theta$ increases with x (§ 71), and, as may be seen from any figure, the concave side is above the curve and *the direction of curvature is upward.**

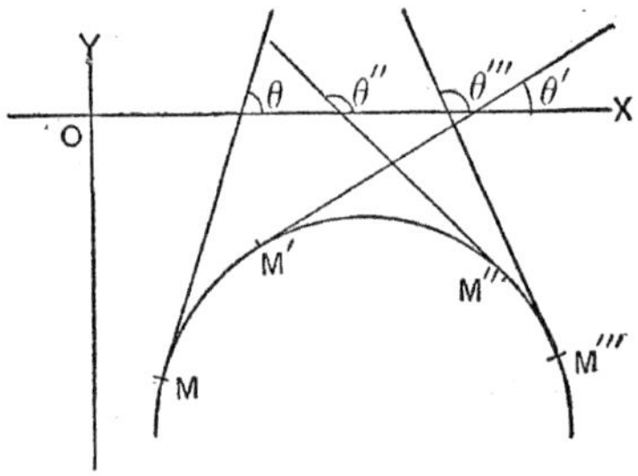

When d^2y/dx^2 *is negative*, $\tan\theta$ is a decreasing function of x, and *the direction of curvature is downward.*

---

* The direction of curvature at a point is along the normal towards the concave side.

In a corresponding manner it may be shown that positive values of d^2x/dy^2 indicate concavity towards the right, and negative values concavity towards the left.

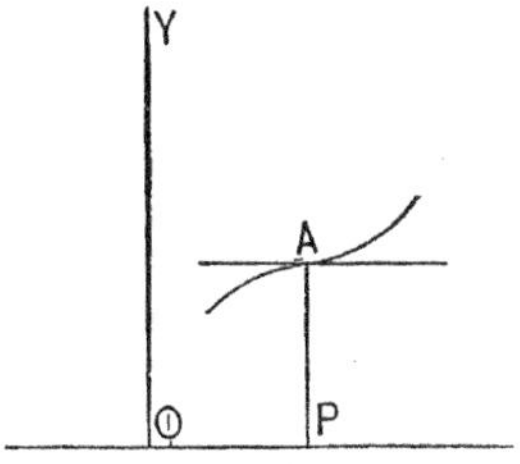

To illustrate, let $y = 2 + (x - 2)^3$.

Then $d^2y/dx^2 = 6(x - 2)$,

and $d^2x/dy^2 = 2/9(2 - x)^5$.

Hence, when $x < 2$, the direction of curvature is downward and to the right. When $x > 2$, the curvature is upward and to the left.

Show that the direction of curvature of $y = e^x$ is always upward and to the left.

160. Polar System.—Representing by p the perpendicular distance PQ from the pole P to the tangent at any

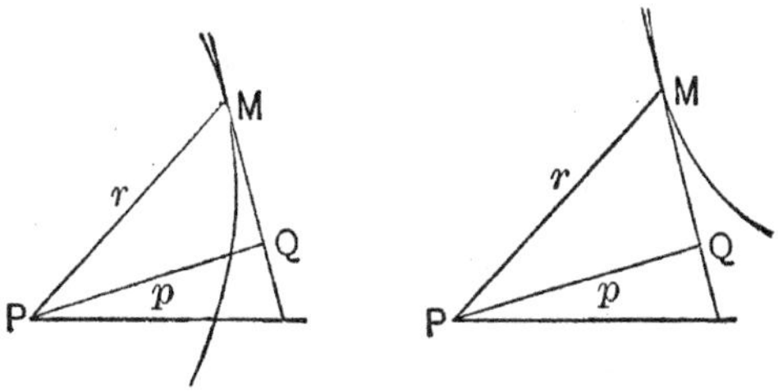

point M on a curve, it is apparent (when the tangent does not pass through P) that the curve is concave towards the pole when p is an increasing function of r, and that it curves away from the pole when p is a decreasing function of r.

Hence, positive values of dp/dr indicate concavity towards the pole and negative values the reverse.

From (1) (§ 150) we have $1/p^2 = u^2 + (du/d\theta)^2$, from which $-dp/p^3 = (u + d^2u/d\theta^2)du$; but $du = -dr/r^2$.

Hence, $dp/dr = (p^3/r^2)(u + d^2u/d\theta^2)$, which, since p is always positive, changes its sign only with $(u + d^2u/d\theta^2)$.

EXAMPLES.

1. Let $r = a\theta^{-\frac{1}{2}}$.

Placing $1/r = u$, we have $u = \theta^{\frac{1}{2}}/a$, $du/d\theta = \theta^{-\frac{1}{2}}/2a$, and $d^2u/d\theta^2 = -\theta^{-\frac{3}{2}}/4a$.

Hence, $$u + d^2u/d\theta^2 = (\theta^{\frac{1}{2}} - \theta^{-\frac{3}{2}}/4)/a.$$

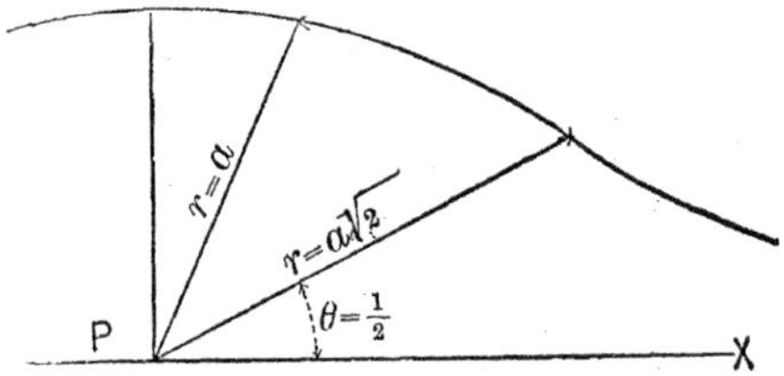

Therefore, when $\theta < \frac{1}{2}$, dp/dr is negative, and the concavity is away from the pole.

When $\theta = \frac{1}{2}$, whence $r = a\sqrt{2}$, $dp/dr = 0$.

When $\theta > \frac{1}{2}$, dp/dr is positive, and the curve is concave towards the pole.

2. Show that $r = p/(1 - \cos\theta)$ is always concave towards the pole.

3. Show that the direction of curvature of $r = a^\theta$ is always towards the pole.

SINGULAR POINTS.

Points of any curve which, independent of coördinates, possess some unusual property are as a class called *singular points.*

161. A Point of Inflexion is one at which, as the *variable increases,* a curve changes its concavity from one side of the curve to the other. The corresponding tangent intersects the curve, and the direction of curvature is reversed.

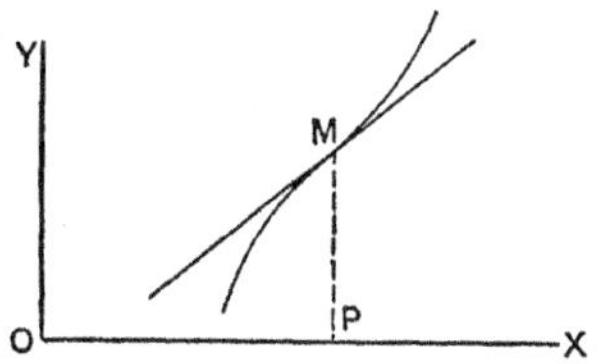

Let $y = fx$ be the equation of a curve. It follows from § 159 that c is the abscissa of a point of inflexion, if as x increases $f''x$ changes its sign in passing through $f''c$.

The real roots of the equations

$$f''x = 0 \quad \text{and} \quad f''x = \infty$$

are therefore *critical* values, and may be tested by a method similar to that described in § 135 in the case of $f'x = 0$ and $f'x = \infty$.

Hence, *the general method* for any *critical* value as c is to determine whether, as h vanishes from any definite value, $f''(c - h)$ and $f''(c + h)$ ultimately have and retain different signs.

When $f''(x)$, $f'''(x)$, etc., are continuous for values of x adjacent to *critical* values, those derived from $f''(x) = 0$ may be examined, as in the case of $f'x = 0$ (§ 136), as follows:

Having $f''(c) = 0$, *substitute* c *for* x *in* $f'''(x), f^{\text{iv}}(x)$, *etc., in order, until a result other than* 0 *is obtained. If the corresponding derivative is of an odd order,* c *is the abscissa of a point of inflexion, otherwise not.*

If a result ∞ is obtained, the general method should be employed.

When $f'''x$ is complex, the general method is usually preferable.

It should be observed that at a point of inflexion the slope of a curve is either a maximum or a minimum, and that the determination of such points is the same as finding those at which $f'(x)$ is a maximum or a minimum.

EXAMPLES.

1. $x^3 - 2x^2y - 2x^2 = 8y.$

$$\frac{dy}{dx} = \frac{x(x^3 + 12x - 16)}{2(x^2 + 4)^2}.$$

$$\frac{d^2y}{dx^2} = \frac{-4(x^3 - 6x^2 - 12x + 8)}{(x^2 + 4)^3} = 0 \text{ gives}$$

$$x = -2, \quad x = 2(2 - \sqrt{3}) = 0.54 -, \quad x = 2(2 + \sqrt{3}) = 7.5.$$

Applying the method § 138, we have

$$\left(\frac{d^3y}{dx^3}\right)_{x=-2} = \frac{-4}{(x^2+4)^3}[x - 0.54][x - 7.5] = -0.2.$$

$$\left(\frac{d^3y}{dx^3}\right)_{x=.54} = \frac{-4}{(x^2+4)^3}[x + 2][x - 7.5] = +0.9.$$

$$\left(\frac{d^3y}{dx^3}\right)_{x=7.5} = \frac{-4}{(x^2+4)^3}[x + 2][x - 0.54] = -.0013.$$

Hence, $(-2, -1)$, $(0.54 -, -0.05)$, and $(7.5 -, 2.6 -)$ are points of inflexion.

2. $y = 2 - 3(x - 2)^{3/5}.$

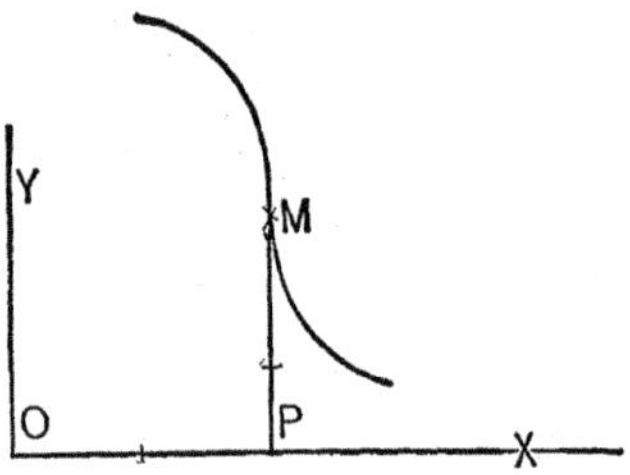

$$\frac{dy}{dx} = \frac{-9}{5(x - 2)^{2/5}}.$$

$$\frac{d^2y}{dx^2} = \frac{18}{25(x - 2)^{7/5}}.$$

$$\frac{18}{25(x - 2)^{7/5}} = 0$$ has no real finite roots.

$$\frac{18}{25(x - 2)^{7/5}} = \infty \text{ gives } x = 2.$$

$(d^2y/dx^2)_{x<2}$ is negative, $(d^2y/dx^2)_{x>2}$ is positive.

Hence, $x = 2 = y$ is a point of inflexion.

3. $y = a\sqrt{(a - x)/x}$.	$x = 3a/4$
4. $a^2y = (x - b)^3$.	$x = b$.
5. $y(x - 2) = (x - 1)(x - 3)$.	$x = 2$.
6. $a^2y = x^3$.	$x = 0 = y$.
7. $y^2(x - a) = x^3 + ax^2$.	$x = -2a$.
8. $y = x^2/a + a[(x - a)/a]^{3/5}$.	$x = a$.
9. $y = x^2 \log(1 - x)$.	$x = 0 = y$.
10. $x^2y = 4a^2(2a - y)$.	$x = \pm 2a/\sqrt{3}$.
11. $y = e^{1/x}$.	$x = -1/2$.
12. $x^3 + y^3 = a^3$.	$x = a, \quad x = 0$.
13. $\log y = \sqrt[3]{x}$.	$x = 8$.
14. $y = x^2 \tan x$.	$x = 0 = y$.
15. $y = a \sin(x/b)$.	$x = 0, b\pi$, etc.
16. $y = (a - x)^{5/3} + ax$.	$x = a$.
17. $y = x^2 - x^{5/2}$.	$x = 64/225$.
18. $y = e^{x^{1/3}}$	$x = 8$.
19. $y = x^3/(a^2 + x^2)$.	$x = 0, x = \pm a\sqrt{3}$.
20. $y = a \tan(x/b)$.	$x = 0 = y$.
21. $y = x^2(x + a)/[a(x - a)]$.	$x = -a(\sqrt[3]{2} - 1)$.
22. $y^3 = ax^2 - x^3$.	$x = a$.
23. $y = 2 + (x - 2)^3$.	$x = 2$.
24. $y = axy + by^2 + cx^3$.	$x = 0 = y$.
25. $y(a^4 - b^4) = x(x - a)^4 - xb^4$.	$x = 2a/5, \quad x = a$
26. $y = a^2x/(x^2 + a^2)$.	$x = 0, \quad x = \pm a\sqrt{3}$.
27. $y = x^2/a - x^3y/a^2$.	$x = \pm a/\sqrt{3}$.
28. $y = x \cos(x/a)$.	$x = 0 = y$.
29. $y = x + 36x^2 - 2x^3 - x^4$.	$x = 2, \quad x = -3$.
30. $y = a\sqrt{x/(2a - x)}$.	$x = a/2$.
31. $axy = x^3 - a^3$.	$x = a$.
32. $a^2y = x^3/3 - ax^2 + 2a^3$.	$x = a$.
33. $xy = a^2 \log(x/a)$.	$x = ae^{3/2}$.
34 $(ay - x^2)^2 = bx^3$.	$x = 9b/64$.
35. $a^3y = x^2(a^2 - x^2)$.	$x = \pm a/\sqrt{6}$.
36. $y(x^2 + a^2) = a^2(a - x)$.	
37. $x^2y^2 = a^2(x^2 - y^2)$.	$x = 0$.

38. $y = 2a\sqrt{(2a - x)/x}$. $\quad x = 3a/2, \quad y = \pm 2a/\sqrt{3}$.

39. $y = be^{-(x/a)^n}$. $\quad x = a\sqrt{(n-1)/n}$.

40. $a^2y = 3bx^2 - x^3$. $\quad x = b$

41. $y = xe^x$. $\quad x = -2$.

42. $a\sqrt{x} = (x - a)\sqrt{y}$. $\quad x = -2a$.

43. $y = e^{-1/x}$. $\quad x = 1/2$.

44. $a^2y = x(a^2 - x^2)$. $\quad x = 0 = y$.

162. Polar Coördinates.—Having $r = f(\theta)$, it follows from § 160 that $r = c$ corresponds to a point of inflexion if $dp/dr = (u + d^2u/d\theta^2)p^3/r^2$ changes its sign in passing through $(dp/dr)_{r=c}$.

Hence, the real roots of the equations $dp/dr = 0$ and $dp/dr = \infty$, or, what is equivalent, $u + d^2u/d\theta^2 = 0$ and $u + d^2u/d\theta^2 = \infty$, are *critical* values which may be tested by methods similar to those indicated in §§ 135 and 136.

It should be observed that p, corresponding to a point of inflexion, is a maximum or a minimum.

EXAMPLES

1. $r = a\theta^2/(\theta^2 - 1) = 1/u$.

Whence $du/d\theta = 2/(a\theta^3), \quad d^2u/d\theta^2 = -6/(a\theta^4)$.

$$u + d^2u/d\theta^2 = (\theta^4 - \theta^2 - 6)/(a\theta^4) = 0$$

gives $\theta = \pm\sqrt{3}$, and changes sign as θ passes through either. Hence, $\theta = \pm\sqrt{3}$, $r = 3a/2$, are points of inflexion.

2. $r = (a + a\theta)/\theta = 1/u$.

Hence, $du/d\theta = a/(a + a\theta)^2$,

$d^2u/d\theta^2 = -2a^2/(a + a\theta)^3$.

$$u + \frac{d^2u}{d\theta^2} = \frac{a^2(\theta^3 + 2\theta^2 + \theta - 2)}{(a + a\theta)^3} = 0$$

gives $\theta = 0.7 -$, and dp/dr changes from $-$ to $+$ as θ passes through $0.7 -$.

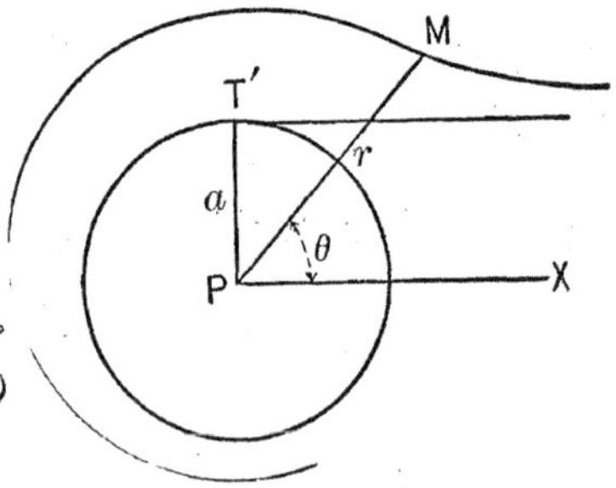

3. $r = a\theta^{-1/2}$. $\quad \theta = 1/2, \quad r = a\sqrt{2}$.

163. **A Multiple Point** is a point common to two or more branches of a curve, and is *double* or *triple*, etc., according to the number of branches. They are classed into *Points of Intersection*, *Shooting Points*, and *Points of Tangency*.

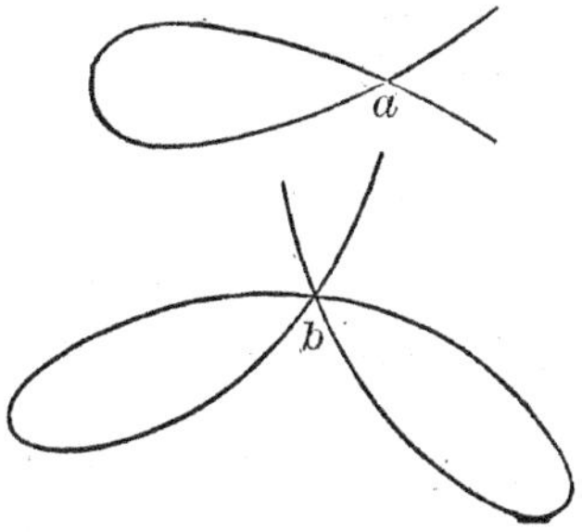

A Multiple Point of Intersection is one through which the branches pass and have different tangents. *a* is a double and *b* is a triple point of intersection. A double point of intersection is also called a *node*.

A Shooting Point is a multiple point at which the branches terminate with different tangents.

A Salient Point is a double shooting point.

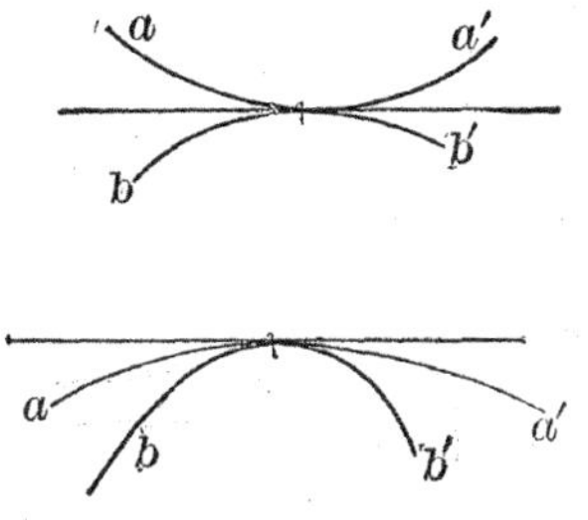

A Multiple Point of Tangency* is one at which the branches have a common tangent.

A Cusp is a double multiple point of tangency at which both branches terminate.

When in the vicinity of a cusp the branches are on opposite sides of the common tangent (Fig. *c*), it is of the

---

* Sometimes called points of osculation or double cusps.

first species, or a *keratoid* (horn) cusp; when on the same

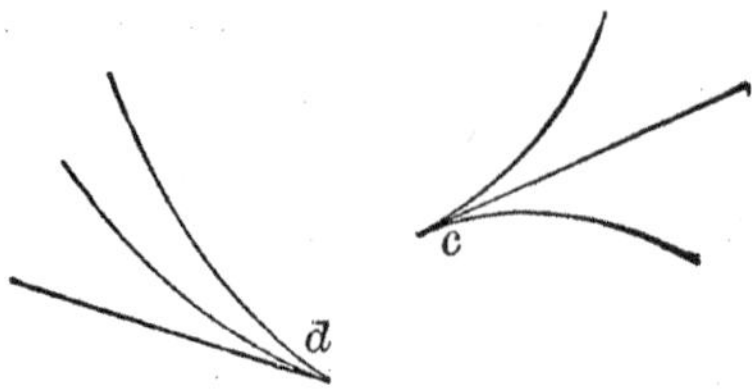

side (Fig. *d*), it is of the second species, or a *ramphoid* (beak) cusp.

A Conjugate Point is an isolated real point of a curve. Thus, the origin $x = 0 = y$ is a real point of the curve whose equation is $y = \pm x\sqrt{x - 2}$, but y is imaginary for $\pm x < 2$. Hence the origin is a conjugate point.

Let $y = F(x)$ be the equation of any curve.

From § 124,

$$F(c + h) = F(c) + F'(c)h + F''(c)h^2/2 + \text{etc.}$$

When (c, d) is a conjugate point, $F(c + h)$ is imaginary for values of h near zero, while $F(c)$ and h are real. Hence, one or more of the expressions $F'(c)$, $F''(c)$, etc., must be imaginary. This important characteristic of a conjugate point is frequently used in testing critical points. Thus, (c, d) is a conjugate point provided $F(c)$ is real, and $F^n(c)$ is imaginary for any entire value of n.

In the example above we find $F'(0) = \pm\sqrt{-2}$.

Since the ordinates of points of a curve adjacent to a conjugate point are imaginary, the number of such ordinates for each point is even. It follows that a conjugate point is a multiple point in the immediate vicinity of which the branches are imaginary. The tangents corresponding to a conjugate point may be real or imaginary, coincident or separate.

Having the equation of any curve with *two or more* branches, if *either* variable, as y, has but one real value, d, corresponding to any real value of the other, as $x = c$, (c, d) is a critical point.

If $(dy/dx)_{(c, d)}$ has two or more values, (c, d) is a multiple point of intersection or tangency according as the several values of $(dy/dx)_{(c, d)}$ are unequal or equal.

When the values of y for points adjacent to and on both sides of (c, d) are imaginary, (c, d) is a conjugate point.

EXAMPLES.

1. $y = 3 \pm (x - 4)\sqrt{x - 2}$.

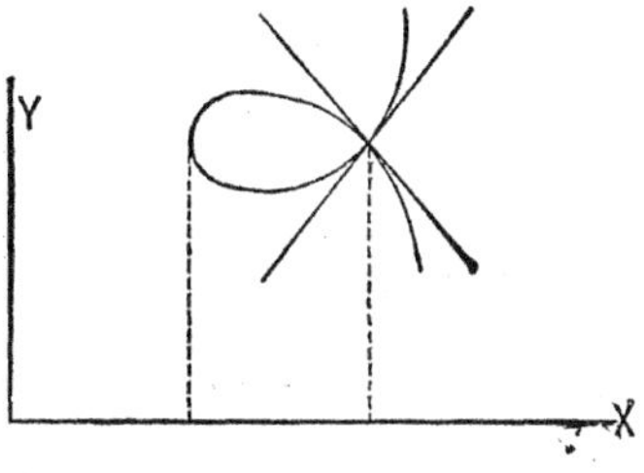

$x < 2$, y is imaginary. $x = 2$, $y = 3$.

$2 < x < 4$, y has two real values. $x = 4$, $y = 3$.

$x > 4$, y has two real values.

Hence, $(2, 3)$ and $(4, 3)$ are critical.

$dy/dx = \pm\sqrt{x - 2} \pm (x - 4)/(2\sqrt{x - 2})$.

$(dy/dx)_{(2, 3)} = \infty$. $(2, 3)$ is not a multiple point.

$(dy/dx)_{(4, 3)} = \pm\sqrt{2}$.

Hence, $(4, 3)$ is a double multiple point of intersection.

2. $y = \pm\sqrt{x^2(x - 2)}/\sqrt{3}$.

$x = 0 = y$. y is imaginary when $x < 0$, or $0 < x < 2$.

Hence, the origin is a conjugate point at which we have

$$(dy/dx)_0 = \mp 2/\sqrt{-6}.$$

3. $x^4 + 2ax^2y = ay^3$.
$x = 0 = y$. $x = 0 \pm h$, y has three values. $(dy/dx)_0 = 0$ and $\pm\sqrt{2}$.

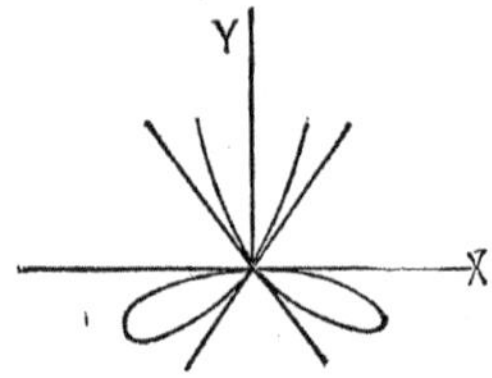

Hence, the origin is a triple point of intersection.

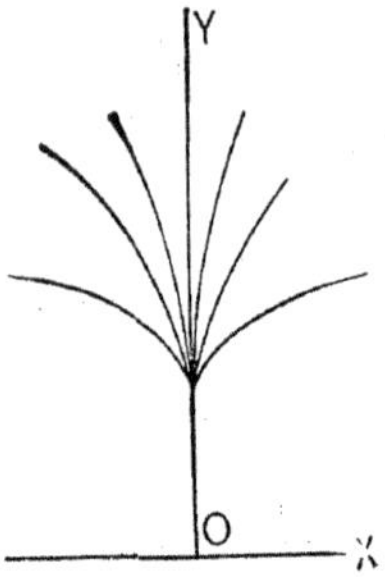

4. $y = 2 + x\tan^{-1}(1/x) = 2 + x\cot^{-1}x$.
$x = 0$, $y = 2$. When $\cot^{-1}x < \pi/2$, x and $\cot^{-1}x$ have the same sign, and $\pm x$ give equal positive value for y.
$(dy/dx)_0 = [\cot^{-1}x - x/(1+x^2)]_0 = \pi/2$ and $3\pi/2$ and $5\pi/2$, etc., or $-\pi/2$ and $-3\pi/2$ and $-5\pi/2$, etc., according as $x \ggg\!\!\to 0$ is positive or negative. Hence, $(0, 2)$ is a shooting point.
$d^2y/dx^2 = -2/(1+x^2)^2$ is negative for $\pm x$; hence, the curve is concave downward.

5. $y = x/(1 + e^{1/x})$.
$x = 0 = y$.

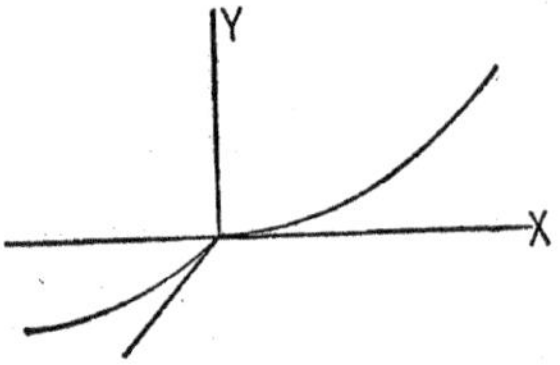

$$\left(\frac{dy}{dx}\right)_0 = \left[\frac{1}{1+e^{1/x}} + \frac{e^{1/x}}{x(1+e^{1/x})^2}\right]$$
$$= 0 \text{ or } 1$$

according as $x \ggg\!\!\to 0$ is positive or negative. Hence, the origin is a salient point.

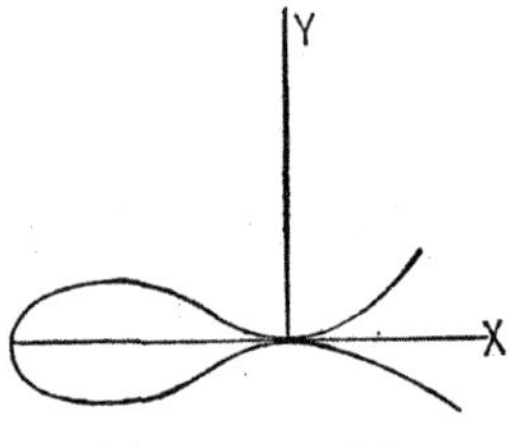

6. $a^2y^2 = cx^4 + x^5$.

$$y = \pm x^2\sqrt{c+x}/a,$$

$x = 0 = y$, and y has double values for values of x between $\pm c$ and 0.

$$\left(\frac{dy}{dx}\right)_0 = \pm\left[\frac{5x^2 + 4cx}{2a\sqrt{x+c}}\right]_0 = \pm 0.$$

Hence the origin is a double point of tangency.

7. $y = 3 \pm (x - 2)^{3/2}$.

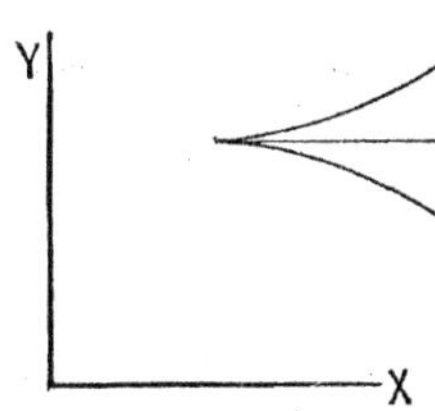

$dy/dx = \pm\, 3\sqrt{x-2}/2$.

$x < 2$, y is imaginary.

$x > 2$, y has double values.

$x = 2$, $y = 3$, and $(dy/dx)_{(2,3)} = \pm 0$.

Hence, $(2, 3)$ is a cusp.

The branches are on opposite sides of the common tangent $y = 3$, and the cusp is of the first species.

8. $(2x - y)^2 = (x - 3)^5$.

$x < 3$ makes y imaginary, $x = 3$ gives $y = 6$, and $x > 3$ gives real double values to y.

$(dy/dx)_{(3,6)} = 2 \pm 0$. Hence, $(3, 6)$ is a cusp.

Since $y = 2x \pm (x - 3)^{5/2}$, the branches are on opposite sides of the common tangent $y = 2x$, and the cusp is of the first species.

A characteristic of a cusp of the *first species* is a change in direction of curvature from one side of the common tangent to the other; while at one of the *second species* the direction of curvature remains upon the same side of the common tangent. Hence, different signs for d^2y/dx^2 corresponding to the two real values of y in the immediate vicinity of a cusp indicate the first species, and like signs the second species.

Thus, in example (7), $\left(\dfrac{d^2y}{dx^2}\right)_{x>2} = \left(\pm \dfrac{3}{4\sqrt{x-2}}\right)_{x>2}$ has values with different signs.

In some cases it is preferable to consider y as the independent variable.

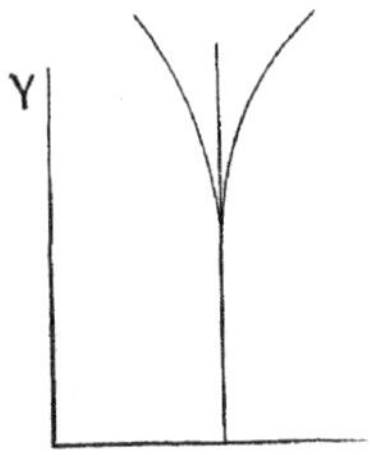

9. $y = 3 + (x - 2)^{2/3}$. y has but one value for each value of x. $x = 2, y = 3$. $dy/dx = 2/[3(x - 2)^{1/3}]$ is negative when $x > 2$, is ∞ when $x = 2$, and is positive when $x < 2$. (2, 3) is evidently a cusp of the first species as in figure.

Solving with respect to x, we have

$$x = 2 \pm (y - 3)^{3/2}, \quad (dx/dy)_{(3, 2)} = (\pm 3\sqrt{y-3}/2)_{(3, 2)} = \pm 0.$$

$y < 3$, x is imaginary. $y = 3$, $x = 2$. $y > 3$ gives double value for x. Hence, (3, 2) is a cusp.

$$\left(\frac{d^2x}{dy^2}\right)_{y>3} = \left(\pm \frac{3}{4\sqrt{y - 3}}\right)_{y>3}$$ has values with different signs, therefore the cusp is of the first species.

Points corresponding to maximum or minimum ordinates at which $dy/dx = \infty$ are cusps.

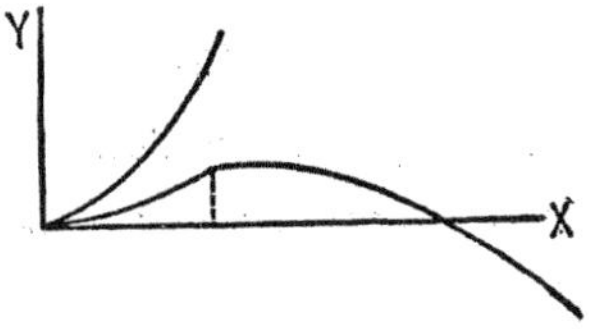

10. $y = x^2 \pm x^{5/2}$.

$$\frac{dy}{dx} = 2x \pm 5x^{3/2}/2.$$

$$\frac{d^2y}{dx^2} = 2 \pm 15x^{1/2}/4.$$

$x = 0 = y$. $x < 0$, y is imaginary.

$x > 0$, y has two real values.

$(dy/dx)_0 = \pm 0$. Hence, the origin is a cusp at which the axis X is the common tangent.

For $0 < x < 1$ both values of y are positive; therefore the cusp is of the second species. This is also indicated by the fact that both values of $(d^2y/dx^2)_{0<x<64/225}$ are positive.

Examine the following curves for multiple points:

11. $y^2 = x^3/(2a - x)$. (0, 0) is a cusp of 1st species.

12. $(y - x)^2 = x^3$. (0, 0) is a cusp.

13. $y = \pm (x^2\sqrt{x^2 - 4})/4$. $x = 0 = y$ is a conjugate point.

14. $(4y - x^3)^2 = (x-4)^5(x - 3)^6$. $(3, 27/4)$ is a conjugate point.

15. $y^2 = 2x^2 + x^3$. $(0, 0)$ is a double point of intersection.

16. $x^{2/3} + y^{2/3} = a^{2/3}$. $(0, \pm a)$ and $(\pm a, 0)$ are cusps of 1st species.

17. $y^2 = x^4 - x^6$. $(0, 0)$ is a double point of tangency.

18. $(y^2 - a^2)^3 = x^4(2x + 3a)^2$. $(-3a/2, \pm a)$ are cusps of 1st species.

19. $y^2(a^2 - x^2) = x^4$. $(0, 0)$ is a double point of tangency.

20. $y = \pm x\sqrt{1 - x^2}$. $(0, 0)$ is a double point of intersection.

21. $(x - a)^5 = (y - x)^2$. (a, a) is a cusp of 1st species.

22. $ay^2 = x^3$. $(0, 0)$ is a cusp of 1st species.

23. $y = a + x + bx^2 \pm cx^{5/2}$. $(0, a)$ is a cusp of 2d species.

24. $y^2 = a^2x^2 - x^4$. $(0, 0)$ is a double point of intersection.

25. $(y - b - cx^2)^2 = (x - a)^5$. $(a, b + ca^2)$ is a cusp of 2d species.

26. $y = \pm x\left[\sqrt{a^2 + x^2}/\sqrt{a^2 - x^2}\right]$. $(0, 0)$ is a double point of intersection.

27. $(y - b)^2 = (x - a)^5$. (a, b) is a cusp of 1st species.

28. $(xy + 1)^2 + (x - 1)^3(x - 2) = 0$. $(1, -1)$ is a cusp.

29. $y^3 + x^3 = 2ax^2$ $(0, 0)$ is a cusp of 1st species.

164. Let $u = f(x, y) = 0$ be the equation in a rational integral form of any algebraic curve, then (2) (§ 111),

$$\frac{dy}{dx} = -\frac{\partial u}{\partial x}\Big/\frac{\partial u}{\partial y}. \quad . \quad . \quad . \quad . \quad . \qquad (1)$$

At a multiple point dy/dx has two or more equal or unequal values.

Since $\partial u/\partial x$ and $\partial u/\partial y$ are rational integral functions, each can have but one value for any set of values of x and y.

Hence, equation (1), two or more values of dy/dx require

$$\partial u/dx = 0 \quad \text{and} \quad \partial u/\partial y = 0. \quad . \quad . \quad . \quad . \qquad (2)$$

Any set of real roots of these equations as (c, d), which also satisfy $f(x, y) = 0$, are therefore critical for multiple points.

$(dy/dx)_{(c,\,d)}$ may be evaluated as in § 117, otherwise (3) (§ 111) gives

$$\frac{\partial^2 u}{\partial x^2} + 2\frac{\partial^2 u}{\partial x\,\partial y}\frac{dy}{dx} + \frac{\partial^2 u}{\partial y^2}\left(\frac{dy}{dx}\right)^2 = 0, \quad . \quad . \quad . \qquad (3)$$

from which the two values of $(dy/dx)_{(c,\,d)}$ may be found.

If in (3) $\partial^2 u/\partial x^2$, $\partial^2 u/\partial x\,\partial y$ and $\partial^2 u/\partial y^2$ vanish for (c, d), (dy/dx) is indeterminate. Then (7) (§ 111),

$$\frac{\partial^3 u}{\partial x^3} + 3\frac{\partial^3 u}{\partial x^2\partial y}\frac{dy}{dx} + 3\frac{\partial^3 u}{\partial x\partial y^2}\left(\frac{dy}{dx}\right)^2 + \frac{\partial^3 u}{\partial y^3}\left(\frac{dy}{dx}\right)^3 = 0, \qquad (4)$$

gives three values for $(dy/dx)_{(c,\,d)}$.

It follows that any algebraic curve whose equation in a rational integral form contains no term of a degree less than the second, with respect to the variables, has a multiple point at the origin.

EXAMPLES.

1. $u = y^2 - x^2(1 - x^2) = 0$.

$\partial u/\partial y = 2y = 0$, $\partial u/\partial x = -2x(1 - x^2) + 2x^3 = 0$, give $x = y = 0$.

Hence, the origin is critical. (3) gives $(dy/dx)_0 = \pm 1$. y has double values for $+1 > x > -1$, hence the origin is a double multiple point of intersection.

$$\left.\begin{aligned} &2.\ u = x^4 + x^2y - y^3 = 0, \\ &\partial u/\partial x = 4x^3 + 2xy = 0, \\ &\partial u/\partial y = x^2 - 3y^2 = 0, \end{aligned}\right\} \text{give } x = 0 = y.$$

$$\partial^2 u/\partial x^2 = 12x^2 + 2y,\ \partial^2 u/\partial x\partial y = 2x,\ \partial^2 u/\partial y^2 = -6y,$$

$$\partial^3 u/\partial x^3 = 24x,\quad \partial^3 u/\partial x^2\partial y = 2,\quad \partial^3 u/\partial x\,\partial y^2 = 0,$$
$$\partial^3 u/\partial y^3 = -6.$$

From (4), $(dy/dx)_0 = 0$, and ± 1. Hence, the origin is a triple point.

$$\left.\begin{aligned} &3.\ u = (4y - 3x)^2 - (x-2)^3/2 = 0, \\ &\partial u/\partial x = -24y + 24x - 3x^2/2 - 6 = 0, \\ &\partial u/\partial y = 32y - 24x = 0, \end{aligned}\right\} \text{give } x = 2,\ y = 3/2.$$

$$\partial^2 u/\partial x^2 = -3x + 24,\quad \partial^2 u/\partial x\,\partial y = -24,\quad \partial^2 u/\partial y^2 = 32.$$

Hence, $$(dy/dx)_{(2,\,3/2)} = 3/4 \pm 0,$$

since $$y = \left[3x \pm \sqrt{(x-2)^3/2}\right]/4.$$

$x < 2$, y is imaginary. $x > 2$, y has double values.

Since $x > 2$ gives one value of y greater and the other less than $3x/4$, the two branches are on opposite sides of the common tangent $y = 3x/4$. Hence $(2, 3/2)$ is a cusp of the first species.

4. $u = y^2 - x^3 - 2ax^2 - a^2x = 0.$

$$\partial u/\partial x = -3x^2 - 4ax - a^2 = 0.$$

$$\partial u/\partial y = 2y = 0.$$

Hence, $(-a, 0)$ is critical. From (1),

$$\left(\frac{dy}{dx}\right)_{(-a,0)} = \left(\frac{3x^2 + 4ax + a^2}{2y}\right)_{(-a,0)} = \frac{0}{0}.$$

(§ 117) $$\left(\frac{d^2y}{dx^2}\right)_{(-a,0)} = \left(\frac{6x + 4a}{2dy/dx}\right)_{(-a,0)} = \pm\sqrt{-a}.$$

Hence, $(-a, 0)$ is a conjugate point.

5. $u = y^2 - 2x^2y - x^4y + 2x^4 = 0.$

$\partial u/\partial x = -4xy - 4x^3y + 8x^3 = 0.$

$\partial u/\partial y = 2y - 2x^2 - x^4 = 0.$

Hence $x = 0 = y$ is a critical point.

From the equation of the curve,

$$y = (x^2 + x^4/2) \pm x^2\sqrt{-4 + 4x^2 + x^4}/2,$$

and is imaginary when x is near zero. Hence the origin is a conjugate point.

Examine the following curves for multiple points :

6. $x^3 - 3axy + y^3 = 0.$ — $(0, 0)$ is a double point of intersection.

7. $x^4 - 2ay^3 - 3a^2y^2 - 2a^2x^2 + a^4 = 0.$ — $(0, -a)$ and $(\pm a, 0)$ are double points of intersection.

8. $x^4 - 2ax^2y - axy^2 + a^2y^2 = 0.$ — $(0, 0)$ is a cusp of 2d species.

9. $y^4 - axy^2 = -x^4.$ — $(0, 0)$ is a cusp of 1st species.

10. $(x^2 + y^2)^2 = a^2(x^2 - y^2).$ — $(0, 0)$ is a double point of intersection.

11. $ay^2 + bx^2 = x^3.$ — $(0, 0)$ is a conjugate point.

12. $ay^2 - x^3 + 4ax^2 - 5a^2x + 2a^3 = 0.$ — $(a, 0)$ is a conjugate point.

13. $x^4 - ax^2y + axy^2 + a^2y^2 = 0.$ — $(0, 0)$ is a conjugate point.

14. $y^2 = x(x + a)^2.$ — $(-a, 0)$ is a conjugate point.

15. $(y-2)^2 = (x-1)^4(x-3)$. (1, 2) is a conjugate point.

16. $y^2(x^2-a^2) = x^4$. (0, 0) is a conjugate point.

17. $x^2 + x^2y^2 - 6ax^2y + a^2y^2 = 0$. (0, 0) is a double point of tangency.

18. $a^3y^2 - 2abx^2y = x^5$. (0, 0) is a double point of tangency.

19. $x^4 - axy^2 = ay^3$. (0, 0) is a triple point and a cusp.

165. A Terminating Point (*point d'arrêt*) is one at which a single branch of a curve terminates.

EXAMPLES.

1. $y = x \log x$.

$x = 0 = y$. y is real when $x > 0$, and is imaginary when $x < 0$. Hence, the origin is a terminating point.

2. $y = e^{-1/x}$.

As $+x \doteq 0$, $y \doteq 0$, and as $-x \doteq 0$, $y \doteq \infty$. Hence, the origin is a terminating point for the right-hand branch.

3. $x^2 \log x + y = x^2y$. (0, 0) is a terminating point.

CHAPTER XV.

CURVATURE OF CURVES.

PLANE CURVES.

166. The Total Curvature of an Arc of Any Curve *is the angle which measures the change in direction of the motion of the generating point while generating the arc.*

Let $MM' = \delta s$ be the length of any varying arc not including a singular point, of any curve. At M and M', respectively, draw the tangents MT and $M'T'$. Each tangent indicates the direction of the motion of the generating point corresponding to its point of tangency. The angle TRT', denoted by $\delta\psi$, included between the tangents at the ends of the arc, measures the change in direction of the motion of the generating point while generating the arc δs, and is its total curvature.

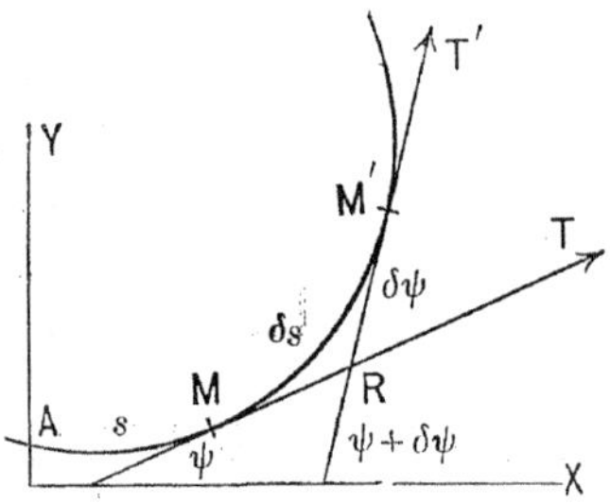

If the extremities of an arc coincide, forming a closed curve without singular points, the corresponding tangents coincide, but the total curvature is 2π and not zero.

167. The Rate of Curvature of a Curve at a Point *is the rate of change, at the point, of its direction regarded as a function of its length.* Thus, in the preceding figure, let ψ

represent the angle which the tangent MT makes with X. It determines the direction, with respect to X, of the motion of the generating point at M, and, regarding ψ as a function of the length of any varying arc of the curve, as $AM = s$, $\dfrac{d\psi}{ds} = \underset{\delta s \ggg\to 0}{\text{limit}} \left[\dfrac{\delta\psi}{\delta s}\right]$ is the *rate* of curvature of the curve s at M. (§ 70.)

168. Rate of Curvature of a Circle at a Point.—Let C be the centre and r the radius of any circle. Then

$$\frac{d\psi}{ds} = \underset{\delta s \ggg\to 0}{\text{limit}} \left[\frac{\delta\psi}{\delta s}\right] = \text{lim.} \left[\frac{\delta\psi}{r\delta\psi}\right] = \frac{1}{r}.$$

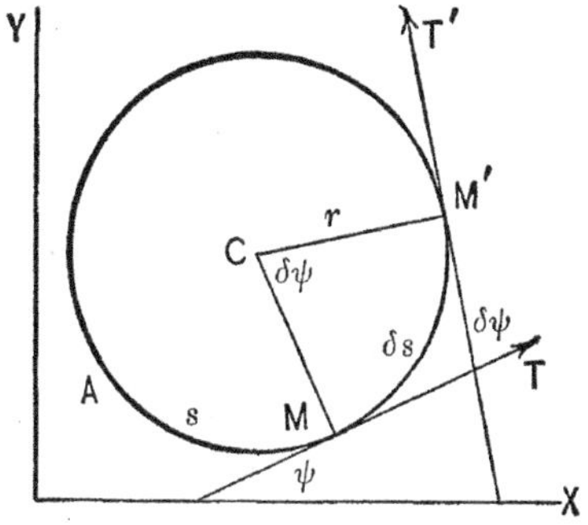

Hence, in any circle the *rate* of curvature is the same at all points, and at any point is equal to the reciprocal of its radius.

169. Circle and Radius of Curvature.—When the radius of a varying circle decreases *continuously*, the *rate* of curvature of the circle at any point increases *continuously*. Hence, a circle may always be assumed having at all points the same *rate* of curvature as that of any given curve at any assumed point.

Such a circle tangent to the curve at the point assumed, and having the same direction of curvature, is called *the circle of curvature of the point*, and its radius and centre are called, respectively, *the radius* and *centre of curvature.* It follows that a radius of curvature is normal to the curve.

Any chord of a circle of curvature which passes through the point of tangency is called a *chord of curvature.*

170. Curvature of a Curve at a Point.—Representing the radius of curvature at any point of any curve by ρ, we have

$$d\psi/ds = 1/\rho. \quad . \quad . \quad . \quad . \quad . \quad . \quad (1)$$

That is, the *rate* of curvature of any curve at any point is equal to the reciprocal of the corresponding radius of curvature, and the *rates* of curvature at different points are inversely as the corresponding radii of curvature.

$d\psi/ds$, corresponding to any point of a curve, multiplied by the unit of length of s is (§ 68) the change that the corresponding value of ψ would undergo were it to retain its rate at the point over the unit of length of s. In other words, $d\psi/ds$ multiplied by the unit of s is the *total curvature of a unit of length of the corresponding circle of curvature,* and it is generally called the *curvature of the curve at the point* or the *curvature of the corresponding circle of curvature.* Its numerical value is the same as that of the correponding *rate* of curvature, and for reasons similar to those given in § 95 it is generally used instead of the *rate.*

It is important not to confound this *curvature of a circle,* which measures the so-called *curvature of the curve at a point,* with the *total* curvature of an arc described in § 166.

ρ and s must be expressed in terms of the same unit of length, and at any point where $\rho =$ unit of length the *rate* of curvature is unity, and the corresponding *curvature of the curve* is *a radian,* which is therefore the unit of curvature.

EXPRESSIONS FOR RADIUS OF CURVATURE.

171. From (1) (§ 170), $\rho = ds/d\psi$, ... (1)

which enables us to determine the rate of curvature at any point from the equation* of the curve in terms of s and ψ.

Thus, having $s = c \tan \psi$; for a catenary,

$$1/\rho = (\cos^2 \psi)/c = c/(s^2 + c^2).$$

2. $s = c\psi$ .... $\rho = c$.

3. $s = a \sin \psi$ .... $\rho = a \cos \psi$.

172. Let $y = f(x)$ be the equation of any plane curve.

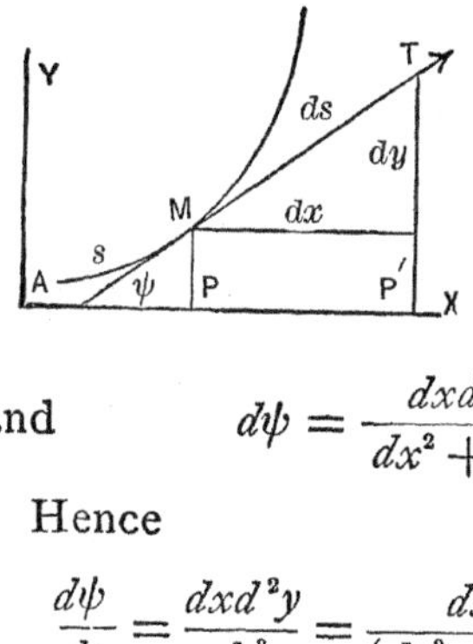

Having, as before, $AM = s$, let $PP' = dx$. Then

$$dx = \cos \psi ds, \quad dy = \sin \psi ds,$$

$$\psi = \tan^{-1} \frac{dy}{dx},$$

and $$d\psi = \frac{dxd^2y}{dx^2 + dy^2} = \frac{dxd^2y}{ds^2}.$$

Hence

$$\frac{d\psi}{ds} = \frac{dxd^2y}{ds^3} = \frac{dxd^2y}{(dx^2 + dy^2)^{3/2}} = \frac{f''(x)}{\left[1 + \overline{f'(x)}^2\right]^{3/2}} = \frac{1}{\rho},$$

or $$\rho = \left[1 + \overline{f'(x)}^2\right]^{3/2} / f''(x), \quad . \; . \; . \; . \; . \quad (1)$$

from which the radius, and therefore the rate of curvature, may be found from the equation of the curve in rectangular coördinates.

Adopting the positive value of $\left[1 + \overline{f'(x)}^2\right]^{3/2}$, ρ will have the same sign as $f''(x)$, which determines the direction of curvature.

---

* Called the *intrinsic* equation of the curve.

In general, at a point of inflexion (§ 161)

$$f''(x) = 0, \quad \text{or} \quad \infty.$$

Hence, at such a point ρ is generally ∞ or 0. In general, at a multiple-point, $f'(x)$ has two or more different values; hence, ρ has two or more values, one for each branch. At a multiple-point of tangency, $f'(x)$ has but one value, but $f''(x)$ has, in general, a value for each branch; hence, ρ differs for each branch.

Comparing (1) with (3) (§ 139), we see that the centre of curvature corresponding to any point of a given curve coincides with the point $(\bar{x}, \bar{y})$ (§ 139) whose distance from the curve measured along the normal is, in general, neither a maximum nor a minimum, and whose coördinates are (2) (§ 139)

$$\left.\begin{aligned} \bar{x} &= x - [1 + \overline{f'(x)}^2] f'(x)/f''(x); \\ \bar{y} &= y + [1 + \overline{f'(x)}^2]/f''(x). \end{aligned}\right\} \quad . \quad . \quad . \quad (2)$$

It follows (§ 139) that the circle of curvature corresponding to any point of a curve, in general, intersects the curve at their common point of tangency.

This is not the case, however, at any point where the curvature is a maximum or a minimum, which, in general, includes any point in the vicinity of which a curve is symmetrical with respect to the corresponding normal. At such a point the curvature, in general, decreases or increases in both directions; consequently in that vicinity the circle of curvature is interior or exterior to the curve.

To illustrate, take the ellipse

$$a^2y^2 + b^2x^2 = a^2b^2,$$

$$\overline{f'(x)}^2 = b^4x^2/(a^4y^2), \qquad f''(x) = -b^4/(a^2y^3).$$

Hence, $1/\rho = ab/(a^2 + b^2 - x^2 - y^2)^{3/2}$
$= - a^4b^4/(a^4y^2 + b^4x^2)^{3/2}$. . . . . (3)

At the vertices $(\pm a, 0)$ $1/\rho = a/b^2$, a maximum.

At the vertices $(0, \pm b)$ $1/\rho = b/a^2$, a minimum.

Hence, at the vertices of the transverse axis the circle of curvature is within the curve, at those of the conjugate axis it is outside, and at all other points it cuts the ellipse.

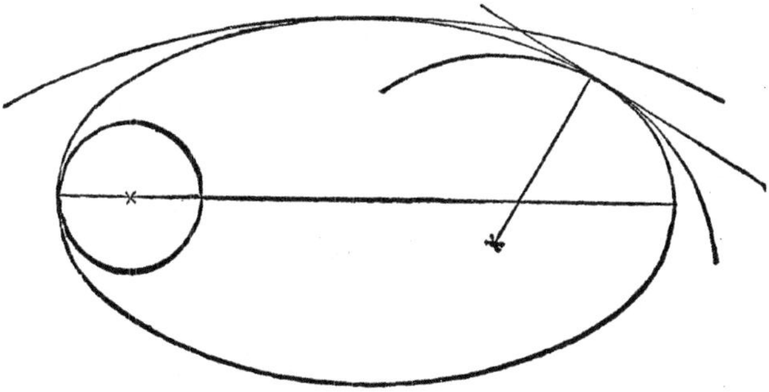

Also (2), $\bar{x} = x - x(a^4y^2 + b^4x^2)/(a^4b^2) = (a^2 - b^2)x^3/a^4;$
$\bar{y} = y - y(a^4y^2 + b^4x^2)/(a^2b^4) = -(a^2 - b^2)y^3/b^4.$ (4)

In (3) put $y^2 = b^2\,(a^2 - x^2)/(a^2$, and $(a^2 - b^2)/a^2 = e^2$, whence $b = a\sqrt{1 - e^2}$. Then

$$1/\rho = \mp a^2\sqrt{1 - e^2}/(a^2 - e^2x^2)^{3/2}. \quad . \; . \; . \quad (5)$$

EXAMPLES.

1. $y = 2x + 3$.

$$\overline{f'(x)}^2 = 4, \qquad f''(x) = 0, \qquad \rho = \infty.$$

2. $x^2 + y^2 = r^2$.

$$\overline{f'(x)}^2 = x^2/(r^2 - x^2), \qquad f''(x) = r^2/(r^2 - x^2)^{3/2},$$

$$\rho = \left(1 + \frac{x^2}{r^2 - x^2}\right)^{3/2} \Big/ \frac{r^2}{(r^2 - x^2)^{3/2}} = r.$$

3. $y^2 = 2px$.

$$\rho = \left(1 + \frac{p}{2x}\right)^{3/2} \Big/ \frac{p^2}{(2px)^{3/2}} = -\frac{(2px + p^2)^{3/2}}{p^2}.$$

$\rho_{x=0} = p =$ one half the parameter.

$\bar{x} = 3x + p, \quad \bar{y} = -y^3/p^2.$

4. $16y^2 + 4x^2 = 64$.

From (3) and (4) (§ 172) we have at $(2, \sqrt{3})$

$$\rho = -5.86, \qquad \bar{x} = 3/8, \qquad \bar{y} = -9\sqrt{3}/4.$$

5. $y^2/b^2 - x^2/a^2 = -1$.

$$\rho = \frac{(a^2x^2 + b^2x^2 - a^4)^{3/2}}{a^4b} = \frac{(a^4y^2 + b^4x^2)^{3/2}}{a^4b^4} = \frac{(e^2x^2 - a^2)^{3/2}}{ab}$$

$$x = \pm a, \qquad y = 0, \quad \text{and} \quad \rho = b^2/a.$$

$$a = b, \qquad \rho = (2x^2 - a^2)^{3/2}/a^2.$$

6. $y^2 = 2px + r^2x^2$.

$$\rho = \left[1 + \left(\frac{p + r^2x}{y}\right)^2\right]^{3/2} \Big/ \left[\frac{r^2y^2 - (p + r^2x)^2}{y^3}\right]$$

$$= \frac{[2px + r^2x^2 + (p + r^2x)^2]^{3/2}}{p^2}.$$

Hence (Example 3, § 149), at any point of a conic, as given, ρ is equal to the cube of the corresponding normal divided by the square of half the parameter.

7. $x = r \text{ vers}^{-1}(y/r) - \sqrt{2ry - y^2}$.

$$\rho = -[1 + (2r/y - 1)]^{3/2}/(r/y^2) = -2\sqrt{2ry}.$$

Hence (Example 6, § 149), at any point of a cycloid ρ is equal to twice the corresponding normal.

$$y = 0 \text{ gives } \rho = 0, \quad \text{and} \quad y = 2r \text{ gives } \rho = 4r.$$

8. $y = 4 - 3(x - 2)^{3/5}$.

$$\rho = \left[1 + \frac{81}{25}(x - 2)^{-4/5}\right]^{3/2} \Big/ \left[\frac{18}{25}(x - 2)^{-7/5}\right]$$

$$= \frac{25}{18}\left[(x - 2)^{14/15} + \frac{81}{25}(x - 2)^{2/15}\right]^{3/2}.$$

(2, 4) is a point of inflexion at which $\rho = 0$. Ex. 2, § 161.

9. $y^2 = 9x$. At $(3, \sqrt{27})$ $\bar{x} = 13.5$, $\bar{y} = -\sqrt{48}$

10. $y^2 = 8x$. $\rho = 2(x+2)^{3/2}/\sqrt{2}$.

11. $x^2/9 + y^2/4 = 1$. $\rho_{x=0} = 9/2$.

12. $xy = m$. $\rho = (x^2 + y^2)^{3/2}/2m$.

13. $y = a(e^{x/a} + e^{-x/a})/2$. $\rho = y^2/a$,

$$\bar{x} = x - y\sqrt{y^2 - a^2}/a, \qquad \bar{y} = 2y.$$

14. $x^{2/3} + y^{2/3} = a^{2/3}$. $\rho = 3\sqrt[3]{axy}$,

$$\bar{x} = x + 3\sqrt[3]{xy^2}, \qquad \bar{y} = y + 3\sqrt[3]{x^2y}.$$

15. $3a^2y = x^3$. $\rho = (a^4 + x^4)^{3/2}/2a^4x$.

16. $y = x^3 - x^2 + 1$. $\rho_{x=1/3} = \infty$ (point of inflexion).

17. $e^{y/a} = \sec(x/a)$. $\rho = a\sec(x/a)$.

18. $y^3 = 6x^2 + x^3$. $\rho = -[y^4 + (4x + x^2)^2]^{3/2}/8x^2y$.

19. $y = x^4 - 4x^3 - 18x^2$. $\rho_{x=0} = -1/36$, $\rho_{x=3} = \infty$.

20. $y = ae^{x/a}$. $\rho = (a^2 + y^2)^{3/2}/ay$.

21. $y^3 = a^2x$. $\bar{x} = (a^4 + 15y^4)/6ya^2$, $\bar{y} = (a^4y - 9y^5)/2a^4$,

$$\rho = (9y^4 + a^4)^{3/2}/6a^4y.$$

22. $3ay^2 = 2x^3$. $\rho = \pm(2a + 3x)^{3/2}\sqrt{x}/a\sqrt{3}$.

23. $x = \sec 2y$. $\rho = (2x^2 - 1)^2/4x$.

24. $y = \log_a x$. $\rho = (M_a^2 + x^2)^{3/2}/M_a x$,

$\rho_{\substack{a=e\\x=1}} = 2\sqrt{2}$.

25. $y = x^3 - x^2 + 1$. $\rho_{x=0} = -1/2$, $\rho_{y=\min.} = 1/2$.

26. $\sqrt{x} + \sqrt{y} = \sqrt{a}$. $\rho = 2(x+y)^{3/2}/\sqrt{a}$.

27. $ay^2 = x^3$. $\rho = (4a + 9x)^{3/2}x^{1/2}/6a$.

28. $y^2 = x^3/(2a - x)$. $\rho = a(8a - 3x)^{3/2}x^{1/2}/3(2a - x)^2$.

29. $a^2y = bx^3 + cx^2y$. $\rho_0 = \infty$.

30. $y = \log\sec x$. $\rho = \sec x$.

31. In a parabola show that any radius of curvature is twice the part of the normal intercepted between the curve and the directrix.

32. Applying (5) to a meridian of the earth, we have, since

$$l = \text{latitude} = \psi - \pi/2,$$

$$\tan^2 l = \cot^2 \psi = 1/\overline{f'(x)}^2.$$

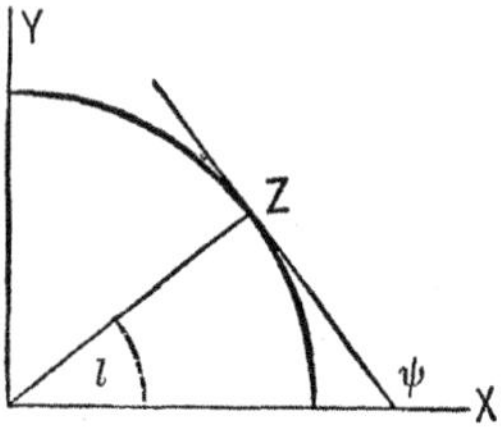

For an ellipse we have

$$\tan^2 l = a^2(a^2 - x^2)/b^2x^2 = (a^2 - x^2)/(1 - e^2)x^2.$$

Hence, $$x^2 = \frac{a^2}{1 + (1 - e^2)\tan^2 l} = \frac{a^2}{\sec^2 l - e^2 \tan^2 l},$$

and $$a^2 - e^2x^2 = \frac{a^2(\sec^2 l - e^2 \sec^2 l)}{\sec^2 l - e^2 \tan^2 l} = \frac{a^2(1 - e^2)}{1 - e^2 \sin^2 l},\text{*}$$

which, substituted in (5), gives

$$1/\rho = (1 - e^2 \sin^2 l)^{3/2}/a(1 - e^2). \quad . \quad . \quad (5)$$

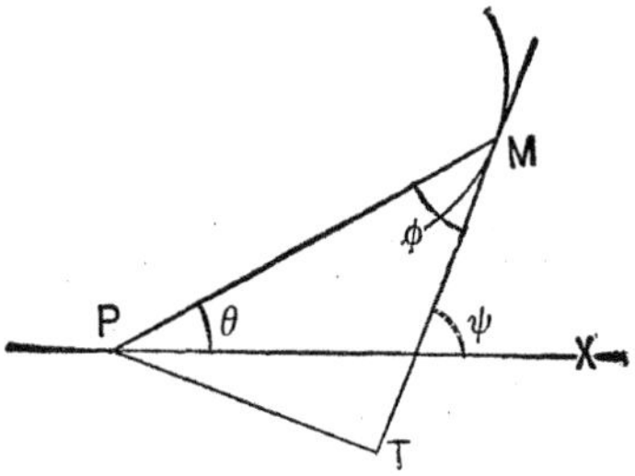

173. In Polar Coördinates.—Differentiating,

$$x = r \cos \theta, \qquad y = r \sin \theta,$$

and substituting in (1) (§ 172), as in Example 11 (§ 115), we have

$$\rho = \left[r^2 + \left(\frac{dr}{d\theta}\right)^2\right]^{3/2} \Big/ \left[r^2 + 2\left(\frac{dr}{d\theta}\right)^2 - r\frac{d^2r}{d\theta^2}\right]. \quad (1)$$

---

* Rice and Johnson's Calculus, p. 355.

Putting $r = 1/u$, whence $dr/d\theta = -(1/u^2)(du/d\theta)$, and $d^2r/d\theta^2 = (2/u^3)(du/d\theta)^2 - (1/u^2)(d^2u/d\theta^2)$, and substituting, we have

$$\rho = \left[u^2 + \left(\frac{du}{d\theta}\right)^2\right]^{3/2} \Big/ \left[u^3\left(u + \frac{d^2u}{d\theta^2}\right)\right]. \quad . \quad . \quad (2)$$

EXAMPLES.

1. $r = a\theta$. $\rho = a(1+\theta^2)^{3/2}/(2+\theta^2) = (a^2+r^2)^{3/2}/(2a^2+r^2)$.
2. $r = a^\theta$. $\rho = r\sqrt{1+\log^2 a}$.
3. $r = a \sin n\theta$. $\rho_{r=0} = na/2$.
4. $r = 2a\cos\theta - a$. $\rho = a(5 - 4\cos\theta)^{3/2}/(9 - 6\cos\theta)$.
5. $r = a(1 - e^2)/(1 - e\cos\theta)$.

$$\rho = a(1 - e^2)(1 - 2e\cos\theta + e^2)^{3/2}/(1 - e\cos\theta)^3.$$

6. $r^2 \cos 2\theta = a^2$. $\rho = -r^3/a^2$.
7. $r = a(2\cos\theta - 1)$. $\rho = a(5 - 4\cos\theta)^{3/2}/(9 - 6\cos\theta)$.
8. $r\cos^2(\theta/2) = p$. $\rho = 2r^{3/2}/\sqrt{p}$.
9. $r = a\sec^2(\theta/2)$. $\rho = 2a\sec^3(\theta/2)$.
10. $r = a\theta^{-\frac{1}{2}}$. $\rho = r(4a^4 + r^4)^{3/2}/2a^2(4a^4 - r^4)$.
11. $r^2\theta = a^2$. $\rho = r(4a^4 + r^4)^{3/2}/2a^2(4a^4 - r^4)$.
12. $r = 4a\sin^2(\theta/2)$.

From which and § 150,

$$dr/(rd\theta) = \cot(\theta/2) = \cot\phi. \quad \therefore \theta/2 = \phi.$$
$$\psi = \theta + \phi = 3\theta/2, \quad \text{and} \quad d\psi = 3d\theta/2.$$

Therefore,

$$\rho = ds/d\psi = (2/3)(ds/d\theta) = (2/3)r\operatorname{cosec}\phi = (8/3)a\sin(\theta/2).$$

13. $r\theta = a$. $\rho = r(a^2 + r^2)^{3/2}/a^3 = a(1+\theta^2)^{3/2}/\theta^4$.
14. $r^2 = a^2\cos 2\theta$. $\rho = a^2/3r$.

174. From § 172 we have

$$dx = \cos\psi ds, \quad \text{and} \quad dy = \sin\psi ds.$$

Differentiating without assuming the independent variable, and writing ds/ρ for $d\psi$, we have

$$d^2x = \cos\psi d^2s - \sin\psi(ds)^2/\rho,$$
$$d^2y = \sin\psi d^2s + \cos\psi(ds)^2/\rho.$$

Squaring and adding, we obtain

$$(d^2x)^2 + (d^2y)^2 = (d^2s)^2 + (ds)^4/\rho^2.$$

Hence, $\rho = ds^2/\sqrt{(d^2x)^2 + (d^2y)^2 - (d^2s)^2}$, . . . (1)

which is a general expression for ρ.

Regarding s as the independent variable $d^2s = 0$, and (1) becomes

$$\rho = d^2s/\sqrt{(d^2x)^2 + (d^2y)^2},$$

from which $1/\rho^2 = (d^2x/ds^2)^2 + (d^2y/ds^2)^2$, ... (2)

which is a convenient form when x and y are given as functions of s.

175. In (1) (§ 172), x is the independent variable. Substituting $(dx\, d^2y - dy\, d^2x)/dx^3$ for d^2y/dx^2, we have (§ 115, Example 11)

$$\rho = (dx^2 + dy^2)^{3/2}/(dx\, d^2y - dy\, d^2x), \quad . \quad . \quad (1)$$

in which neither x nor y is independent. This form is convenient when x and y are given as functions of a third variable. Thus, having

1. $x = a(\phi - \sin\phi)$, $\qquad y = a(1 - \cos\phi)$,

we have, as in Example 12 (§ 115), $\rho = -4a \sin(\phi/2)$.

2. $\begin{cases} x = c \sin 2\theta(1 + \cos 2\theta). \\ y = c \cos 2\theta(1 - \cos 2\theta). \end{cases}$ $\qquad \rho = 4c \cos 3\theta.$

3. $\begin{cases} x = a \cos\theta. \\ y = b \sin\theta. \end{cases}$ $\qquad \rho = (a^4y^2 + b^4x^2)^{3/2}/(a^4b^4).$

4. $\begin{cases} x = a(2\cos\theta + \cos 2\theta). \\ y = a(2\sin\theta - \sin 2\theta). \end{cases}$ $\qquad \rho = -8a \sin(3\theta/2).$

5. $\begin{cases} x = (a + b)\cos\theta - b\cos\dfrac{a + b}{b}\theta. \\ y = (a + b)\sin\theta - b\sin\dfrac{a + b}{b}\theta. \end{cases}$ $\qquad \rho = \dfrac{4b(a + b)}{a + 2b}\sin\dfrac{a}{2b}\theta.$

176. ρ in Terms of r and p.—From § 150.

$$p = r^2 / \sqrt{r^2 + (dr/d\theta)^2}.$$

Hence,

$$\frac{dp}{dr} = \frac{2r\sqrt{r^2 + \left(\frac{dr}{d\theta}\right)^2} - r^2\left(r + \frac{d^2r}{d\theta^2}\right) \Big/ \left[\sqrt{r^2 + \left(\frac{dr}{d\theta}\right)^2}\,\right]}{r^2 + \left(\frac{dr}{d\theta}\right)^2}$$

$$= \left[r^3 + 2r\left(\frac{dr}{d\theta}\right)^2 - r^2\frac{d^2r}{d\theta^2}\right] \Big/ \left[r^2 + \frac{dr^2}{d\theta^2}\right]^{3/2}.$$

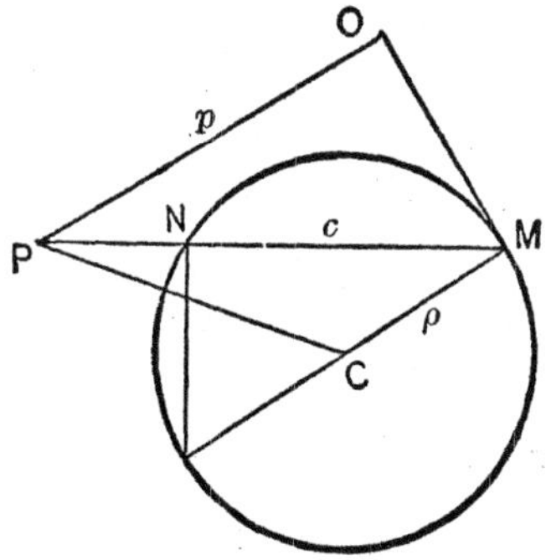

Comparing with (1) (§ 173), we have

$$\rho = r\,dr/dp. \quad . \quad . \quad (1)$$

Let C be the centre, and $CM = \rho$, the radius of curvature corresponding to M. Let $c = MN$ represent the chord of curvature which passes through P, and $PM = r$. Then

$$c = 2\rho \cos PMC = 2\rho \sin OMP = 2\rho p/r = 2p\,dr/dp.$$

EXAMPLES.

1. $p = ar.$	$\rho = r/a, \quad c = 2r.$
2. $ar = p^2.$	$\rho = 2p^3/a^2 = 2(r^3/a)^{1/2}, \quad c = 2p^2/a = 2r.$
3. $r^2 = ap$	$\rho = a/2, \quad c = r.$
4. $a^2 + b^2 - r^2 = a^2b^2/p^2.$	$\rho = a^2b^2/p^3.$
5. $r^3 = a^2p.$	$\rho = a^2/3r, \quad c = 2r/3.$
6. $r^3 = 2ap^2.$	$c = 4r/3.$
7. $r^{m+1} = a^m p.$	$\rho = r^2/(m+1)p, \quad c = 2r/(m+1).$
8. $p^2 = r^2 - a^2.$	$\rho = p, \quad c = 2p^2/r.$

9. $r^2 = a^2 \cos 2\theta$.

Putting $r^{-1} = u$, we have $a^2u^2 = \sec 2\theta$.

Hence, $du/d\theta = u \tan 2\theta$, $u^2 + du^2/d\theta^2 = a^4u^6$, $1/p^2 = a^4/r^6$, and $dp/dr = 3r^2/a^2$. Therefore, $\rho = a^2/3r$, and $c = 2r/3$.

10. $r = 2a(1 - \cos\theta)$. $\rho = 8a \sin(\theta/2)/3 = 4\sqrt{ar}/3$, $c = 4r/3$.

11. $r = a(1 + \cos\theta)$. $\rho = 2\sqrt{2ar}/3$, $c = 4r/3$.

12. $r = a\theta$.

$$dr/d\theta = a; \text{ also (§ 150), } dr/d\theta = r\sqrt{r^2 - p^2}/p.$$

Hence, $r\sqrt{r^2 - p^2}/p = a$, or $r^4 - p^2r^2 = a^2p^2$.

From which $rdr/dp = p(r^2 + a^2)/2(2r^2 - p^2)$.

But $p^2 = r^4/(r^2 + a^2)$. Therefore, $\rho = (r^2 + a^2)^{3/2}/2(r^2 + 2a^2)$.

The chord of curvature through the pole

$$= 2pdr/dp = 2p^2(r^2 + a^2)/2r(2r^2 - p^2) = r(r^2 + a^2)/(r^2 + 2a^2).$$

Hence, $\rho = c$ when $(\pi^2 + a^2)^{3/2} = 2r(r^2 + a^2)$ or $r = a/\sqrt{3}$.

CURVES OF DOUBLE CURVATURE.

177.* Let $MM' = \delta s$ be the length of any varying arc not including a singular point, of any curve of double curvature. At M and M' respectively draw the tangents MT and $M'T'$, and through the origin draw the two right lines OE and OE' parallel to them respectively. Upon each lay off a length l, and join the extremities E and E' by the right line EE', forming an isosceles triangle in which the angle EOE', designated by $\delta\psi$, measures the *total* curvature of the arc δs. (§ 166.)

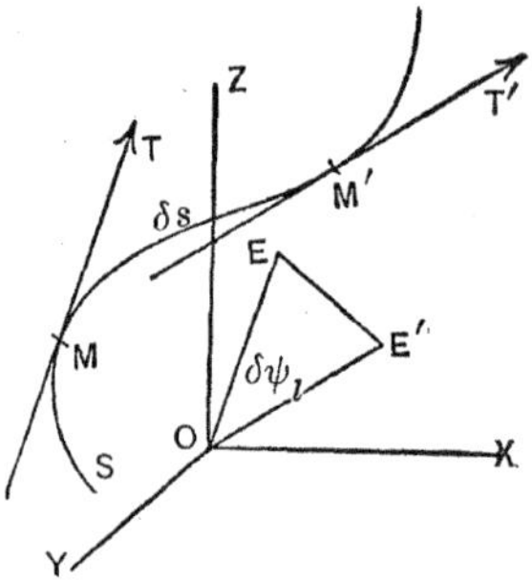

* Modification of method in Calcul Différentiel, par J. Bertrand, page 614.

The *rate* of curvature of the curve at M is (§ 167)

$$\frac{d\psi}{ds} = \underset{\delta s \ggg\!\!\to 0}{\text{limit}} \left[\frac{\delta\psi}{\delta s}\right]. \quad . \quad . \quad . \quad . \quad . \qquad (1)$$

Let α, β, γ, and $\alpha + \delta\alpha$, $\beta + \delta\beta$, $\gamma + \delta\gamma$, represent the angles which the tangents MT and $M'T'$ make respectively with the coördinate axes.

From the triangle EOE' we have

$$EE'/l = 2 \sin (\delta\psi/2); \text{ hence,}$$

$$\underset{\delta\psi \ggg\!\!\to 0}{\text{limit}} \left[\frac{EE'}{l} \Big/ \delta\psi\right] = \text{limit} \left[\sin \left(\frac{\delta\psi}{2}\right) \Big/ \frac{\delta\psi}{2}\right] = 1. \quad (2)$$

The coördinates of E and E' are respectively $l \cos \alpha$, $l \cos \beta$, $l \cos \gamma$, and $l \cos (\alpha + \delta\alpha)$, $l \cos (\beta + \delta\beta)$, $l \cos (\gamma + \delta\gamma)$. Substituting in formula

$$D = \sqrt{[(x'' - x')^2 + (y'' - y')^2 + (z'' - z')^2]},$$

and dividing by l, we have

$$EE'/l = \sqrt{([\cos (\alpha + \delta\alpha) - \cos \alpha]^2}$$
$$+ [\cos (\beta + \delta\beta) - \cos \beta]^2 + [\cos (\gamma + \delta\gamma) - \cos \gamma]^2).$$

Dividing by δs, taking the limit as $\delta s \ggg\!\!\to 0$, whence $\delta\psi \ggg\!\!\to 0$, substituting $\delta\psi$ for EE'/l, (2), we have (1) and § 170 for the rate of curvature at M,

$$\underset{\delta s \ggg\!\!\to 0}{\text{limit}} \left[\frac{\delta\psi}{\delta s}\right] = \sqrt{\left(\frac{d \cos \alpha}{ds}\right)^2 + \left(\frac{d \cos \beta}{ds}\right)^2 + \left(\frac{d \cos \gamma}{ds}\right)^2}$$

$$= \frac{d\psi}{ds} = \frac{1}{\rho}. \quad . \quad . \quad . \quad . \quad . \quad . \quad . \quad . \quad . \quad . \qquad (3)$$

From § 88,

$$\cos \alpha = dx/ds, \quad \cos \beta = dy/ds, \quad \cos \gamma = dz/ds.$$

Differentiating and substituting in (3), we have

$$\frac{1}{\rho} = \sqrt{\left(\frac{d^2x}{ds^2}\right)^2 + \left(\frac{d^2y}{ds^2}\right)^2 + \left(\frac{d^2z}{ds^2}\right)^2}, \quad . \quad . \quad . \quad (4)$$

in which s is the independent variable.

In order to obtain a more general expression for $1/\rho$, place (§ 115)

$$\frac{d^2x}{ds^2} = \frac{ds\, d^2x - dx\, d^2s}{ds^3}, \quad \frac{d^2y}{ds^2} = \frac{ds\, d^2y - dy\, d^2s}{ds^3}, \text{ etc.,}$$

giving

$$\frac{1}{\rho} = \sqrt{\frac{(ds\, d^2x - dx\, d^2s)^2 + (ds\, d^2y - dy\, d^2s)^2 + (ds\, d^2z - dz\, d^2s)^2}{ds^6}}$$

which may be written

$$\frac{1}{\rho} = \sqrt{\frac{ds^2[(d^2x)^2 + (d^2y)^2 + (d^2z)^2] - 2ds\,d^2s(dx\,d^2x + dy\,d^2y + dz\,d^2z) + (d^2s)^2(dx^2 + dy^2 + dz^2)}{ds^6}}.$$

From § 88, $\qquad ds^2 = dx^2 + dy^2 + dz^2.$

Whence, $\qquad ds\, d^2s = dx\, d^2x + dy\, d^2y + dz\, d^2z.$

Therefore,

$$\frac{1}{\rho} = \sqrt{\frac{(d^2x)^2 + (d^2y)^2 + (d^2z)^2 - (d^2s)^2}{ds^4}}, \quad . \quad (5)$$

which is a general expression for the rate of curvature at any point of any curve.

If the curve is of single curvature, its plane may be taken as that of XY, z will be zero, and (5) reduces to (1) (§ 174).

CHAPTER XVI.

INVOLUTES AND EVOLUTES.

178. Each point of any given curve as $MM'M''$, has, in general, a centre of curvature.

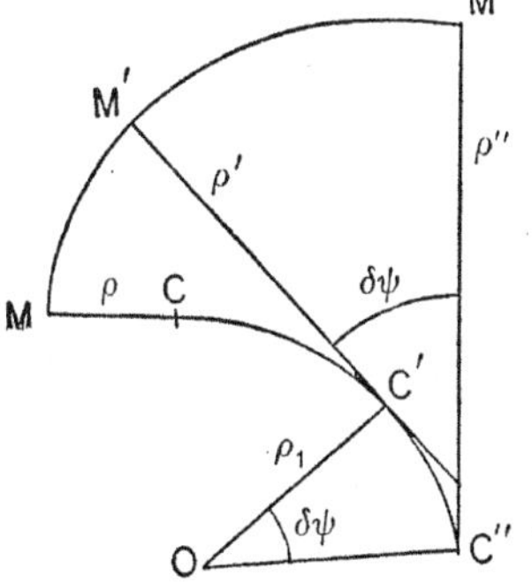

The locus of the centre of curvature of any given curve is called its **evolute**. Thus, $CC'C''$ is the evolute of $MM'M''$.

The given curve $MM'M''$ is called an **involute** of its evolute.

179. Coördinates of the Centre of Curvature.—Let C be the centre, and $CM = \rho$ the radius of curvature corresponding to M, whose coördinates are x, y. Then

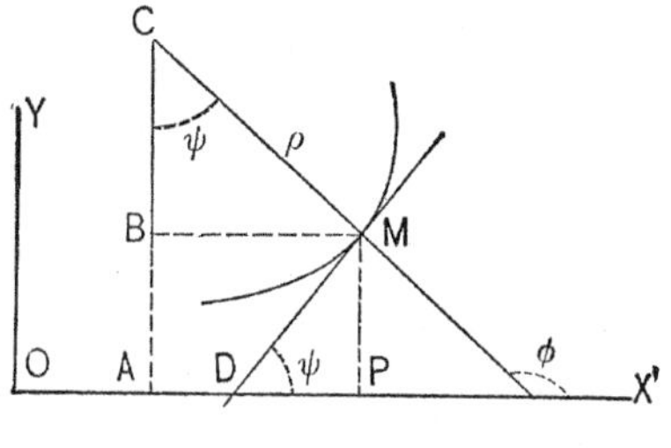

$$\left.\begin{aligned} OA &= OP - AP, \quad \text{or} \quad \bar{x} = x - \rho \sin \psi, \\ AC &= AB + BC, \quad \text{or} \quad \bar{y} = y + \rho \cos \psi, \end{aligned}\right\} \quad . \quad . \quad (1)$$

which correspond with (2), § 172, since (§ 170)

$$\rho = \frac{ds}{d\psi}, \quad \text{and (§ 172)} \quad \frac{dx}{d\psi} = \left[1 + \left(\frac{dy}{dx}\right)^2\right] \Big/ \frac{d^2y}{dx^2},$$

$$dx/ds = \cos\psi, \quad dy/ds = \sin\psi, \quad \text{giving}$$

$$\rho \sin\psi = \frac{ds}{d\psi}\frac{dy}{ds} = \frac{dy}{d\psi} = \frac{dy}{dx}\frac{dx}{d\psi} = \frac{dy}{dx}\left[1 + \left(\frac{dy}{dx}\right)^2\right] \Big/ \frac{d^2y}{dx^2};$$

$$\rho \cos\psi = \frac{ds}{d\psi}\frac{dx}{ds} = \frac{dx}{d\psi} = \left[1 + \left(\frac{dy}{dx}\right)^2\right] \Big/ \frac{d^2y}{dx^2}.$$

180. Differentiating (1), § 179, with respect to s, we have

$$\left.\begin{aligned} d\bar{x}/ds &= dx/ds - \rho\cos\psi d\psi/ds - \sin\psi d\rho/ds, \\ d\bar{y}/ds &= dy/ds - \rho\sin\psi d\psi/ds + \cos\psi d\rho/ds. \end{aligned}\right\} \quad (1)$$

But (§§ 170, 172)

$$d\psi/ds = 1/\rho, \quad \cos\psi = dx/ds, \quad \sin\psi = dy/ds.$$

Therefore,

$$\left.\begin{aligned} d\bar{x}/ds &= -\sin\psi d\rho/ds, \\ d\bar{y}/ds &= \cos\psi d\rho/ds. \end{aligned}\right\} \quad . \quad . \quad . \quad . \quad (2)$$

Hence, by division, $d\bar{y}/d\bar{x} = -\cot\psi$.

Represent the angle which a tangent to the evolute at $(\bar{x}, \bar{y})$ makes with X by ψ_1; then,

$$d\bar{y}/d\bar{x} = \tan\psi_1 = -\cot\psi = -dx/dy, \text{ or } \psi_1 = \psi + \pi/2.$$

Therefore, the tangent to the evolute at $(\bar{x}, \bar{y})$ is normal to the involute at (x, y); or, in other words, the radius of curvature of any curve at any point is tangent to the evolute at the corresponding centre of curvature. An evolute must therefore be drawn tangent to all radii of curvature of the involute. An evolute is therefore the *limit* of the locus of points of intersection of adjacent normals to the involute, as the number of normals corresponding to any definite portion of the involute is increased without limit.

It follows that a radius of curvature which is unlimited in length is an asymptote to the evolute. Hence, in general, the normal at a point of inflexion is an asymptote to the evolute.

Also, when in the vicinity of any point a curve is symmetrical with respect to the normal at the point, the corresponding point of the evolute is a cusp.

181. Squaring both members of (2), § 180, and adding each to each, we have

$$(\overline{dx}^2 + \overline{dy}^2)/ds^2 = d\rho^2/ds^2. \quad \text{Hence,} \ d\rho = \pm\sqrt{\overline{dx}^2 + \overline{dy}^2}.$$

Let s represent the length of a varying portion of the evolute, then (§ 87) $ds = \sqrt{\overline{dx}^2 + \overline{dy}^2}$. Hence, $d\rho = ds$, and (§ 74) $\rho = s \pm a$, in which a is a constant.

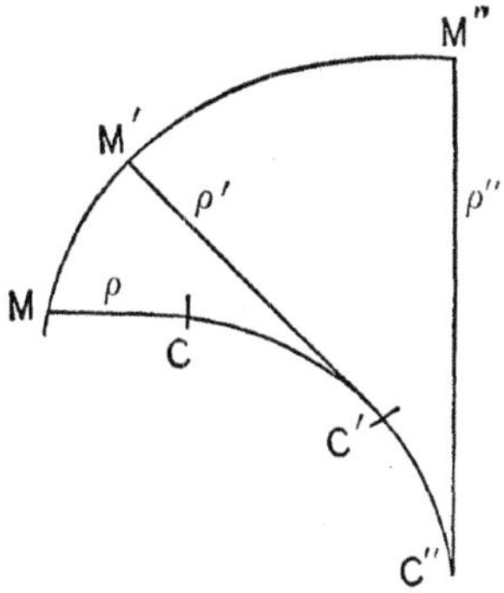

Let $C'M' = \rho'$, and $C''M'' = \rho''$, be the radii of curvature corresponding to M' and M'' respectively. Measuring the evolute from C, and denoting the lengths of the arcs CC' and CC'' by s_1 and s_2 respectively, we have

$$\rho' = s_1 + a; \qquad \rho'' = s_2 + a. \quad . \quad . \quad . \quad (1)$$

Hence,

$$\rho'' - \rho' = s_2 - s_1 = \text{arc } C'C'', \quad . \quad . \quad . \quad (2)$$

that is, in general, *the difference between any two radii of curvature of an involute is equal to the arc of the evolute between the corresponding centres of curvature.*

Exceptions exist when the arc of the evolute includes a singular point or is discontinuous.

Measuring the evolute from C, we have

$$\text{arc } CC' = \rho' - \rho, \quad \text{or} \quad \rho' = \text{arc } CC' + \rho;$$

also, $$\text{arc } CC'' = \rho'' - \rho, \quad \text{or} \quad \rho'' = \text{arc } CC'' + \rho.$$

Hence, (1), $$a = \rho.$$

Similarly it may be shown that the constant a in equation (1) is, in general, equal to the radius of curvature which passes through the point of the evolute from which it is measured.

It follows that as a right line rolls tangentially upon, or as a string is unwound from, any curve, each point describes an involute to the given curve as an evolute. Hence, while an involute has but one evolute, each evolute has an unlimited number of involutes. Any two involutes corresponding to the same evolute are separated by a constant distance measured along the normals, and are called parallel curves.

182. In general, at cusps of the first species and at points of inflexion ρ changes sign. Hence, at such points we have $\rho = 0$, or ∞, and the evolutes pass through these points or have infinite branches to which the corresponding normals are asymptotes.

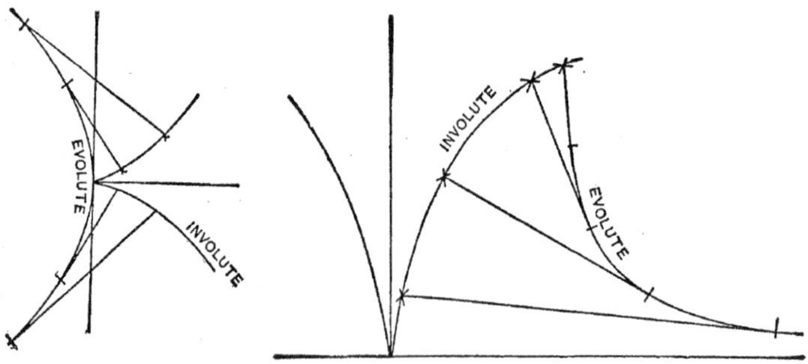

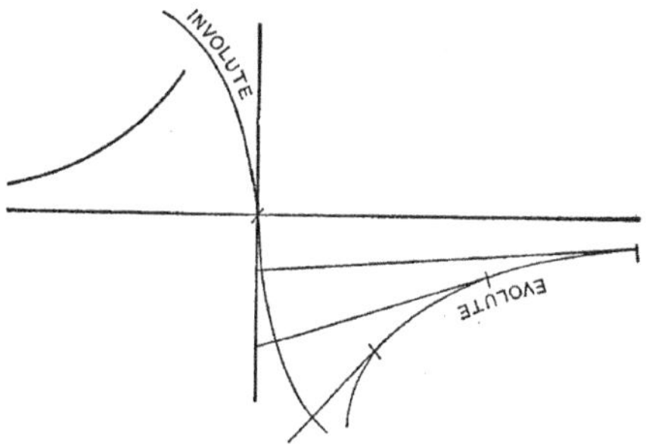

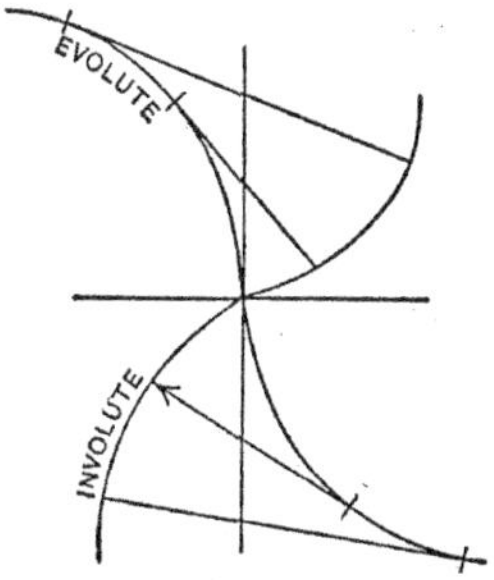

At a cusp of the 2d species ρ does not change its sign, and the corresponding point of the evolute will, in general, be a point of inflexion.

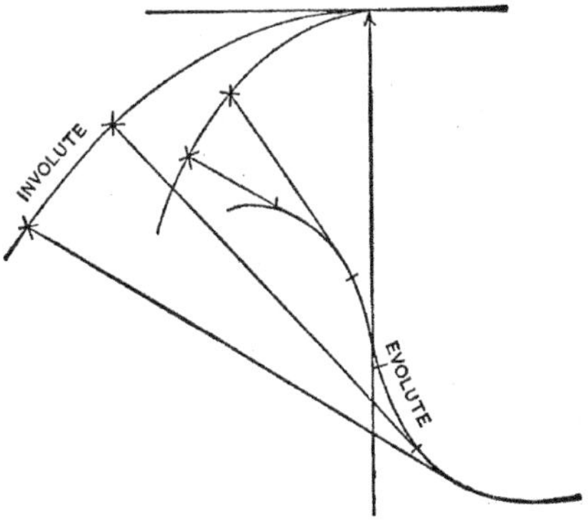

183. Radius of Curvature of an Evolute.—The angle $\delta\psi$, between any two tangents as those at M' and M'', is equal to that between the corresponding radii of curvature to the involute, and since these radii of curvature are tangents to the evolute, the angle which they make with each other is equal to $C''OC'$, included between the corresponding radii of curvature of the evolute.

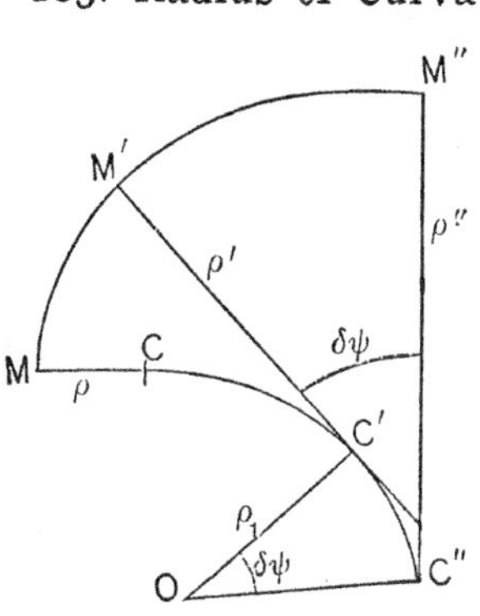

Let $s_1 = CC'$, $\delta s_1 = C'C''$, and ρ_1 = radius of curvature of evolute at C'.

Then (§ 170) $$\rho_1 = \underset{\delta s_1 \Rightarrow 0}{\text{limit}} \left[\frac{\delta s_1}{\delta\psi}\right].$$

But δs_1 corresponds to $\delta s = M'M''$ of the involute, and vanishes with it. Also (§ 181), $\delta s_1 = \rho'' - \rho' = \delta\rho$.

Hence, $$\rho_1 = \underset{\delta s \Rightarrow 0}{\text{limit}} \left[\frac{\delta\rho}{\delta\psi}\right] = \frac{d\rho}{d\psi} = \frac{d^2 s}{d\psi^2}, \quad . \quad . \quad . \quad (1)$$

in which s is the length of an arc of the involute measured up to M' and ψ is the angle which the tangent at M' makes with a fixed right line.

184. Equation of the Evolute.—Let

$$y = f(x) \quad . \quad . \quad . \quad . \quad . \quad . \quad . \quad . \quad (1)$$

be the equation of any given plane curve. The coördinates of the centre of curvature for any point (x, y) of the curve are (1), § 172,

$$\left.\begin{aligned} \bar{x} &= x - [1 + \overline{f'(x)}^2] f'(x)/f''(x); \\ \bar{y} &= y + (1 + \overline{f'(x)}^2)/f''(x). \end{aligned}\right\} \quad . \quad . \quad (2)$$

Expressions for $f'(x)$ and $f''(x)$ in terms of x and y, obtained by differentiating (1), substituted in (2) give $\bar{x}$ and $\bar{y}$ in terms of x and y. Combining these equations with (1), eliminating x and y, we have $\bar{y} = F(\bar{x})$ for the evolute of $y = f(x)$.

EXAMPLES.

1. $y^2 = 2px$.

$$dy/dx = p/y, \qquad d^2y/dx^2 = -p^2/y^3.$$

Substituting in (2), we have

$$\bar{x} = x - \left[1 + \frac{p^2}{y^2}\right]\frac{p}{y} \Big/ \left(-\frac{p^2}{y^3}\right) = x + \frac{y^2}{p} + p. \quad \therefore x = \frac{\bar{x} - p}{3},$$

$$\bar{y} = y + \left(1 + \frac{p^2}{y^2}\right) \Big/ \left(-\frac{p^2}{y^3}\right) = -\frac{y^3}{p^2}. \quad \therefore y^2 = (p^2\bar{y})^{2/3}.$$

Combining with $y^2 = 2px$, and eliminating x and y we have

$$\bar{y}^2 = 8(\bar{x} - p)^3/27p$$

for the evolute of the parabola.

$C'CC''$ is the evolute of the parabola $C''OC'$.

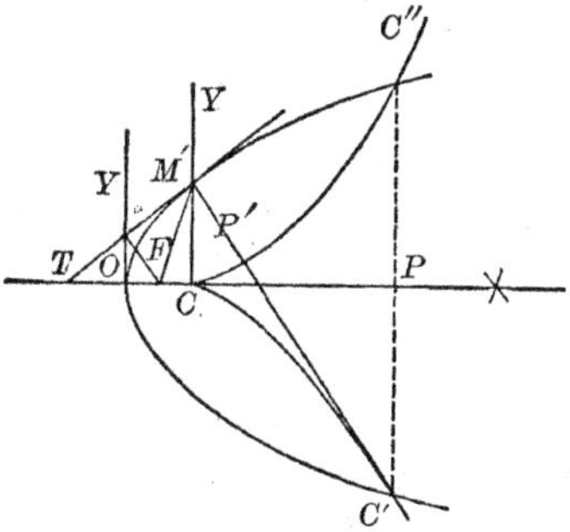

$$\bar{y} = 0$$

gives $\bar{x} = p = OC$.

$$\bar{x} = 4p = OP$$

gives

$$\bar{y} = \pm p\sqrt{8} = \pm PC''$$

for points common to the parabola and its evolute.

The arc $CC' = (3\sqrt{3} - 1)p$.

Transferring the origin to C, the axes remaining parallel, we have, denoting the new coördinates by x and y,

$$\bar{x} = p + x, \qquad \bar{y} = y.$$

Hence, we have $y^2 = 8x^3/(27p)$ for the evolute. The branch CC' belongs to OC'' and CC'' to OC'.

Let $r = FM'$ represent the focal distance of any point as M', and let $l = FY$ represent the perpendicular from the focus to the tangent TM'.

Then $\overline{FY}^2 = FM' \times FO$, or $l^2 = pr/2$.

Hence, $2ldl = pdr/2$, or $dr/dl = 4l/p$.

From § 176, $c =$ chord of curvature through F

$= 2ldr/dl$.

Hence, $c = 8l^2/p = 8pr/2p = 4r = 4FM'$.

That is, in a common parabola the chord of curvature through the focus is equal to four times the focal distance of the point of tangency.

2. $a^2y^2 + b^2x^2 = a^2b^2$.

$$dy/dx = -b^2x/a^2y, \qquad d^2y/dx^2 = -b^4/a^2y^3.$$

Substituting in (2) and reducing, we have

$$\bar{x} = (a^2 - b^2)x^3/a^4; \qquad \therefore x = (a^4\bar{x}/[a^2 - b^2])^{1/3}.$$

$$\bar{y} = -(a^2 - b^2)y^3/b^4; \qquad \therefore y = -(b^4\bar{y}/[a^2 - b^2])^{1/3}.$$

Combining with $a^2y^2 + b^2x^2 = a^2b^2$, we have

$$(a\bar{x})^{2/3} + (b\bar{y})^{2/3} = (a^2 - b^2)^{2/3}$$

for the evolute of the ellipse.

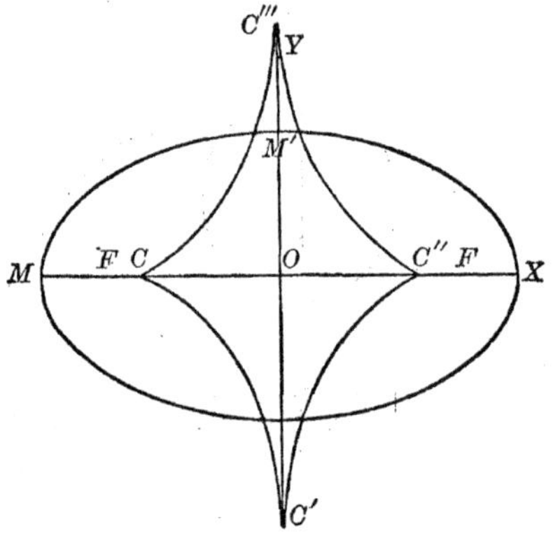

$CC'C''C'''$ is the evolute of the ellipse $MM'X$; $\bar{x} = 0$ gives $\bar{y} = (a^2 - b^2)/b = OC'''$.

When $e^2 = (a^2 - b^2)/a^2 = 1/2$, we have $a^2 = 2b^2$, and $OC''' = b$, in which case the vertices C''' and C' are on the curve. They are without (as in the figure) or within the ellipse according as e^2 is greater or less than $1/2$.

Measuring the evolute from C, we have $\rho = MC = b^2/a$, and $\rho' = M'C' = a^2/b$.

Hence, arc $CC' = a^2/b - b^2/a = (a^3 - b^3)/ab$.

$$\text{Axis } CC'' = 2a - 2b^2/a = 2(a^2 - b^2)/a.$$

$$\text{Axis } C'C''' = 2b + 2(a^2/b - 2b) = 2(a^2 - b^2)/b.$$

Hence, *the axes of the evolute are inversely as the corresponding axes of the ellipse.*

3. $x = r \text{ vers}^{-1}(y/r) - \sqrt{2ry - y^2}$.

$dy/dx = (2r/y - 1)^{1/2}, \qquad d^2y/dx^2 = - r/y^2$.

Substituting in (2), reducing and combining with the given equation, we have

$$\bar{x} = r \text{ vers}^{-1}(-\bar{y}/r) + \sqrt{-2r\bar{y} - \bar{y}^2} \quad . \quad . \quad (a)$$

for the evolute of the cycloid.

Produce BA, making $AO' = BA = 2r$.

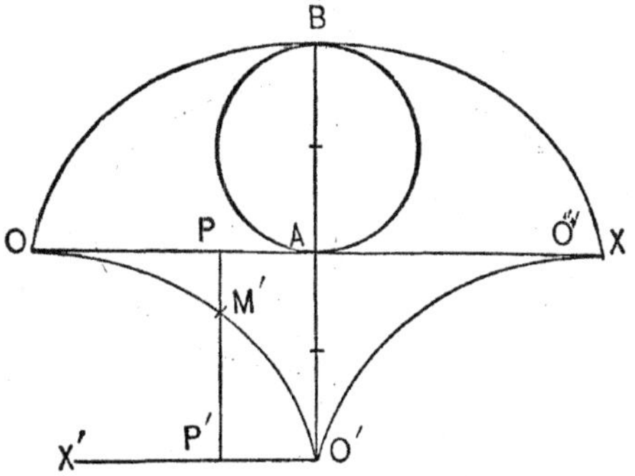

$OM'O'O''$ is the form of an evolute of a cycloid. The branch OO' belongs to OB and $O'O''$ to BO''.

Transferring the origin to O', taking $O'X'$ and $O'A$ as the new coördinate axes, denoting the new coördinates by x and y, we have for any point of the branch OO', as M',

$$\bar{x} = OA - O'P' = \pi r - x, \; \bar{y} = AO' + P'M' = -2r + y,$$

which substituted in (a) give

$$x = \pi r - r \text{ vers}^{-1}[(2r - y)/r] - \sqrt{2ry - y^2}.$$

But $\quad \pi r - r \text{ vers}^{-1}[(2r - y)/r] = r \text{ vers}^{-1}(y/r)$.

Hence, $x = r \text{ vers}^{-1}(y/r) - \sqrt{2ry - y^2}$

is the equation of $O'M'O$ referred to the new axes. It is of the same form and contains the same constants as the equation of the cycloid; hence, the evolute of a cycloid is an equal cycloid.

At $\quad O, \rho = 0,\quad$ and at $\quad B, \rho' = O'B = 4r.$

Hence, $\quad \text{arc } OM'O' = \text{arc } OB = \rho' - \rho = 4r.$

Therefore, arc $OBO'' = 8r$, that is, *the length of one branch of a cycloid is equal to four times that of the diameter of its generating circle.*

4. $x^2 + y^2 = R^2.$ $\qquad \bar{x} = \bar{y} = 0.$

5. $a^2y^2 - b^2x^2 = -a^2b^2.$ $\qquad (a\bar{x})^{2/3} - (b\bar{y})^{2/3} = (a^2 + b^2)^{2/3}.$

6. $x^2 = 4ay.$ $\qquad \bar{x}^2 = 4(\bar{y} - 2a)^3/27a.$

7. $x^{2/3} + y^{2/3} = a^{2/3}.$ $\qquad (\bar{x} + \bar{y})^{2/3} + (\bar{x} - \bar{y})^{2/3} = 2a^{2/3}.$

8. $2xy = a^2.$ $\qquad (\bar{x} + \bar{y})^{2/3} - (\bar{x} - \bar{y})^{2/3} = 2a^{2/3}.$

185. Equation of Evolute in Polar Coördinates.—Let C be the centre of curvature corresponding to M of a curve referred to P as a pole, and PX as a fixed right line. Draw PT and PT_1 perpendicular to MT and MT_1 respectively.

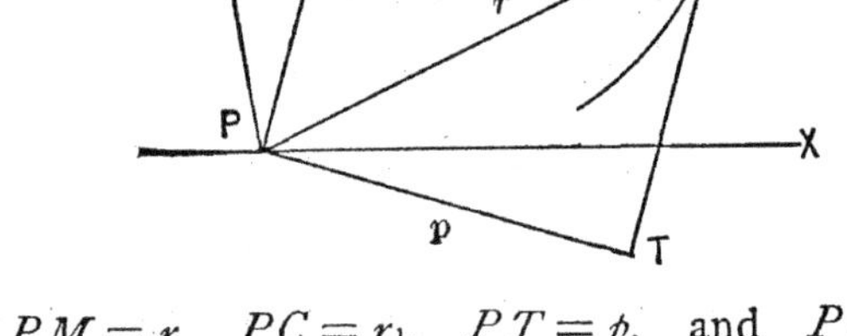

Let $\quad PM = r,\quad PC = r_1,\quad PT = p,\quad$ and $\quad PT_1 = p_1.$

Then $\qquad r_1^2 = \rho^2 + r^2 - 2r\rho \cos PMC.$

But $r \cos PMC = r \sin PMT = p.$

Hence, $r_1^2 = \rho^2 + r^2 - 2\rho p.$ . . . . . . (1)

Also, $p_1^2 = \overline{TM}^2 = r^2 - p^2;$ . . . . . . (2)

and (1), § 176, $\rho = r\, dr/dp.$ . . . . (3)

From the equation of the curve find p in terms of r, giving

$$p = F(r). \quad \ldots \ldots \quad (4)$$

By eliminating r, ρ, and p, there results an equation between r_1 and p_1 for the evolute. Thus, having $r = a^\theta$, then $p = cr$, in which $c = 1/\sqrt{1 + \log^2 a}$. (Ex. 3, § 150.)

$$dp = c dr, \quad dr/dp = 1/c.$$

Hence, $\rho = r/c, \quad p_1^2 = r^2 - c^2r^2 = r^2(1 - c^2),$

and $r_1^2 = r^2/c^2 + r^2 - 2r^2 = r^2(1 - c^2)/c^2 = p_1^2/c^2,$

or $p_1^2 = c^2r_1^2$ and $p_1 = cr_1,$

for the evolute, which is a logarithmic spiral similar to the given curve.

186. Having ρ in terms of ψ, equation (1), § 183, enables us to express ρ_1 in terms of ψ_1. Thus,

$$r = 4a \sin^2 (\theta/2).$$

Example 12, § 173. When $\psi = 3\theta/2$,

$$\rho = 8a \sin (\theta/2)/3.$$

Hence, $\rho = 8a \sin (\psi/3)/3.$

(1), § 183,

$$\rho_1 = d\rho/d\psi = 8a \cos (\psi/3)/9 = 8a \sin (\pi/2 + \psi/3)/9.$$

Let ψ_1 represent the angle which ρ makes with X.

Then $\psi = \psi_1 - \pi/2$, and $\pi/2 + \psi/3 = (\psi_1 + \pi)/3$.

Hence, $$\rho_1 = 8a \sin [(\psi_1 + \pi)/3]/9.$$

187. Equation of an Involute.—Combine the equations

$$\bar{y} = F(\bar{x}), \quad . \quad . \quad . \quad . \quad . \quad . \quad . \quad (1)$$

$$d\bar{y}/d\bar{x} = - dx/dy, \quad . \quad . \quad . \quad . \quad . \quad (2)$$

$$(x - \bar{x}) + (y - \bar{y})dy/dx = 0, \quad . \quad . \quad . \quad . \quad (3)$$

eliminating $\bar{x}$ and $\bar{y}$, and there will result in general a differential equation involving x and y.

188. Involute of a Circle.—Let a = radius of circle, and let A be the initial position of generating point. Let TB = circum. AT be any position of the tangent rolling upon circum. AT. Take origin and pole at O. Let

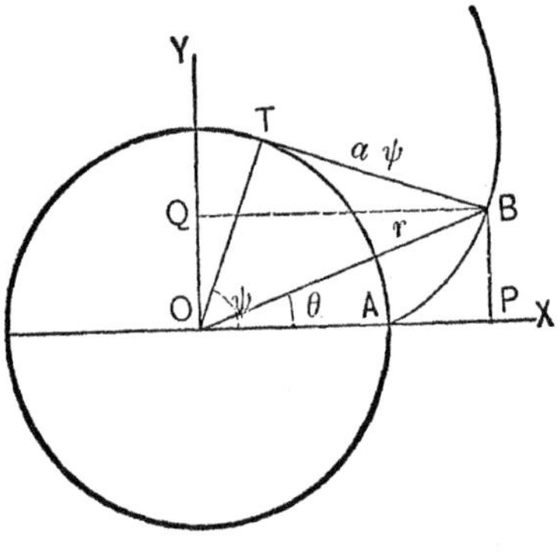

ψ = angle AOT, and let θ = angle AOB, which radius vector r makes with X. Then tangent $TB = a\psi$, and we have, for the rectangular coördinates of any point, as B, by projecting the two lines OT and TB upon X and Y respectively,

$$\left.\begin{aligned} x = OP = a \cos \psi + a\psi \sin \psi, \\ y = OQ = a \sin \psi - a\psi \cos \psi. \end{aligned}\right\} \quad . \quad . \quad (1)$$

In polar coördinates we have, from the right triangle OTB,

$$\sqrt{r^2 - a^2} = a\psi.$$

But $\theta = \psi -$ angle $BOT = \psi - \sec^{-1}(r/a)$.

Hence, $a\psi = a\theta + a\sec^{-1}(r/a)$,

and $\sqrt{r^2 - a^2} = a\theta + a\sec^{-1}(r/a)^{-1}$. . . . (2)

CHAPTER XVII.

ORDERS OF CONTACT OF CURVES AND OSCULATING LINES.

189. Let $y = f(x)$ and $y = \phi(x)$ be the equations of any two given lines, as BB and DD, which have in common a

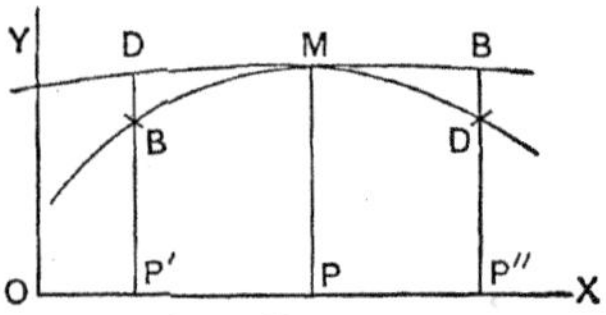

Fig. 1

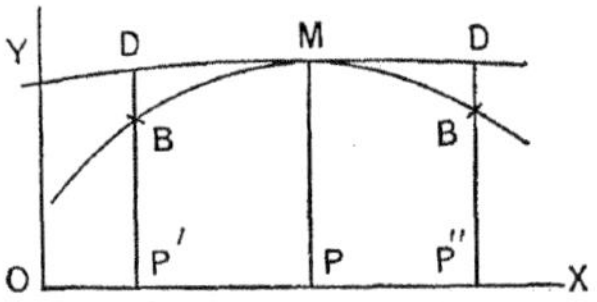

Fig. 2

point M, whose abscissa is $OP = a$, and whose ordinate is

$$PM = b = f(a) = \phi(a).$$

Increase a by an infinitesimal $h = PP''$, and we have for the corresponding ordinates, $P''B$ and $P''D$, designated by y' and y'' respectively (§ 124),

$$y' = f(a+h) = f(a) + f'(a)h + f''(a)h^2/\underline{|2} + \ldots \quad (1)$$

$$y'' = \phi(a+h) = \phi(a) + \phi'(a)h + \phi''(a)h^2/\underline{|2} + \ldots \quad (2)$$

Subtracting (2) from (1), member from member, we have

$$y' - y'' = DB = [f'(a) - \phi'(a)]h + [f''(a) - \phi''(a)]h^2/\underline{|2} + \ldots \quad (3)$$

When $f'(a) = \phi'(a)$,

$$y' - y'' = DB = [f''(a) - \phi''(a)]h^2/\underline{|2} + [f'''(a) - \phi'''(a)]h^3/\underline{|3} + \ldots \quad (4)$$

the lines are tangent to each other, and are said to have a contact of at least the *first order*.

When, *also*, $f''(a) = \phi''(a)$,

$$y' - y'' = DB = [f'''(a) - \phi'''(a)]h^3/\underline{|3} + \dots \quad (5)$$

the lines have a contact of at least *the second order*.

When, *also*, $f'''(a) = \phi'''(a)$,

$$y' - y'' = DB = [f^{iv}(a) - \phi^{iv}(a)]h^4/\underline{|4} + \dots \quad (6)$$

the lines have a contact of at least *the third order*; and, in general, the order of contact of any two lines having a point in common is denoted by the *greatest number*, beginning with the first, of successive derivatives of their ordinates corresponding to the common point, which are, respectively, equal each to each.

Denoting the order of contact of any two lines by n, we have also $f^n(a) = \phi^n(a)$, making with $f(a) = \phi(a)$, $n + 1$ conditions, and giving

$$y' - y'' = DB = [f^{n+1}(a) - \phi^{n+1}(a)]h^{n+1}/\underline{|n+1} + \dots \quad (7)$$

From which we see that when $(n + 1)$ is odd, the sign of $y' - y'' = DB$ will change when the sign of h changes, that is, if D is above B when h is negative and vanishing, it will be below B immediately after h becomes positive. The lines will then intersect, as shown in Fig. 1. When $(n + 1)$ is even, the sign of $y' - y'' = DB$ will not change with that of h, and the lines will not intersect. (See Fig. 2.)

Hence, *when the order of contact is even, lines intersect at the common point, and when it is odd, they do not.*

To illustrate, take the two equations

$$y^2 = 4x, \quad . \quad . \quad . \quad (1) \quad \text{and} \quad (x - 5)^2 + (y + 2)^2 = 32. \quad (2)$$

Combining, we find that the point (1, 2) is common.

(1) gives $f'(1) = 1$, $f''(1) = -1/2$, and $f'''(1) = 3/4$.

(2) gives $\phi'(1) = 1$, $\phi''(1) = -1/2$, and $\phi'''(1) = 3/8$.

Hence, the circle (2) has a contact of the second order with the parabola (1), and intersects it at the point (1, 2).

Determine the common points, and order of contact at each, in the following pairs of lines:

1. $\begin{cases} y^2 = 4x. \\ y = x + 1. \end{cases}$ Ans. (1, 2) in common. Contact of 1st order.

2. $\begin{cases} y = x^3. \\ y = 3x^2 - 3x + 1. \end{cases}$ Ans. (1, 1) 2d order.

3. $\begin{cases} 4y = x^2 - 4. \\ y^2 + x^2 = 2y + 3. \end{cases}$ Ans. (0, − 1) 3d order.

Two lines having at a common point a contact of the nth order with a third line, have a contact with each other of at least the nth order.

190. Osculating Lines.—The line of any species of line, which at a given point of a given line has the highest possible order of contact, is called *an osculatrix* or *osculating line.* Thus, the circle which at the given point has the highest possible order of contact is called *an osculating circle.* The parabola of highest contact is called *an osculating parabola.*

To determine the equation of an osculatrix at a given point of a given line, assume the general equation of the species of line in its reduced form.

The problem then is to determine such values for the arbitrary constants contained therein as will cause the required line to have the highest possible order of contact.

Since the osculatrix must pass through the given point, substitute its coördinates in the general equation, giving one equation between the required quantities, and diminishing the number of arbitrary constants by unity.

From the general equation of the species and the equation of the given line determine expressions for the successive derivatives of the ordinates to include those whose order is denoted by the number, less unity, of the constants in the reduced form of the general equation of the species.

Substitute the coördinates of the given point in each, and place the results corresponding to derivatives of the same order equal to each other. The resulting equations with the one before obtained will equal in number the required quantities, which in general may be determined. Their values substituted in the general equation will give the required *osculatrix*. The order of contact will in general be denoted by the number, less unity, of arbitrary constants entering the general equation, but in exceptional cases it may be higher.

EXAMPLES.

1. Find the equation of the osculating right line to the parabola $y^2 = 9x$ at the point (1, 3).

In $y = ax + b$ substitute the coördinates (1, 3), giving

$$3 = a + b. \quad . \quad . \quad . \quad . \quad . \quad . \quad . \quad (1)$$

From $y = ax + b$ we find $\phi'(x) = a$.

From $y^2 = 9x$ we find $f'(x) = 9/2y$.

Substituting the coördinates (1, 3) in each, and placing the results equal to each other, we have $a = 9/6 = 3/2$, which in (1) gives $b = 3/2$. Hence, $y = 3x/2 + 3/2$ is the equation of the required line, which is tangent to the parabola.

Since the general equation of a right line contains but two arbitrary constants, it cannot, in general, have a contact of an order higher than the first with a plane curve.

An exception exists at a point of inflexion where, in general, for both the curve and the right line, we have, denoting the abscissa of the point by a, $f''(a) = 0$. The contact is, therefore, at least of the second order.

At a point of inflexion the direction of curvature changes, and $y' - y'' = DB$ (Fig. 1, § 189) changes sign with h. Equation (7), § 189, shows that this occurs only when $n + 1$ is odd. Hence, the order of contact is even and the tangent intersects the curve.

2. Find the equation of the osculating circle to the parabola $y^2 = 4x$ at the point (1, 2).

$(x - a)^2 + (y - b)^2 = R^2$ is the general equation of the circle. Substituting the coördinates (1, 2), we have

$$(1 - a)^2 + (2 - b)^2 = R^2. \quad . \quad . \quad . \quad . \quad (1)$$

Differentiating the general equation of the circle, we find

$$\phi'(x) = - (x - a)/y - b, \quad \text{and} \quad \phi''(x) = - R^2/(y - b)^3.$$

From $y^2 = 4x$ we obtain

$$f'(x) = 2/y, \quad \text{and} \quad f''(x) = - 4/y^3.$$

Substituting the coördinates (1, 2) in each, and placing the results corresponding to derivatives of the same order equal to each other, we have

$$- (1 - a)/(2 - b) = 1 \quad \text{and} \quad - R^2/(2 - b)^3 = -4/2^3,$$

which with (1) give

$$a = 5, \quad b = - 2, \quad \text{and} \quad R^2 = 32.$$

Hence, $(x - 5)^2 + (y + 2)^2 = 32$ is the required equation.

191. Osculating Circle at any point (x', y') of any plane curve whose equation is $y = f(x)$.

Substituting (x', y') in $(x - a)^2 + (y - b)^2 = R^2$, we have

$$(x' - a)^2 + (y' - b)^2 = R^2. \quad . \quad . \quad . \quad . \quad (a)$$

From $(x - a)^2 + (y - b)^2 = R^2$ we obtain

$$\phi'(x) = -\frac{x-a}{y-b}, \quad \text{and} \quad \phi''(x) = -\left[1 + \left(\frac{x-a}{y-b}\right)^2\right] \Big/ (y-b);$$

and from $y = f(x)$ we derive expressions for $f'(x)$ and $f''(x)$.

Substituting (x', y') in each, and placing the results corresponding to the derivatives of the same order equal to each other, we have

$$-(x' - a)/(y' - b) = f'(x'),$$

$$\text{and} \quad -[1 + \overline{f'(x')}^2]/(y' - b) = f''(x'),$$

which with (a) give, omitting the primes,

$$R = [1 + \overline{f'(x)}^2]^{3/2}/f''(x), \quad . \quad . \quad . \quad . \quad . \quad (1)$$

$$a = x - [1 + \overline{f'(x)}^2]f'(x)/f''(x), \quad . \quad . \quad . \quad (2)$$

$$b = y + [1 + \overline{f'(x)}^2]/f''(x). \quad . \quad . \quad . \quad . \quad . \quad (3)$$

Comparing these with (1) and (2), § 172, we see that *the osculatory circle at any point of a plane curve is the circle of curvature.*

EXAMPLES.

1. Find the equation and radius of the osculating circle to the curve $4(y + 1) = x^2$, at $(0, -1)$.

Ans. $y^2 + x^2 = 2y + 3$; radius $= 2$.

2. Find the radius of the osculating circle to the parabola $y^2 = 9x$, at $(3, \sqrt{27})$. Ans. 16.04.

3. Find the equation and radius of the osculating circle to the parabola $y^2 = 16x$ at (1, 4).

Ans. $(x - 11)^2 + (y + 1)^2 = 125$; radius $= 5\sqrt{5}$.

192. In general, an osculating circle has a contact of the second order with any plane curve, but § 172 shows that at a point where the curvature of a curve is a maximum or a minimum, the circle of curvature, and therefore the osculating circle, does not intersect the curve. The order of contact is therefore odd, and of a degree higher than the second. That the contact in such cases is at least of the third order may be shown as follows:

From (1), § 172, and (1), § 191, we have

$$R = \rho = [1 + \overline{f'(x)}^2]^{3/2}/f''(x);$$

and in order that ρ may be a maximum or a minimum,

$$\frac{d\rho}{dx} = 0 \text{ gives} \qquad 3\overline{f''(x)}^2 f'(x) - f'''(x)[1 + \overline{f'(x)}^2] = 0,$$

whence $\quad f'''(x) = 3\overline{f''(x)}^2 f'(x)/[1 + \overline{f'(x)}^2].$

From § 114 we have for a circle

$$\frac{d^3y}{dx^3} = 3\left(\frac{d^2y}{dx^2}\right)^2 \frac{dy}{dx} \Big/ \left[1 + \left(\frac{dy}{dx}\right)^2\right].$$

Hence, *when the radius of curvature is a maximum or a minimum, the circle of curvature has a contact with the curve of at least the third order.*

It follows that the order of contact of an osculating circle at a vertex of a conic is odd, higher than the second, and the circle does not intersect the conic.

CHAPTER XVIII.

ENVELOPES.

193. In

$$u = f(x, y, a) = 0 \quad . \quad . \quad . \quad . \quad . \quad . \quad (1)$$

let a be an arbitrary constant. By giving all possible values to a, (1) will represent a series of lines, all of the same kind or family, unlimited in number, and, in general, intersecting each other in order. a is then called a **variable parameter.**

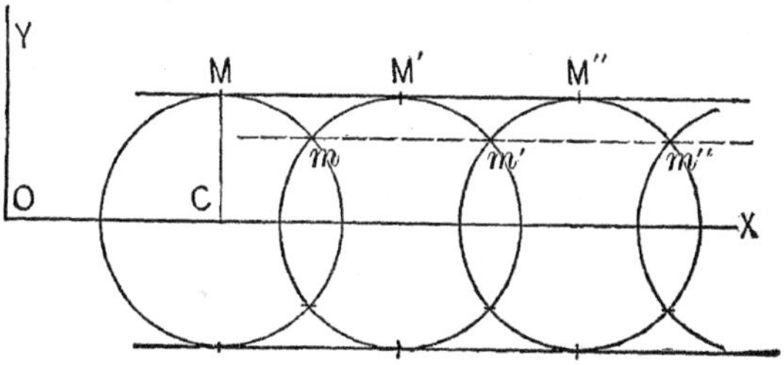

To illustrate, let

$$u = (x - a)^2 + y^2 - 9 = 0.$$

By giving different values to a, the equation may represent a series of circles having the same radius, their centres on X, and intersecting each other in order in points as m, m', etc.

In general, any value of a in (1) corresponds to a determinate particular line, and $a + h$ to another line of the same kind having for its equation

$$u' = f(x, y, a + h) = 0. \quad . \quad . \quad . \quad . \quad (2)$$

This second line, which may be regarded as a second state of the first, will, in general, as h vanishes, ultimately intersect the first, in points m, m', etc., the coördinates of which will satisfy both (1) and (2). If (1) and (2) be combined and a eliminated, the resulting equation will be that of the *locus* of the points m, m', etc., of intersection of all of the series of lines represented by (1), each with its second state. If in this resulting equation h be made equal to zero, we will obtain the equation of the *limit of the above locus*, and this *limit* is called *an envelope* of the series of lines. In the case of the circles, the right line $MM'M''$ is the limit of the locus $mm'm''$, and is an envelope of the circles.

In combining (1) and (2) so as to eliminate a, complicated expressions frequently arise which may sometimes be avoided by the following method of Calculus.

Since the coördinates of the points m, m', etc., common to each of the lines of the series represented by (1), and its second state in order, satisfy both (1) and (2), they will satisfy the equation

$$(u' - u)/h = [f(x, y, a + h) - f(x, y, a)]/h = 0. \quad (3)$$

As h vanishes, the points m, m', etc., approach limiting positions M, M', etc., and the coördinates of the points M, M', etc., will satisfy both (1) and the equation

$$\partial u/\partial a = \partial f(x, y, a)/\partial a = 0, \quad . \quad . \quad . \quad (4)$$

which (3) approaches as h vanishes.

If, therefore, (1) and (4) be combined so as to eliminate a, the resulting equation will be that of the *locus of the limiting positions* of the points of intersection of the series of lines represented by (1), each with its second state. This locus is the same as the *limit of the locus* of the points of intersection, etc., before obtained, and is, therefore, an envelope of the series.

Hence, an envelope of any series of lines determined by giving all possible values to a variable parameter in an equation involving two variables only may be defined as *the limit of the locus*, or the *locus of the limiting positions* of points of intersection of the series of lines, each with its second state, under the law that the difference in position between each second state and its primitive vanishes.

To obtain the equation of an envelope of a series of lines given by an equation with a variable parameter, we have the following rule:

Combine the given equation with its differential equation taken with respect to the variable parameter, and eliminate the parameter.

From (1), regarding a as constant, we obtain

$$du/dx = \partial u/\partial x + (\partial u/\partial y)(dy/dx) = 0$$

for the differential equation of *each of the lines of the series.*

From (1) and (4), $a = \phi(x)$, which substituted in (1) gives the equation of an envelope. Hence, differentiating (1) regarding $a = \phi(x)$, we have

$$\frac{du}{dx} = \frac{\partial u}{\partial x} + \frac{\partial u}{\partial y}\frac{dy}{dx} + \frac{\partial u}{\partial a}\frac{da}{dx} = 0;$$

or, since $\partial u/\partial a = 0$,

$$du/dx = \partial u/\partial x + (\partial u/\partial y)(dy/dx) = 0,$$

for the differential equation of *an envelope.* Each line of the series, and an envelope, have, therefore, the same differential equation, and dy/dx at any point common to any line of the series, and an envelope will be the same for both. *An envelope is therefore tangent to all of the lines of the series.*

EXAMPLES.

1. Find the equation of an envelope of the series of circles given by the equation

$$u = (x - a)^2 + y^2 - 9 = 0, \quad . \quad . \quad . \quad . \quad (1)$$

when a is a variable parameter.

$$\partial u/\partial a = -2(x - a) = 0. \quad . \quad . \quad . \quad . \quad (2)$$

Combining (1) and (2) so as to eliminate a, we have for the required envelope

$$y = \pm 3, \qquad x = 0/0.$$

Hence, the right lines $MM'M''$ and $MM_1M_{,,}$ (see figure, § 193) are the envelopes.

2. Find the envelope of the curves given by

$$y = x \tan \theta - x^2/(4h \cos^2 \theta), \text{ as } \theta \text{ varies.}$$

$$\partial u/\partial \theta = x/\cos^2 \theta - x^2 \sin \theta/2h \cos^3 \theta = 0.$$

Hence, $\tan \theta = 2h/x$, $1 + \tan^2 \theta = (x^2 + 4h^2)/x^2 = \sec^2 \theta$,

and $\cos^2 \theta = x^2/(x^2 + 4h^2)$, $4h \cos^2 \theta = 4hx^2/(x^2 + 4h^2)$. Substituting in given equation, we find $y = h - x^2/(4h)$ for the required envelope.

3. Find the envelope of a right line of a given length c which moves with its ends on the coördinate axes.

Let θ equal the angle which the right line makes with X. Then intercepts are, respectively, $c \cos \theta$ and $c \sin \theta$, giving

$$u = x \sec \theta + y \operatorname{cosec} \theta - c = 0 \quad . \quad . \quad . \quad (1)$$

for the line, in which θ is the variable parameter.

$$\partial u/\partial \theta = x \sec \theta \tan \theta - y \operatorname{cosec} \theta \cot \theta = 0. \quad . \quad (2)$$

Combining (1) and (2), we find

$$x = c \cos^3 \theta, \qquad y = c \sin^3 \theta;$$

and eliminating θ, we have for the required envelope

$$x^{2/3} + y^{2/3} = c^{2/3}.$$

Otherwise, let a and b represent the intercepts, respectively, giving

$$x/a + y/b = 1 \quad . \quad . \quad . \quad (3) \quad \text{and} \quad a^2 + b^2 = c^2. \quad . \quad . \quad (4)$$

Regarding b as the variable parameter, we may by means

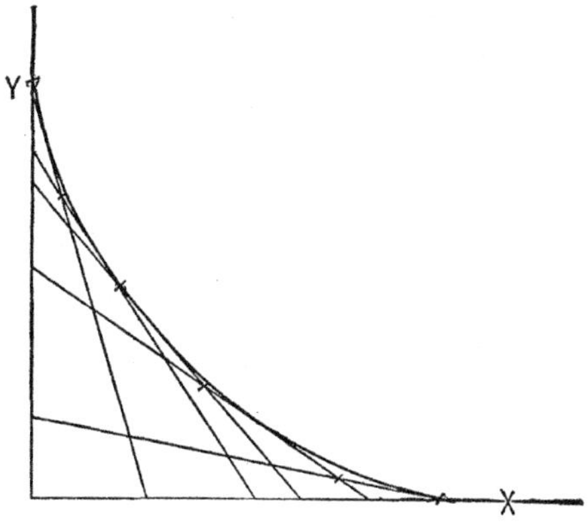

of (4) eliminate a from (3), and proceed as before ; or, since a is a function of b, we have

$$x\partial a/a^2 + y\partial b/b^2 = 0. \quad . \quad . \quad . \quad . \quad (5)$$

$$a\partial a + b\partial b = 0. \quad . \quad . \quad . \quad . \quad . \quad . \quad (6)$$

Combining (3), (4), (5), (6), eliminating $\partial a/\partial b$, a and b, we have

$$(a^2 + b^2)y = b^3; \qquad \therefore\ b = (c^2y)^{1/3}, \quad \text{and} \quad a = (c^2x)^{1/3}.$$

Substituting these expressions in (3), we have for the envelope

$$x^{2/3} + y^{2/3} = c^{2/3}.$$

In a similar manner, when there are n parameters and $n - 1$ equations of condition between them, we may differentiate the n given equations, regarding $n-1$ of the parameters as functions of the variable parameter. Then, by combining the n differential equations with the given equation of the series, the parameters may be eliminated and the envelope determined.

4. Find the envelope of a series of concentric and coaxal ellipses having the same area.

The given equations are

$$a^2y^2 + b^2x^2 = a^2b^2, \quad \text{and} \quad ab = c^2.$$

$$xy = c^2/2 \text{ is the envelope.}$$

5. Find the envelope of a right line moving so that its perpendicular distance from the origin remains constant.

$$u = f(x, y, \alpha) = x \cos \alpha + y \sin \alpha - p = 0,$$

$$\partial u/\partial a = -x \sin \alpha + y \cos \alpha = 0,$$

and $\quad x^2 + y^2 = p^2$ is the envelope.

6. Find the envelope of the hypothenuse of a right triangle moving so that the area of the triangle remains constant.

Let a = constant area, b = base = variable parameter, and c = altitude = $2a/b$. Then $2a/b^2$ = tan of angle hypothenuse makes with base, and taking the coördinate axes to coincide with the legs, we have

$$y = 2ax/b^2 + 2a/b \text{ for the hypothenuse,}$$

and $\quad xy = a/2$ for the required envelope.

7. Find the envelope of a right line moving so that the sum of its intercepts on the coördinate axes is constant.

Let a and b represent the intercepts upon X and Y respectively, and let $c = a + b$; then the equation of the line is

$$x/a + y/(c - a) = 1,$$

and a is the variable parameter.

$$\text{Ans. } x^2 + y^2 - 2xy - 2cx - 2cy + c^2 = 0$$

$$\text{or } \sqrt{x} + \sqrt{y} = \sqrt{c}.$$

8. Find the envelope of a right line moving so that the product of its distances from two fixed points is constant.

Take X to coincide with right line joining the two points, and the origin to be at its middle point. Let $(a, 0)$ and $(-a, 0)$ be the two fixed points, and

$$x \cos \alpha + y \sin \alpha = p \quad . \quad . \quad . \quad . \quad . \quad (1)$$

the equation of the line.

The distances from the fixed points to the line are, respectively,

$$a \cos \alpha - p \quad \text{and} \quad -a \cos \alpha - p.$$

Hence, $\quad p^2 - a^2 \cos^2 \alpha = c^2 = \text{constant}. \quad . \quad . \quad . \quad (2)$

From (1) and (2), regarding α as the variable parameter, we find for the envelope

$$x^2/(c^2 + a^2) + y^2/c^2 = 1.$$

9. Find the envelope of the right lines whose equation is

$$y - y' = 2b(x - x')/(1 - b^2), \quad . \quad . \quad . \quad . \quad . \quad (1)$$

when y' is the variable parameter, and we have

$$y'^2 = 2px' \quad \text{and} \quad b = -y'/p.$$

Eliminating b and x', (1) becomes

$$yy'^2 - p^2y + p^2y' - 2py'x = 0. \quad . \quad . \quad . \quad (2)$$

Hence, $\quad \partial u/\partial y' = 2yy' + p^2 - 2px = 0,$

and $\quad y' = (2px - p^2)/2y. \quad . \quad . \quad . \quad . \quad . \quad . \quad (3)$

Combining (3) and (2), eliminating y', we have

$$x = \pm \sqrt{-y^2} + p/2$$

for the envelope, which is a point at the focus of the parabola, and is the *Caustic* of rays of light reflected from the concave side of a parabola, the incident rays being parallel to the axis which coincides with X.

10. Find the envelope of the right lines whose equation is

$$y - y' = 2b(x - x')/(1 - b^2), \quad . \quad . \quad . \quad . \quad (1)$$

when x' is the variable parameter, and we have

$$y'^2 + x'^2 = a^2 \quad \text{and} \quad b = y'/x'.$$

Eliminating b, (1) becomes

$$x - x' = (y'/x' - x'/y')(y' - y)/2. \quad . \quad . \quad (2)$$

Differentiating and reducing, we obtain

$$\partial u/\partial x' = y' - a^{2/3}y^{1/3} = 0, \quad \text{or} \quad y' = a^{2/3}y^{1/3}. \quad . \quad (3)$$

Substituting, in (2), $y'x'^2/a^2$ for $y' - y$, we have, after combination with (3) and reduction,

$$x' = 2a^{2/3}x/(a^{2/3} + 2y^{2/3}).$$

Squaring the expressions for y' and x', and substituting in $y'^2 + x'^2 = a^2$, we have for the envelope

$$a^{4/3}y^{2/3} + 4a^{4/3}x^2/(a^{2/3} + 2y^{2/3})^2 = a^2,$$

which is the equation of the *Caustic* of rays of light reflected from a circle, the incident rays being parallel to X.

11. Find the envelope of the polar line to the ellipse $9y^2 + 4x^2 = 36$ as the pole moves along the right line $y = 2x + 1$.

Let x'' and y'' be the coördinates of the pole, giving $y'' = 2x'' + 1$.

Then $9yy'' + 4xx'' = 36$ is the equation of the polar line Substituting $2x'' + 1$ for y'', we have

$$u = 18yx'' + 9y + 4xx'' - 36 = 0,$$

in which x'' is the variable parameter.

$$\partial u/\partial x'' = 18y + 4x = 0.$$

Combining the last two equations, we have $0x'' + 9y = 36$, or $y = 4$, $x = -18$ for the required envelope.

Lines.	Variable Parameters.	Envelopes.
12. $y = ax + b/a.$	$a.$	$y^2 = 4bx.$
13. $ax - y = x^2(1 + a^2)/2p.$	$a.$	$x^2 + 2py = p^2.$
14. $x^2/a + y^2/(a - h) = 1.$	$a.$	$(x \pm \sqrt{h})^2 + y^2 = 0.$
15. $x^2 + y^2 = r^2.$	$r.$	
16. $y^2 = ax - a^2.$	$a.$	$y = \pm x/2.$
17. $x \cos 3\phi + y \sin 3\phi = a(\cos 2\phi)^{3/2}.$	$\phi.$	$(x^2 + y^2)^2 = a^2(x^2 - y^2).$
18. $\begin{cases} ny + mx = a^2. \\ n^2 + (m - d)^2 = r^2. \end{cases}$	$\begin{cases} m. \\ n = f(m). \end{cases}$	$r^2y^2 + (r^2 - d^2)x^2. + 2da^2x = a^4.$
19. $cy^2 = a^2x - a^3.$	$a.$	$y^2 = 4x^3/27c.$
20. $\begin{cases} (x - a)^2 + (y - b)^2 = r^2. \\ a^2 + b^2 = c^2. \end{cases}$	$\begin{cases} a. \\ b = f(a). \end{cases}$	$x^2 + y^2 = (c \pm r)^2.$
21. $y = 2ax + a^4.$	$a.$	$16y^3 + 27x^4 = 0.$
22. $y^2 + (x - a)^2 - 2pa = 0.$	$a.$	$y^2 = p(p + 2x).$
23. $y^2 = 2px.$	$p.$	$x = 0.$
24. $\begin{cases} (x - a)^2 + y^2 = r^2. \\ r = ca. \end{cases}$	$a.$	$y = \pm cx/(1 - c^2).$
25. $y^2 = ax - a^2.$	$a.$	$4y^2 = x^2.$
26. $y = ax + a^4.$	$m.$	$256y^3 + 27x^4 = 0.$
27. $\dfrac{a^2 \cos\theta}{x} - \dfrac{b^2 \sin\theta}{y} = \dfrac{c^2}{a}$	$\theta.$	$\dfrac{a^4}{x^2} + \dfrac{b^4}{y^2} = \dfrac{c^4}{a^2}.$
28. $y = mx + \sqrt{a^2m^2 + b^2}.$	$m.$	$a^2y^2 + b^2x^2 = a^2b^2.$
29. $\begin{cases} (x/a)^2 + (y/b)^2 = 1. \\ a + b = c. \end{cases}$	$a.$	$x^{2/3} + y^{2/3} = c^{2/3}.$

194. *The envelope of the normals of any given curve is its evolute.*

This follows from the definitions of envelopes and evolutes; otherwise, the equation of a normal to a curve $y = f(x)$, at (x', y'), is (§ 148)

$$x - x' + (y - y')f'(x') = 0, \quad . \quad . \quad . \quad (1)$$

in which $y' = f(x')$, and x' may be taken as the variable parameter. Differentiating with respect to x', we have

$$-1 - \overline{f'(x')}^2 + (y - y')f''(x') = 0. \quad . \quad . \quad (2)$$

Combining (1) and (2), we find

$$\left.\begin{aligned} x &= x' - [1 + \overline{f'(x')}^2]f'(x'/f''(x'), \\ y &= y' + [1 + \overline{f'(x')}^2]/f''(x'), \end{aligned}\right\} \quad . \quad . \quad (3)$$

for the limiting position of the intersection of the normal at (x', y') and its second state; which is therefore the point of tangency of the envelope to the normal at (x', y'). Comparing (3) with (1), § 172, we see that this point is the corresponding centre of curvature of the given curve. Hence, the envelope of the normals is the evolute of the curve.

Combining (3) and $y' = f(x')$, x' may be eliminated and the equation of the envelope obtained.

EXAMPLES.

1. Find the envelope of the normals to the parabola $y^2 = 4ax$.

Here $$y' = f(x') = 2a^{1/2}x'^{1/2},$$

$$f'(x') = a^{1/2}/x'^{1/2}, \qquad f''(x') = -a^{1/2}/(2x'^{3/2}).$$

Substituting in (3) and eliminating y', we have

$$x = x' - \frac{(1 + a/x')(a^{1/2}/x'^{1/2})}{-a^{1/2}/(2x')^{3/2}} = 3x' + 2a,$$

$$y = y' + \frac{1 + a/x'}{-a^{1/2}/2x'^{3/2}} = -2x'^{3/2}/a^{1/2},$$

from which, eliminating x', we obtain for the required envelope $ay^2 = 4(x - 2a)^3/27$. (See Example 1, § 184.)

2. Find the envelope of the normals to an ellipse by the above method.

Otherwise, the equation of the normal to an ellipse at a point whose eccentric angle is denoted by θ is

$$u = ax \sec \theta - by \operatorname{cosec} \theta - a^2 + b^2 = 0.$$

Regarding θ as the variable parameter,

$$\partial u/\partial \theta = ax \sec \theta \tan \theta + by \operatorname{cosec} \theta \cot \theta = 0.$$

Combining and eliminating θ, we have for the required envelope $$(ax)^{2/3} + (by)^{2/3} = (a^2 - b^2)^{2/3}.$$

CHAPTER XIX.

CURVE TRACING.

195. The foregoing principles, with those from Analytic Geometry, enable us, in general, to trace curves from their equations with great accuracy.

RECTANGULAR COÖRDINATES.

No fixed rule or directions apply in all cases, but, in general, it is desirable to determine—

From the equation of curve.

1°. Symmetry with respect to the coördinate axes.

2°. Limiting coördinates and asymptotes parallel to them.

3°. Points on the coördinate axes.

4°. Terminating points.

In general, requiring the first derivative of the ordinate.

5°. Direction of curve at points on the coördinate axes.

6°. Asymptotes oblique to coördinate axes.

7°. Multiple points.

In general, requiring the second derivative of the ordinate.

8°. Character of cusps.

9°. Maximum and minimum ordinates.

10°. Direction of curvature and points of inflexion.

EXAMPLES.

1. $y = a^2x/(a^2 + x^2)$.

Each value of x from $-\infty$ to $+\infty$ gives a real value for y. The curve is therefore unlimited in both directions with X, and is limited in both directions of Y.

As $x \Rightarrow \pm \infty$, $y \Rightarrow \pm 0$. Hence, X is an asymptote in both directions.

$x = 0$ gives $y = 0$, and $y = 0$ gives $x = 0$, or $\pm \infty$.

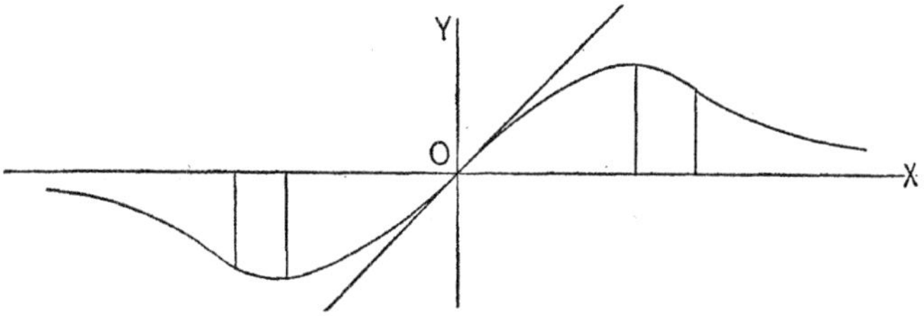

Hence, (0, 0) is the only point at which the curve cuts the coördinate axes.

$f'(x) = a^2(a^2 - x^2)/(a^2 + x^2)^2$.

$f'(0) = 1$. Hence, the direction of the curve at the origin makes an angle of 45° with X.

$f'(x) = 0$ gives $x = \pm a$ and $\pm \infty$.

$f''(x) = 2a^2x(x^2 - 3a^2)/(a^2 + x^2)^3$.

$f''(a)$ is negative; $\therefore y = a/2$ is a maximum.

$f''(-a)$ is positive; $\therefore y = -a/2$ is a minimum.

$f''(x) = 0$ gives $x = 0$ and $\pm a\sqrt{3}$.

$f''(x)$ is negative for values of x from $-\infty$ to $-a\sqrt{3}$, and the curve is concave downward.

For values of x from $-a\sqrt{3}$ to 0 $f''(x)$ is positive, and the concave side is above.

As x varies from 0 to $a\sqrt{3}$, $f''(x)$ is again negative, and the concavity is downward.

Values of x from $a\sqrt{3}$ to $+\infty$ make $f''(x)$ positive, and the curve is concave upward.

It follows that

$$(-a\sqrt{3},\ -a\sqrt{3}/4),\ (0, 0),\quad \text{and}\quad (a\sqrt{3},\ a\sqrt{3}/4)$$

are points of inflexion.

2. $x^3 - 2x^2y - 2x^2 = 8y$; $\therefore\ y = x^2(x-2)/2(x^2+4)$.

$$x = -\infty = y. \qquad x = 0 = y. \qquad x = \infty = y.$$

The curve is unlimited in both directions along X and Y.

$y = 0$ gives $x = 0$ and 2. Hence, the curve cuts X at the points $(0, 0)$ and $(2, 0)$.

$f'(x) = x(x^3 + 12x - 16)/2(x^2+4)^2$.

$f'(0) = 0$. Hence, at the origin X is a tangent.

$f'(2) = 1/4$. Therefore, at the point $(2, 0)$ the curve makes $\tan^{-1}(1/4)$ with X.

$f'(x) = 0$ gives $x = 0$ and 1.19 nearly.

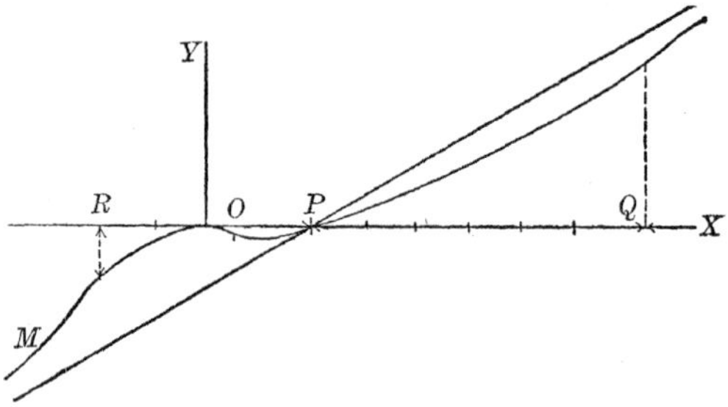

Expanding the expression for y, we have

$y = x/2 - 1 + (4 - 2x)/(x^2+4)$, in which, as $x \to \infty$,

$$y \to (x/2 - 1).$$

Hence, $y = x/2 - 1$ is an asymptote.

$f''(x) = -4(x^3 - 6x^2 - 12x + 8)/(x^2+4)^3$.

$f''(0)$ is negative; hence, $y = 0$ is a maximum.

$f''(1.19) = 0$ is positive; hence, $y = -0.11$ is a minimum.

$f''(x) = 0$ gives $x = -2$, $4 - 2\sqrt{3} = 0.54$, and $4 + 2\sqrt{3} = 7.5$.

Values of x from $-\infty$ to -2 make $f''(x)$ positive, and the corresponding part of the curve is concave upward. As x varies from -2 to 0.54, $f''(x)$ is negative, and the concavity is downward. When $0.54 < x < 7.5$, $f''(x)$ is positive, and the concave side is above the curve; but $x > 7.5$ makes $f''(x)$ negative, and the concavity is downward. It follows that $(-2, -1)$, $(0.54, -0.05)$, and $(7.5, 2.6)$ are points of inflexion.

3. $y = a^2x/(x-a)^2$.

$$f'(x) = -a^2(a+x)/(x-a)^3.$$

$$f''(x) = 2a^2(x+2a)/(x-a)^4.$$

The curve is unlimited in both directions of X, and in the positive direction of Y.

$y = 0$ gives $x = 0$ and $\pm\infty$.

X is an asymptote in both directions, and since $y \ggg\!\!\rightarrow \infty$ as $x \ggg\!\!\rightarrow a$, $x = a$ is an asymptote.

$f'(0) = 1$. Hence, at the origin the curve makes $\tan^{-1} 1$ with X.

$f'(x) = 0$ gives $x = -a$, and $f''(-a)$ is positive. Hence, $y = -a/4$ is a minimum.

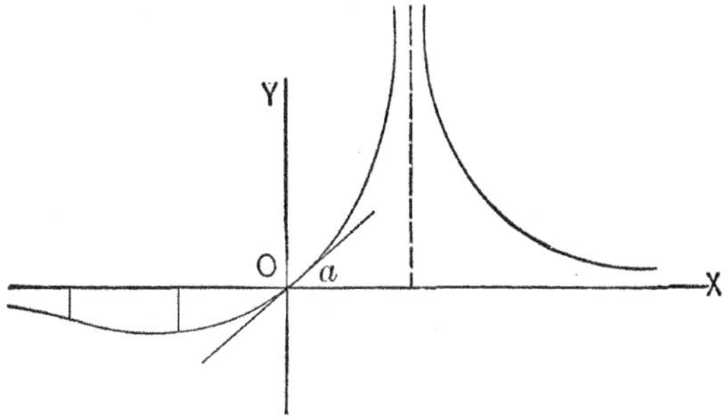

To the left of the point of inflexion $(-2a, -2a/9)$

the concave side is below, and to the right it is above, the curve.

4. $y^3 = 2ax^2 - x^3$. $\therefore$ $y = x^{2/3}(2a - x)^{1/3}$.

$f'(x) = (4ax - 3x^2)/3y^2$.

$f''(x) = -8a^2/9x^{4/3}(2a-x)^{5/3}$.

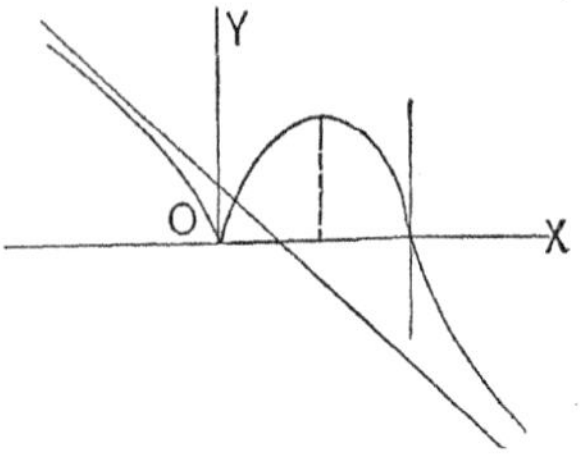

The curve is unlimited in both directions along X and Y. It cuts X at (0, 0) and $(2a, 0)$. Y is tangent to both branches at the origin, which is a cusp of the first species; and the tangent at $(2a, 0)$ is perpendicular to X. $y = -x + 2a/3$ is the equation of an asymptote in both directions. $y = a\sqrt[3]{32}/3$, corresponding to $x = 4a/3$, is a maximum. $(2a, 0)$ is a point of inflexion to the left of which the curve is concave downward, and to the right of which it is concave upward.

5. $y = x(3 - x)^4/16$.

$f'(x) = (3 - x)^3(3 - 5x)/16$.

$f''(x) = (3 - x)^2(5x - 6)/4$.

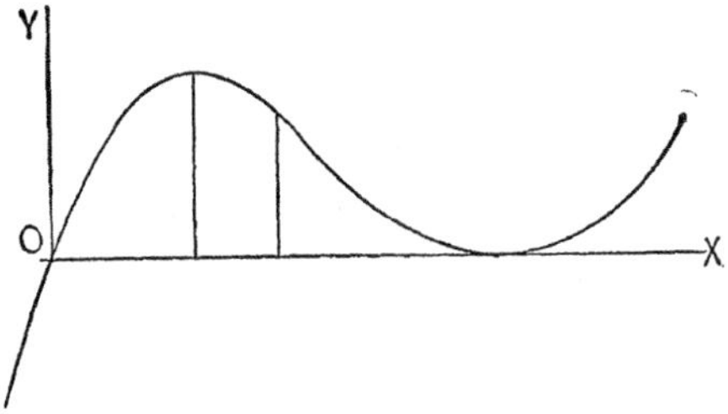

As $x \rightarrow \pm\infty$, $y \rightarrow \pm\infty$. $x = 0 = y$. The curve is unlimited in the directions of X and Y. (0, 0) and (3, 0) are points on X. $f'(0) = 5.06$, and $f'(3) = 0$. $f'(x) = 0$ gives $x = 3/5$ and 3. $f''(3/5)$ is negative; hence, $y = 1.24$ is a maximum. $f'(x)$ changes sign in passing through

$f'(3) = 0$ (§ 135), and $y = 0$ is a minimum. (1.2, 0.79) is a point of inflexion to the left of which the curve is concave downward, and to the right of which the concavity is upward.

6. $y^2 = x^5 + x^4$.

$$f'(x) = \pm x(5x + 4)/2\sqrt{x+1}.$$

$$f''(x) = \pm (15x^2 + 24x + 8)/4(x+1)^{3/2}.$$

The curve is symmetrical with respect to X.

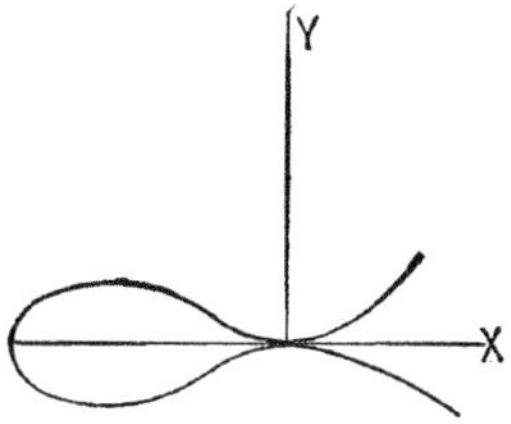

Values of $x < -1$ give imaginary expressions for y. $x = -1$ gives $y = \pm 0$. $x > -1$ gives two values for y equal with opposite signs. As $x \ggg\!\!\to \infty$, $y \ggg\!\!\to \pm\infty$. The curve is, therefore, limited in the direction of X negative by the ordinate corresponding to $x = -1$, and is unlimited in the other directions along X and Y.

$(-1, \pm 0)$ and $(0, \pm 0)$ are points on X.

$f'(-1) = \pm\infty$, and $f'(0) = \pm 0$. Hence, at $(-1, \pm 0)$ the tangent is parallel to Y, and X is tangent to both branches at the origin, which is a multiple point of tangency. (Example 6, page 281.)

$f'(x) = 0$ gives $x = 0$ or $-4/5$. $f''(0)$ is positive for the upper and negative for the lower branch. Hence, the zero ordinates at the origin are, respectively, a minimum and a maximum. $f''(-4/5)$ is negative for the upper and positive for the lower branch. Hence, the corresponding ordinates are, respectively, a maximum and a minimum.

$f''(x) = 0$ gives $x = (-12 \pm \sqrt{24})/15$. Points of both branches corresponding to the upper sign are points of inflexion, and the direction of curvature is as indicated in the figure.

7. $y = \pm x\sqrt{a^2 + x^2}/\sqrt{a^2 - x^2}$.

$$f'(x) = \pm \frac{a^4 + 2a^2x^2 - x^4}{(a^2 - x^2)^{3/2}(a^2 + x^2)^{1/2}}.$$

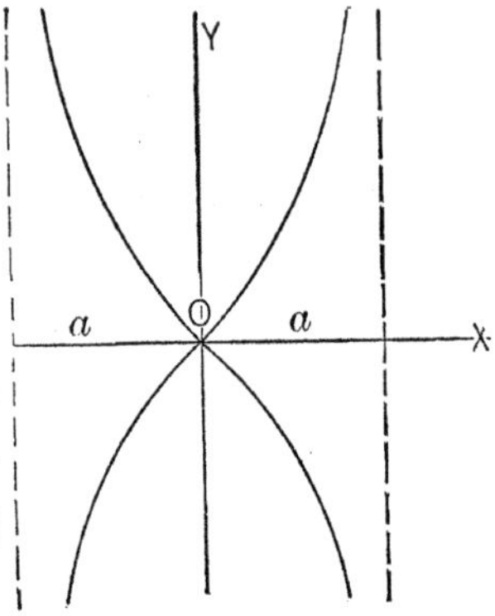

The curve is symmetrical with respect to X and Y. It is limited in both directions of X by the asymptotes $x = \pm a$, and is unlimited in both directions along Y.

$$f'(0) = \pm 1, \quad \text{and} \quad f'(\pm a) = \pm \infty.$$

Both branches pass through the origin, one inclined at an angle of $45°$, and the other at an angle of $135°$, with X. $f'(x)$ is an increasing function for the branches above X, which are, therefore, concave upward, and a decreasing function for those below X, which are concave downward.

8. $y = 1/e^{1/x}$.

$$f'(x) = 1/x^2e^{1/x}.$$

$$f''(x) = (1 - 2x)/x^4e^{1/x}.$$

As $x \rightarrow \mp \infty$, $y \rightarrow 1$. As $-x \rightarrow 0$, $y \rightarrow \infty$. As $+x \rightarrow 0$, $y \rightarrow 0$.

The curve is limited by X in the direction of Y negative, and is discontinuous at the origin which is a terminating point for the right-hand branch. As $-x \ggg\!\to 0$, $f'(x) \ggg\!\to \infty$; and as $+x \ggg\!\to 0$, $f'(x) \ggg\!\to 0$. Hence, X is a tangent at the origin. Y is an asymptote to the left-hand branch, and $y = 1$ is an asymptote to both branches. As x varies con-

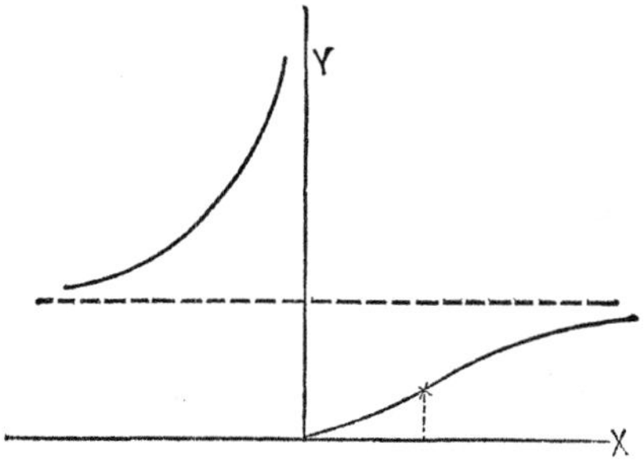

tinuously, $f'(x)$ does not change sign, and there are no maximum or minimum ordinates. Corresponding to $x = 1/2$, there is a point of inflexion, to the left of which the curve is concave upward, and to the right of which the curvature is downward.

9. **The Logarithmic Curve.**

$$x = e^y, \quad \therefore y = \log x.$$

$$f'(x) = 1/x. \quad f''(x) = -1/x^2.$$

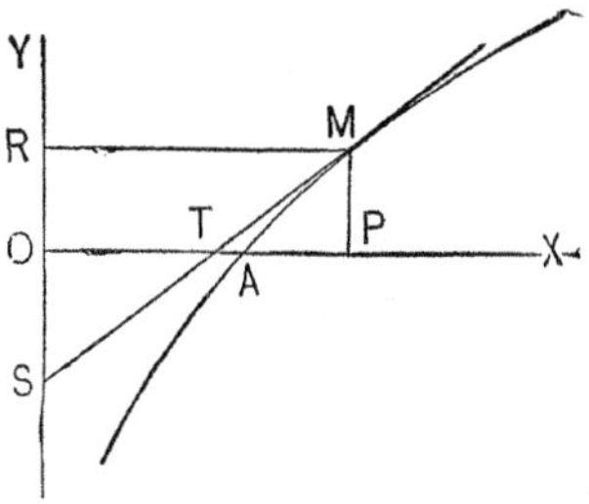

As $x \ggg\!\to 0$, $y \ggg\!\to -\infty$. $x = 1$, $y = 0$. As $x \ggg\!\to \infty$, $y \ggg\!\to \infty$. The curve is limited in the direction of X negative by Y, which is an asymptote, and is unlimited in the other directions along X and Y. $f'(1) = 1$. Hence, at $(1, 0)$ the curve makes $\tan^{-1} 1$ with X. $f''(x)$ is negative for all points of the curve, and the concave side is below. RS, the subtangent on Y, is (§ 149) $xdy/dx = 1 = OA$.

10. $y^2 = ax^2 + bx^3$.

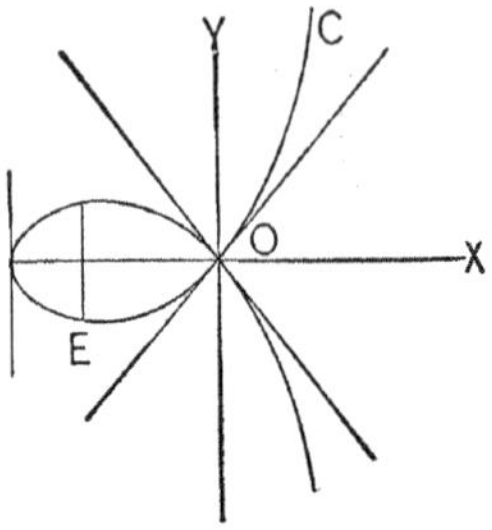

$f'(x) = \pm (a + 3bx/2)/\sqrt{a + bx}$.

$f''(x) = \pm (4ab + 3b^2x)/4(a + bx)^{3/2}$.

The curve is symmetrical with respect to X.

As $x \rightarrow \infty$, $y \rightarrow \pm \infty$. $x = 0 = y$. $x = -a/b$, $y = \pm 0$. $x < -a/b$, y is imaginary.

The curve is limited in the direction of X negative by the ordinate corresponding to $x = -a/b$, and is unlimited in the other directions along X and Y.

$f'(-a/b) = \pm \infty$. $f'(0) = \pm \sqrt{a}$. The origin is, therefore, a double multiple point.

$f'(x) = 0$ gives $x = -2a/3b$, for which $y = \pm 2a \sqrt{3a/9b}$ are maximum and minimum ordinates.

For $x > -a/b$ the first value of $f''(x)$ is positive, and the second negative. The branch $BEOC$ is, therefore, concave upward, and the other is concave downward.

POLAR COÖRDINATES: EXAMPLES.

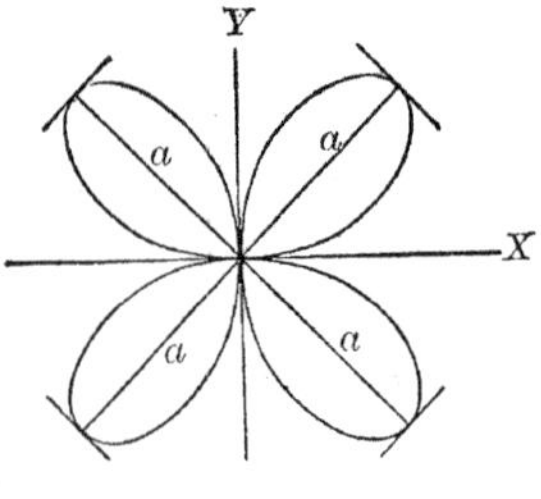

1. $r = a \sin 2\theta$.

$$dr/d\theta = 2a \cos 2\theta.$$

As θ varies from 0 to $\pi/4$, r changes from 0 to a, and as θ varies from $\pi/4$ to $\pi/2$, r changes from a to 0; completing a loop in the first angle. As θ varies from $\pi/2$ to π, r is negative, and changes from 0, through $-a$, to 0, forming a loop in the fourth angle. As θ varies from π to $3\pi/2$, r is positive and changes from 0, through a, to 0, forming a loop in the third angle. As θ varies from $3\pi/2$ to 2π, r is negative and changes from 0, through $-a$, to 0, forming a loop in the second angle.

As θ passes through $\pi/4$, $dr/d\theta$ changes from $+$ to $-$. Hence, $r = a$ is a maximum. As θ passes through $3\pi/4$, $dr/d\theta$ changes from $-$ to $+$, and $r = -a$ is a minimum. As θ passes through $5\pi/4$, $dr/d\theta$ changes from $+$ to $-$, and $r = a$ is a maximum. As θ passes through $7\pi/4$, $dr/d\theta$ changes from $-$ to $+$, and $r = -a$ is a minimum.

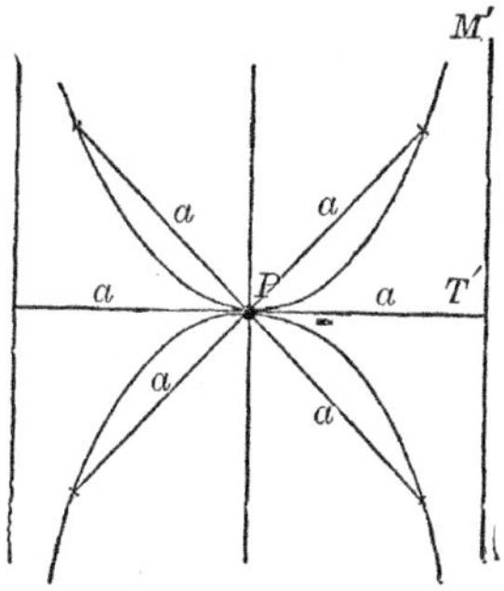

2. $r = a \tan \theta$.

$$dr/d\theta = a/\cos^2 \theta.$$

r is always an increasing function of θ.

Values of θ from 0 to $\pi/2$ give the branch PM'. As θ varies from $\pi/2$ to π, r is negative and increases from $-\infty$ to 0, giving the branch in the fourth angle. Values of θ from π to $3\pi/2$ determine the branch in the third angle, and the branch in the second angle is due to negative values of r corresponding to values of θ from $3\pi/2$ to 2π.

$$\text{The subtangent} = r^2 d\theta/dr = a \sin^2 \theta.$$

$\theta = \pi/2$ or $3\pi/2$ gives $r = \infty$, and the subtangent $= a$. Hence (§ 157), $r \cos \theta = \pm a$ are asymptotes.

3. **The Spiral of Archimedes,** $r = a\theta$.

Estimating from the pole P, where $r = 0 = \theta$, r increases directly with θ. Denoting the value of r after one revolution by

$$r_1 = PA = 2\pi a,$$

we have $a = r_1/2\pi$,

and $r = r_1\theta/2\pi$.

Hence,

$PO = r_1/4$, $PO' = r_1/2$,

$PO'' = r_1 3/4$, etc.

$$dr = r_1 d\theta/2\pi;$$

$$\therefore d\theta/dr = 2\pi/r_1.$$

Subtangent $= r^2 d\theta/dr$

$= r_1\theta^2/2\pi$. (§ 150.)

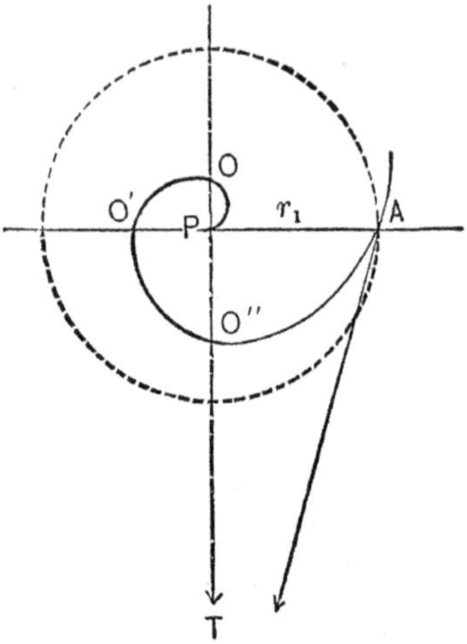

Hence, subtangents are to each other as the squares of the corresponding radii. $\theta = m2\pi$ gives subt $= m^2 2\pi r_1$.

$$\text{Subnormal} = dr/d\theta = r_1/2\pi.$$

4. **The Parabolic Spiral,** $r^2 = a^2\theta$.

$r_1 = a\sqrt{2\pi}$; $\therefore a = r_1/\sqrt{2\pi}$, and $r^2 = r_1^2\theta/2\pi$.

$dr/d\theta = a^2/2r =$ subnormal.

Subtangent $= 2r^3/a^2$.

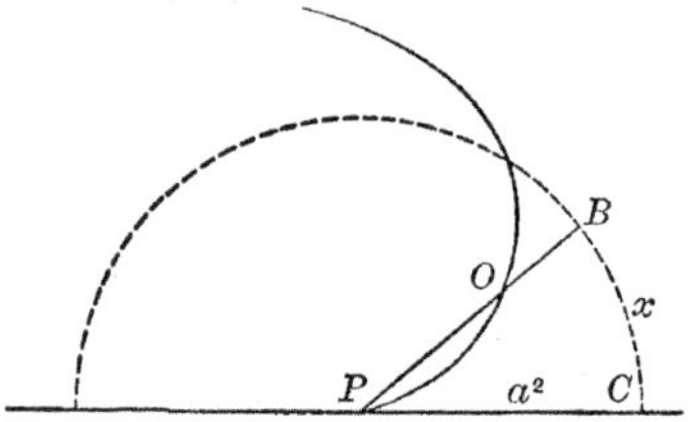

This spiral may be constructed by first constructing the parabola $y^2 = x$, and the circle CB with centre at P and radius $= a^2$. Then lay off from C the arc CB equal to an assumed abscissa of the parabola, and upon the radius PB lay off from P PO equal to the corresponding ordinate. O will be a point of the curve, since $r = PO = y = \sqrt{x} = \sqrt{a^2\theta}$.

5. **The Hyperbolic Spiral,** $r = a/\theta$.

$$r_1 = a/2\pi; \quad \therefore a = 2\pi r_1 \quad \text{and} \quad r = 2\pi r_1/\theta.$$

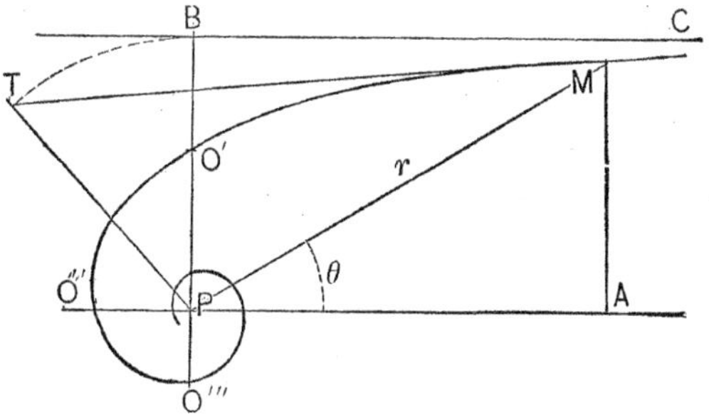

At the pole $r = 0$ and $\theta = \infty$. r varies inversely with θ. Hence, $PO' = 4r_1$, $PO'' = 2r_1$, $PO''' = 4r_1/3$, etc.

$dr/d\theta = -a/\theta^2$; subtangent $PT = -a$.

$\theta = 0$ gives $r = \infty$, and the subtangent $PB = -a$. Hence (§ 157), BC, parallel to PA, is an asymptote.

From § 150, tan $PMT = r d\theta/dr = -\theta$, which leads to a construction of the curve by points. Thus, with P as a centre and radius $= a$, describe a circle. Draw any radius

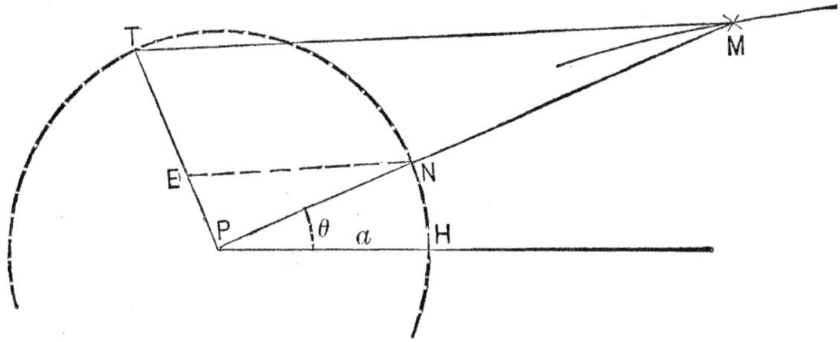

vector as PM, and the corresponding subtangent PT. Lay off $PE =$ arc HN, and draw EN. TM drawn parallel to EN will determine a point M of the curve; for

$$\tan PMT = \tan PNE = PE/a = HN/a = \theta.$$

6. **The Logarithmic Spiral,** $r = a^\theta$.

$d\theta/dr = M_a/r$, whence (§ 150) $\tan PMT = \tan \phi = M_a$, and ϕ is constant.

Also, $\sin \phi = rd\theta/ds$.

Hence, $p = r \sin \phi = r^2 d\theta/ds = cr$,

in which c represents the sine of the constant angle made by the radius vector with the curve.

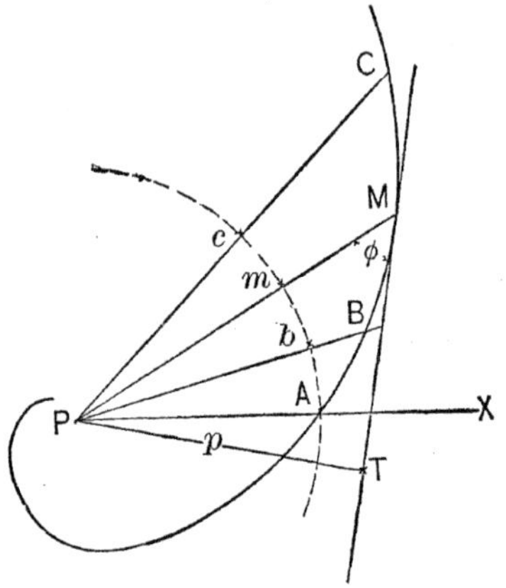

If $\theta = 0$ be increased by equal angles, the corresponding values of r will be in geometrical progression. With any convenient radius, as PA, describe a circle, and lay off equal arcs Ab, bm, mc, etc. Draw the right lines PA, Pb, Pm, etc., and let PA be the initial side of θ. PA will then represent unity. Make $\theta = Ab/PA$, and determine the corresponding value of $r = PB$. PB/PA will be the ratio of the progression, and the distances PM, PC, etc., from P to corresponding points of the curve are readily determined. Since $r = 0$ requires $\theta = -\infty$, the number of spires from A to P is unlimited.

CHAPTER XX.

APPLICATIONS TO SURFACES.

196. $u = F(x, y, z) = 0 \quad . \quad (1) \quad \text{or} \quad z = f(x, y) \quad (2)$
is the general equation of any surface.

Assuming the co-ordinate axes perpendicular to each other, and x and y as the independent variables, we have

$$du = \frac{\partial u}{dx}dx + \frac{\partial u}{dy}dy + \frac{\partial u}{dz}dz = 0, \quad . \quad . \quad . \quad (3)$$

or

$$dz = \frac{\partial z}{dx}dx + \frac{\partial z}{dy}dy, \quad . \quad . \quad . \quad . \quad . \quad . \quad . \quad . \quad . \quad (4)$$

for the general differential equation of the surface.

197. *To find the equation of the tangent plane to any surface at a given point.*

Let $z = \phi(x, y)$ be the equation of the surface and $P(x', y', z')$ be the given point.

Then $$dz = \frac{\partial z'}{dx'}dx + \frac{\partial z'}{dy'}dy, \quad . \quad . \quad . \quad . \quad . \quad . \quad (1)$$

in which $\frac{\partial z'}{dx'}$ and $\frac{\partial z'}{dy'}$ are constant for a fixed point (§ 102).

The equations of the tangent line PQ in the vertical plane PMN are, from the figure,

$$\frac{x - x'}{dx} = \frac{y - y'}{dy} = \frac{z - z'}{dz}. \quad . \quad . \quad . \quad . \quad (2)$$

Equations (1) and (2) are relations existing for any set of values of dx, dy, and dz, corresponding to the point P and the corresponding tangent.

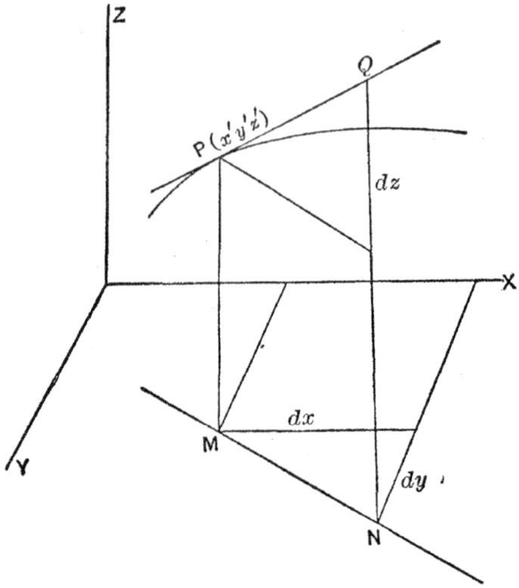

Eliminating dx, dy, and dz, we find the locus of all these tangents to be

$$z - z' = \frac{\partial z'}{dx'}(x - x') + \frac{\partial z'}{dy'}(y - y'). \quad . \quad . \quad (3)$$

If the equation of the surface be in form

$$u = F(x, y, z) = 0,$$

then $$du = \frac{\partial u}{dx'}dx + \frac{\partial u}{dy'}dy + \frac{\partial u}{dz'}dz = 0,$$

in which dx, dy, and dz are the same as in (1).

Combining this equation with (2), the equation of the tangent plane becomes

$$\frac{\partial u}{dx'}(x - x') + \frac{\partial u}{dy'}(y - y') + \frac{\partial u}{dz'}(z - z') = 0. \qquad (4)$$

(For tangent plane to surface of 2d order only. Compare C. Smith's Solid Geo., § 52.) Note. If $\frac{\partial u}{dx'} = \frac{\partial u}{dy'} = \frac{\partial u}{dz'} = 0$, the plane is indeterminate.

EXAMPLES.

1. Find the equation of the tangent plane at the point $(2, 3, \sqrt{23})$ on the surface $x^2 + y^2 + z^2 = 36$.

$$\partial z'/dx' = -x'/z' = -2/\sqrt{23}.$$
$$\partial z'/dy' = -y'/z' = -3\sqrt{23}.$$

Substituting in (3), we have

$$z - \sqrt{23} = (x - 2)(-2/\sqrt{23}) + (y - 3)(-3/\sqrt{23}),$$

or $2x + 3y + \sqrt{23}z = 36$, for the required plane,

or

$$\partial u/dx' = 2x' = 4, \ \partial u/dy' = 2y' = 6, \ \partial u/dz' = 2z' = 2\sqrt{23}$$

Substituting in (4), we obtain same result.

2. Find the equation of the tangent plane at (x', y', z') on the ellipsoid $x^2/a^2 + y^2/b^2 + z^2/c^2 = 1$.

Ans. $xx'/a^2 + yy'/b^2 + zz'/c^2 = 1$.

3. Find the equation of the tangent plane at (x', y', z') on the surface whose equation is

$$mx^2 + ny^2 + pz^2 + m''x + l = 0.$$

Ans. $2mx'x + 2ny'y + 2pz'z + m''(x + x') + 2l = 0$.

4. Find the equation of the tangent plane to the ellipsoid whose equation is

$$4x^2 + 2y^2 + z^2 = 10, \text{ at } (1, -1, 2).$$

Ans. $2x - y + z = 5$.

5. Find the equation of the tangent plane to the ellipsoid

$$x^2/16+y^2/9+z^2/4=1, \text{ at } (3, 1, \sqrt{47/36}).$$

Ans. $3x/16+y/9+(\sqrt{47/36})z/4=1.$

6. Find the equation of the tangent plane at any point of the surface $x^{\frac{2}{3}}+y^{\frac{2}{3}}+z^{\frac{2}{3}}=a^{\frac{2}{3}}$; and show that the sum of the squares of the intercepts on the axes made by a tangent plane is constant.

198. *The normal at any point of a surface.*

Let equation of surface be $z=\phi(x, y)$, and point be (x', y', z'). The normal passes through (x', y', z') and is perpendicular to the plane given by (3), § 197; hence its equations are

$$\frac{x-x'}{\partial z'/dx'}=\frac{y-y'}{\partial z'/dy'}=\frac{z-z'}{-1}. \quad . \quad . \quad . \quad . \quad (1)$$

If equation of surface be $u=F(x, y, z)=0$, equations of normal are

$$\frac{x-x'}{\partial u/dx'}=\frac{y-y'}{\partial u/dy'}=\frac{z-z'}{\partial u/dz'}. \quad . \quad . \quad . \quad . \quad (2)$$

EXAMPLES.

1. Deduce formulas for the direction cosines of the normal given by (1); by (2). And find the cosines of the angles the tangent planes corresponding to (1) and (2) make with XY, XZ, and YZ.

2. Find the tangent of the angle that the tangent plane to $x^2+y^2+z^2=36$ at $(2, 3, \sqrt{23})$ makes with XY.

199. *To find the equations of the tangent line to a given curve at a given point.*

Let equations of curve be

$$F(x, y, z) = 0, \quad . \quad . \quad . \quad . \quad . \quad . \quad . \quad (a)$$
$$\phi(x, y, z) = 0, \quad . \quad . \quad . \quad . \quad . \quad . \quad . \quad (b)$$

and let (x', y', z') be the point of tangency.

The required tangent line lies in the tangent plane to each of the surfaces (a) and (b) at point (x', y', z'); hence its equations are

$$\frac{\partial F}{dx'}(x - x') + \frac{\partial F}{dy'}(y - y') + \frac{\partial F}{dz'}(z - z') = 0,$$

$$\frac{\partial \phi}{dx'}(x - x') + \frac{\partial \phi}{dy'}(y - y') + \frac{\partial \phi}{dz'}(z - z') = 0.$$

If the curve be given by two of its projecting cylinders as

$$f(x, z) = 0, \quad . \quad . \quad . \quad . \quad . \quad . \quad . \quad . \quad (a')$$
$$\psi(y, z) = 0, \quad . \quad . \quad . \quad . \quad . \quad . \quad . \quad . \quad (b')$$

the equations of the tangent become

$$\left.\begin{aligned} \frac{\partial f}{dx'}(x - x') + \frac{\partial f}{dz'}(z - z') = 0, \\ \frac{\partial \psi}{dy'}(y - y') + \frac{\partial \psi}{dz'}(z - z') = 0, \end{aligned}\right\}$$

which (§ 148) are the equations of lines in XZ and YZ respectively tangent to the curves (a') and (b') at the projections of (x', y', z'). The problem is thus reduced to finding the equations of right lines tangent to two of the projections of the given curve at the projections of the given point.

Ex. Show that the curve whose equations are

$$x^2 + y^2 = a^2 \quad \text{and} \quad z = aC \tan^{-1}\frac{y}{x}$$

makes a constant angle with the axis Z.

200. *To find the equation of the normal plane to a curve at a given point.*

Let the equations of the curve be

$$F(x, y, z) = 0 \quad \text{and} \quad \phi(x, y, z) = 0,$$

and (x', y', z') be the given point.

The normal plane is perpendicular to the tangent to the curve at the given point.

Let equation of normal plane be

$$\lambda(x - x') + \mu(y - y') + \nu(z - z') = 0.$$

If this plane be perpendicular to the tangent we have the conditions

$$\begin{cases} \lambda\dfrac{\partial F}{dx'} + \mu\dfrac{\partial F}{dy'} + \nu\dfrac{\partial F}{dz'} = 0, \\ \lambda\dfrac{\partial \phi}{dx'} + \mu\dfrac{\partial \phi}{dy'} + \nu\dfrac{\partial \phi}{dz'} = 0. \end{cases}$$

Eliminating λ, μ, and ν, the required equation is

$$\begin{vmatrix} x - x', & y - y', & z - z' \\ \dfrac{\partial F}{dx'}, & \dfrac{\partial F}{dy'}, & \dfrac{\partial F}{dz'} \\ \dfrac{\partial \phi}{dx'}, & \dfrac{\partial \phi}{dy'}, & \dfrac{\partial \phi}{dz'} \end{vmatrix} = 0.$$

201. The numerical value of the expression

$$dz/\sqrt{dx^2 + dy^2} = \tan QMT$$

measures *the slope* of the surface $z = f(x, y)$, at the point M, along any section, as MB, made by a vertical plane through M, and whose trace on YX makes with X $\tan^{-1} dy/dx = \phi$ (§ 102).

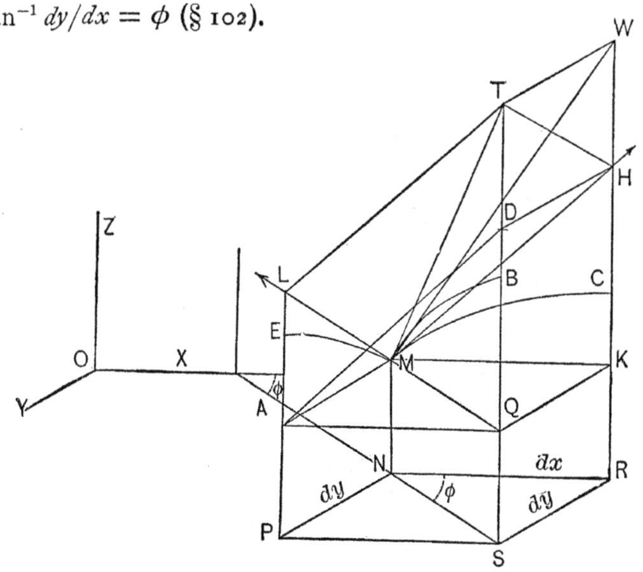

Writing $\quad \tan QMT = \tan s = \dfrac{dz/dx}{\sqrt{1 + (dy/dx)^2}},$

we have, (2), § 102,

$$\tan s = \frac{\partial z/dx + (\partial z/dy)(dy/dx)}{\sqrt{1 + (dy/dx)^2}}.$$

Placing

$$dy/dx = \tan \phi = m, \quad \partial z/dx = p, \quad \text{and} \quad \partial z/dy = q,$$

we have

$$\tan s = (p + mq)/\sqrt{1 + m^2}. \quad . \quad . \quad . \quad . \quad (1)$$

Application.—At the point $(2, 3, \sqrt{23})$ on the surface

$$x^2 + y^2 + z^2 = 36 \quad . \quad . \qquad . \quad . \quad (a)$$

find the slope of the curve cut out by the plane

$$y = 2x - 1, \quad z = 0/0. \quad . \quad . \quad . \quad . \quad . \quad (b)$$

From (a), $\partial z/dx = -x/z$, and $\partial z/dy = -y/z$.

Hence, $p = -2/\sqrt{23}$, and $q = -3/\sqrt{23}$.

From (b), $dy/dx = 2 = m$. Therefore,

$$\tan s = [-2/\sqrt{23} - 6/\sqrt{23}]/\sqrt{1+4} = -0.746 +.$$

Hence, $0.746 +$ is the required slope.

202. At any point, as M, $\tan s$ varies with m. To determine m, in order that the slope shall be a maximum, we place

$$\frac{d \tan s}{dm} = \frac{q - mp}{(1 + m^2)^{3/2}} = 0,$$

whence $q - mp = 0$, or $m = q/p$.

When $\tan s$ is positive, maximum values of $\tan s$ and the slope are the same; but when $\tan s$ is negative, the slope is a maximum when $\tan s$ is a minimum.

Application.—Find the equation of the vertical plane which passes through the point $(2, 3, \sqrt{23})$ on the surface $x^2 + y^2 + z^2 = 36$, and which cuts from the surface the line with the maximum slope.

$y = mx + b$, $z = 0/0$, is the general form of the required equation.

At $(2, 3, \sqrt{23})$ $\quad p = \partial z/dx = -2/\sqrt{23}$,

and $\quad q = \partial z/dy = -3/\sqrt{23}$.

Hence, $m = 3/2$. The trace on XY must pass through $(2, 3)$. Hence, we have

$$3 = 3 + b, \quad \text{or} \quad b = 0, \quad \text{and} \quad y = 3x/2, \quad z = 0/0,$$

is the required plane. The maximum slope is approximately .751.

When $p + mq = 0$, or $m = -p/q$, $\tan s = 0$, and the slope is a minimum, since numerical values only of $\tan s$ are considered.

In the above application $m = -p/q = -2/3$. Hence, $3 = -4/3 + b$, giving $b = 13/3$, and $y = -2x/3 + 13/3$, $z = 0/0$, is the plane which cuts out the curve whose tangent at M is parallel to XY.

The intersection of the surface by the horizontal plane through the given point is a horizontal line, and $F(x, y, c) = 0$, $z = c$, are its equations.

SHORT-TITLE CATALOGUE

OF THE

PUBLICATIONS

OF

JOHN WILEY & SONS,

NEW YORK.

LONDON: CHAPMAN & HALL, LIMITED.

ARRANGED UNDER SUBJECTS.

Descriptive circulars sent on application.
Books marked with an asterisk are sold at *net* prices **only.**
All books are bound in cloth unless otherwise stated.

AGRICULTURE.

Armsby's Manual of Cattle-feeding....................12mo, $1 75
Budd and Hansen's American Horticultural Manual:
Part I.—Propagation, Culture, and Improvement....12mo, 1 50
Part II.—Systematic Pomology. (*In preparation.*)
Downing's Fruits and Fruit-trees of America..............8vo, 5 00
Grotenfelt's Principles of Modern Dairy Practice. (Woll.)..12mo, 2 00
Kemp's Landscape Gardening.........................12mo, 2 50
Maynard's Landscape Gardening as Applied to Home Decoration. 12mo, 1 50
Sanderson's Insects Injurious to Staple Crops............12mo, 1 50
" Insects Injurious to Garden Crops. (*In preparation.*)
" Insects Injuring Fruits. (*In preparation.*)
Stockbridge's Rocks and Soils............................8vo, 2 50
Woll's Handbook for Farmers and Dairymen............16mo, 1 50

ARCHITECTURE.

Baldwin's Steam Heating for Buildings...............12mo, 2 50
Berg's Buildings and Structures of American Railroads....4to, 5 00
Birkmire's Planning and Construction of American Theatres.8vo, 3 00
" Architectural Iron and Steel..................8vo, 3 50
" Compound Riveted Girders as Applied in Buildings. 8vo, 2 00
" Planning and Construction of High Office Buildings. 8vo, 3 50
" Skeleton Construction in Buildings..............8vo, 3 00
Briggs's Modern American School Buildings............8vo, 4 00
Carpenter's Heating and Ventilating of Buildings.........8vo, 4 00
Freitag's Architectural Engineering. 2d Edition, Rewritten.8vo, 3 50
" Fireproofing of Steel Buildings...............8vo, 2 50
Gerhard's Guide to Sanitary House-inspection...........16mo, 1 00
" Theatre Fires and Panics...................12mo, 1 50
Hatfield's American House Carpenter...................8vo, 5 00
Holly's Carpenters' and Joiners' Handbook.............18mo, 75
Kidder's Architect's and Builder's Pocket-book..16mo, morocco, 4 00
Merrill's Stones for Building and Decoration.............8vo, 5 00
Monckton's Stair-building...............................4to, 4 00

Patton's Practical Treatise on Foundations................8vo, 5 00
Siebert and Biggin's Modern Stone-cutting and Masonry..8vo, 1 50
Snow's Principal Species of Wood: Their Characteristic Properties. (*In preparation.*)
Wait's Engineering and Architectural Jurisprudence......8vo, 6 00
Sheep, 6 50
" Law of Operations Preliminary to Construction in Engineering and Architecture.........................8vo, 5 00
Sheep, 5 50
" Law of Contracts..................................8vo, 3 00
Woodbury's Fire Protection of Mills.....................8vo, 2 50
Worcester and Atkinson's Small Hospitals, Establishment and Maintenance, and Suggestions for Hospital Architecture, with Plans for a Small Hospital..................12mo, 1 25
The World's Columbian Exposition of 1893..........Large 4to, 1 00

ARMY AND NAVY.

Bernadou's Smokeless Powder, Nitro-cellulose, and the Theory of the Cellulose Molecule.......................12mo, 2 50
* Bruff's Text-book Ordnance and Gunnery................8vo, 6 00
Chase's Screw Propellers and Marine Propulsion..........8vo, 3 00
Craig's Azimuth...4to, 3 50
Crehore and Squire's Polarizing Photo-chronograph......8vo, 3 00
Cronkhite's Gunnery for Non-commissioned Officers..24mo, mor., 2 00
* Davis's Elements of Law..................................8vo, 2 50
* " Treatise on the Military Law of United States..8vo, 7 00
* Sheep, 7 50
De Brack's Cavalry Outpost Duties. (Carr.)....24mo, morocco, 2 00
Dietz's Soldier's First Aid Handbook...........16mo, morocco, 1 25
* Dredge's Modern French Artillery..........4to, half morocco, 15 00
Durand's Resistance and Populsion of Ships...............8vo, 5 00
* Dyer's Handbook of Light Artillery.................12mo, 3 00
Eissler's Modern High Explosives........................8vo, 4 00
* Fiebeger's Text-book on Field Fortification........Small 8vo, 2 00
Hamilton's The Gunner's Catechism...................18mo, 1 00
* Hoff's Elementary Naval Tactics........................8vo, 1 50
Ingalls's Handbook of Problems in Direct Fire............8vo, 4 00
* " Ballistic Tables..................................8vo, 1 50
Lyons's Treatise on Electromagnetic Phenomena..........8vo, 6 00
* Mahan's Permanent Fortifications. (Mercur.)..8vo, half mor., 7 50
Manual for Courts-martial..................16mo, morocco, 1 50
* Mercur's Attack of Fortified Places....................12mo, 2 00
* " Elements of the Art of War...................8vo, 4 00
Metcalf's Cost of Manufactures—And the Administration of Workshops, Public and Private............8vo, 5 00
* " Ordnance and Gunnery.......................12mo, 5 00
Murray's Infantry Drill Regulations..............18mo, paper, 10
* Phelps's Practical Marine Surveying...................8vo, 2 50
Powell's Army Officer's Examiner........................12mo, 4 00
Sharpe's Art of Subsisting Armies in War......18mo, morocco, 1 50
Walke's Lectures on Explosives...........................8vo, 4 00
* Wheeler's Siege Operations and Military Mining..........8vo, 2 00
Winthrop's Abridgment of Military Law................12mo, 2 50
Woodhull's Notes on Military Hygiene..................16mo, 1 50
Young's Simple Elements of Navigation........16mo, morocco, 1 00
Second Edition, Enlarged and Revised......16mo, mor., 2 00

ASSAYING.

Fletcher's Practical Instructions in Quantitative Assaying with the Blowpipe..........................12mo, morocco, 1 50
Furman's Manual of Practical Assaying................8vo, 3 00
Miller's Manual of Assaying..........................12mo, 1 00
O'Driscoll's Notes on the Treatment of Gold Ores..........8vo, 2 00
Ricketts and Miller's Notes on Assaying................8vo, 3 00
Wilson's Cyanide Processes..........................12mo, 1 50
" Chlorination Process..........................12mo, 1 50

ASTRONOMY.

Craig's Azimuth....................................4to, 3 50
Doolittle's Treatise on Practical Astronomy...............8vo, 4 00
Gore's Elements of Geodesy..............................8vo, 2 50
Hayford's Text-book of Geodetic Astronomy..............8vo, 3 00
Merriman's Elements of Precise Surveying and Geodesy....8vo, 2 50
* Michie and Harlow's Practical Astronomy..............8vo, 3 00
* White's Elements of Theoretical and Descriptive Astronomy. 12mo, 2 00

BOTANY.

Baldwin's Orchids of New England................Small 8vo, 1 50
Davenport's Statistical Methods, with Special Reference to Biological Variation......................16mo, morocco, 1 25
Thomé and Bennett's Structural and Physiological Botany. 16mo, 2 25
Westermaier's Compendium of General Botany. (Schneider.) 8vo, 2 00

CHEMISTRY.

Adriance's Laboratory Calculations and Specific Gravity Tables. 12mo, 1 25
Allen's Tables for Iron Analysis..........................8vo, 3 00
Arnold's Compendium of Chemistry. (Mandel.) (*In preparation.*)
Austen's Notes for Chemical Students..................12mo, 1 50
Bernadou's Smokeless Powder.—Nitro-cellulose, and Theory of the Cellulose Molecule..........................12mo, 2 50
Bolton's Quantitative Analysis..........................8vo, 1 50
Brush and Penfield's Manual of Determinative Mineralogy...8vo, 4 00
Classen's Quantitative Chemical Analysis by Electrolysis. (Herrick—Boltwood.) 8vo, 3 00
Cohn's Indicators and Test-papers.....................12mo, 2 00
Craft's Short Course in Qualitative Chemical Analysis. (Schaeffer.) 12mo, 2 00
Drechsel's Chemical Reactions. (Merrill.)..............12mo, 1 25
Eissler's Modern High Explosives.........................8vo, 4 00
Effront's Enzymes and their Applications. (Prescott.)....8vo, 3 00
Erdmann's Introduction to Chemical Preparations. (Dunlap.) 12mo, 1 25
Fletcher's Practical Instructions in Quantitative Assaying with the Blowpipe..........................12mo, morocco, 1 50
Fowler's Sewage Works Analyses.......................12mo, 2 00
Fresenius's Manual of Qualitative Chemical Analysis. (Wells.) 8vo, 5 00
" Manual of Qualitative Chemical Analysis. Part I. Descriptive. (Wells.)..........................8vo, 3 00
" System of Instruction in Quantitative Chemical Analysis. (Allen.)..............................8vo, 6 00

Fuertes's Water and Public Health......................12mo, 1 50
Furman's Manual of Practical Assaying....................8vo, 3 00
Gill's Gas and Fuel Analysis for Engineers..............12mo, 1 25
Grotenfelt's Principles of Modern Dairy Practice. (Woll.)..12mo, 2 00
Hammarsten's Text-book of Physiological Chemistry. (Mandel.)
8vo, 4 00
Helm's Principles of Mathematical Chemistry. (Morgan.) 12mo, 1 50
Hinds's Inorganic Chemistry.................................8vo, 3 00
* " Laboratory Manual for Students................12mo, 75
Holleman's Text-book of Inorganic Chemistry. (Cooper.)...8vo, 2 50
" " " " Organic " (Walker and Mott.)
(In preparation.)
Hopkins's Oil-chemists' Handbook.........................8vo, 3 00
Keep's Cast Iron..8vo, 2 50
Ladd's Manual of Quantitative Chemical Analysis.........12mo, 1 00
Landauer's Spectrum Analysis. (Tingle.)..................8vo, 3 00
Lassar-Cohn's Practical Urinary Analysis. (Lorenz.) *(In preparation.)*
Leach's The Inspection and Analysis of Food with Special Reference to State Control. *(In preparation.)*
Löb's Electrolysis and Electrosynthesis of Organic Compounds. (Lorenz.) 12mo, 1 00
Mandel's Handbook for Bio-chemical Laboratory......12mo, 1 50
Mason's Water-supply. (Considered Principally from a Sanitary Standpoint.) 3d Edition, Rewritten......8vo, 4 00
" Examination of water. (Chemical and Bacteriological.) 12mo, 1 25
Meyer's Determination of Radicles in Carbon Compounds. (Tingle.) 12mo, 1 00
Miller's Manual of Assaying.................................12mo, 1 00
Mixter's Elementary Text-book of Chemistry............12mo, 1 50
Morgan's Outline of Theory of Solution and its Results..12mo, 1 00
" Elements of Physical Chemistry................12mo, 2 00
Nichols's Water-supply. (Considered mainly from a Chemical and Sanitary Standpoint, 1883.).....................8vo, 2 50
O'Brine's Laboratory Guide in Chemical Analysis..........8vo, 2 00
O'Driscoll's Notes on the Treatment of Gold Ores..........8vo, 2 00
Ost and Kolbeck's Text-book of Chemical Technology. (Lorenz—Bozart.) *(In preparation.)*
* Penfield's Notes on Determinative Mineralogy and Record of Mineral Tests.............................8vo, paper, 0 50
Pinner's Introduction to Organic Chemistry. (Austen.) 12mo, 1 50
Poole's Calorific Power of Fuels.............................8vo, 3 00
* Reisig's Guide to Piece-dyeing..........................8vo, 25 00
Richards and Woodman's Air, Water, and Food from a Sanitary Standpoint 8vo, 2 00
Richards's Cost of Living as Modified by Sanitary Science 12mo, 1 00
" Cost of Food, a Study in Dietaries...........12mo, 1 00
* Richards and Williams's The Dietary Computer..........8vo, 1 50
Ricketts and Russell's Skeleton Notes upon Inorganic Chemistry. (Part I.—Non-metallic Elements.)..8vo, morocco, 75
Ricketts and Miller's Notes on Assaying.................8vo, 3 00
Rideal's Sewage and the Bacterial Purification of Sewage..8vo, 3 50
Ruddiman's Incompatibilities in Prescriptions.............8vo, 2 00
Schimpf's Text-book of Volumetric Analysis............12mo, 2 50
Spencer's Handbook for Chemists of Beet-sugar Houses. 16mo, mor., 3 00
" Handbook for Sugar Manufacturers and their Chemists ....16mo, morocco, 2 00
Stockbridge's Rocks and Soils...............................8vo, 2 50

* Tillman's Elementary Lessons in Heat..................8vo, 1 50
* " Descriptive General Chemistry.................8vo, 3 00
Turneaure and Russell's Public Water-supplies...........8vo, 5 00
Van Deventer's Physical Chemistry for Beginners. (Boltwood.)
12mo, 1 50
Walke's Lectures on Explosives.........................8vo, 4 00
Wells's Laboratory Guide in Qualitative Chemical Analysis..8vo, 1 50
" Short Course in Inorganic Qualitative Chemical Analysis for Engineering Students...............12mo, 1 50
Whipple's Microscopy of Drinking-water..................8vo, 3 50
Wiechmann's Sugar Analysis.........................Small 8vo, 2 50
" Lecture-notes on Theoretical Chemistry....12mo, 3 00
Wilson's Cyanide Processes............................12mo, 1 50
" Chlorination Process............................12mo, 1 50
Wulling's Elementary Course in Inorganic Pharmaceutical and Medical Chemistry.................................12mo, 2 00

CIVIL ENGINEERING.

BRIDGES AND ROOFS. HYDRAULICS. MATERIALS OF ENGINEERING. RAILWAY ENGINEERING.

Baker's Engineers' Surveying Instruments..............12mo, 3 00
Bixby's Graphical Computing Table...Paper, 19½ x 24¼ inches. 25
Davis's Elevation and Stadia Tables......................8vo, 1 00
Folwell's Sewerage. (Designing and Maintenance.).......8vo, 3 00
Freitag's Architectural Engineering. 2d Ed., Rewritten...8vo, 3 50
French and Ives's Stereotomy............................8vo, 2 50
Goodhue's Municipal Improvements.....................12mo, 1 75
Goodrich's Economic Disposal of Towns' Refuse..........8vo, 3 50
Gore's Elements of Geodesy..............................8vo, 2 50
Hayford's Text-book of Geodetic Astronomy..............8vo, 3 00
Howe's Retaining-walls for Earth.......................12mo, 1 25
Johnson's Theory and Practice of Surveying.........Small 8vo, 4 00
" Stadia and Earth-work Tables...................8vo, 1 25
Kiersted's Sewage Disposal.............................12mo, 1 25
Laplace's Philosophical Essay on Probabilities. (Truscott and Emory.)..................................12mo, 2 00
Mahan's Treatise on Civil Engineering. (1873.) (Wood.)..8vo, 5 00
* Mahan's Descriptive Geometry..........................8vo, 1 50
Merriman's Elements of Precise Surveying and Geodesy....8vo, 2 50
Merriman and Brooks's Handbook for Surveyors....16mo, mor., 2 00
Merriman's Elements of Sanitary Engineering............8vo, 2 00
Nugent's Plane Surveying................................8vo, 3 50
Ogden's Sewer Design...................................12mo, 2 00
Patton's Treatise on Civil Engineering........8vo, half leather, 7 50
Reed's Topographical Drawing and Sketching..............4to, 5 00
Rideal's Sewage and the Bacterial Purification of Sewage...8vo, 3 50
Siebert and Biggin's Modern Stone-cutting and Masonry....8vo, 1 50
Smith's Manual of Topographical Drawing. (McMillan.)..8vo, 2 50
* Trautwine's Civil Engineer's Pocket-book....16mo, morocco, 5 00
Wait's Engineering and Architectural Jurisprudence......8vo, 6 00
Sheep, 6 50
" Law of Operations Preliminary to Construction in Engineering and Architecture.....................8vo, 5 00
Sheep, 5 50
Wait's Law of Contracts.................................8vo, 3 00
Warren's Stereotomy—Problems in Stone-cutting........8vo, 2 50
Webb's Problems in the Use and Adjustment of Engineering Instruments ....16mo, morocco, 1 25

* Wheeler's Elementary Course of Civil Engineering........8vo, 4 00
Wilson's Topographic Surveying..........................8vo, 3 50

BRIDGES AND ROOFS.

Boller's Practical Treatise on the Construction of Iron Highway Bridges 8vo, 2 00
* Boller's Thames River Bridge...................4to, paper, 5 00
Burr's Course on the Stresses in Bridges and Roof Trusses, Arched Ribs, and Suspension Bridges..............8vo, 3 50
Du Bois's Mechanics of Engineering. Vol. II.........Small 4to, 10 00
Foster's Treatise on Wooden Trestle Bridges............4to, 5 00
Fowler's Coffer-dam Process for Piers...................8vo, 2 50
Greene's Roof Trusses...................................8vo, 1 25
" Bridge Trusses.......................................8vo, 2 50
" Arches in Wood, Iron, and Stone......................8vo, 2 50
Howe's Treatise on Arches...............................8vo, 4 00
" Design of Simple Roof-trusses in Wood and Steel. 8vo, 2 00
Johnson, Bryan and Turneaure's Theory and Practice in the Designing of Modern Framed Structures......Small 4to, 10 00
Merriman and Jacoby's Text-book on Roofs and Bridges:
Part I.—Stresses in Simple Trusses......................8vo, 2 50
Part II.—Graphic Statics................................8vo, 2 50
Part III.—Bridge Design. Fourth Ed., Rewritten......8vo, 2 50
Part IV.—Higher Structures..............................8vo, 2 50
Morison's Memphis Bridge................................4to, 10 00
Waddell's De Pontibus, a Pocket Book for Bridge Engineers. 16mo, mor., 3 00
" Specifications for Steel Bridges..............12mo, 1 25
Wood's Treatise on the Theory of the Construction of Bridges and Roofs 8vo, 2 00
Wright's Designing of Draw-spans:
Part I.—Plate-girder Draws..............................8vo, 2 50
Part II.—Riveted-truss and Pin-connected Long-span Draws. 8vo, 2 50
Two parts in one volume.................................8vo, 3 50

HYDRAULICS.

Bazin's Experiments upon the Contraction of the Liquid Vein Issuing from an Orifice. (Trautwine.)..............8vo, 2 00
Bovey's Treatise on Hydraulics..........................8vo, 5 00
Church's Mechanics of Engineering.......................8vo, 6 00
" Diagrams of Mean Velocity of Water in Open Channels paper, 1 50
Coffin's Graphical Solution of Hydraulic Problems..16mo, mor., 2 50
Flather's Dynamometers, and the Measurement of Power.12mo, 3 00
Folwell's Water-supply Engineering......................8vo, 4 00
Frizell's Water-power....................................8vo, 5 00
Fuertes's Water and Public Health......................12mo, 1 50
" Water-filtration Works...............................12mo, 2 50
Ganguillet and Kutter's General Formula for the Uniform Flow of Water in Rivers and Other Channels. (Hering and Trautwine.)..8vo, 4 00
Hazen's Filtration of Public Water-supply...............8vo, 3 00
Hazlehurst's Towers and Tanks for Water-works........8vo, 2 50
Herschel's 115 Experiments on the Carrying Capacity of Large, Riveted, Metal Conduits..........................8vo, 2 00

Mason's Water-supply. (Considered Principally from a Sanitary Standpoint.) 8vo, 4 00
Merriman's Treatise on Hydraulics 8vo, 4 00
* Michie's Elements of Analytical Mechanics 8vo, 4 00
Schuyler's Reservoirs for Irrigation, Water-power, and Domestic Water-supply . . . . . . Large 8vo, 5 00
Turneaure and Russell. Public Water-supplies 8vo, 5 00
Wegmann's Design and Construction of Dams 4to, 5 00
" Water-supply of the City of New York from 1658 to 1895 . . . . . . 4to, 10 00
Weisbach's Hydraulics and Hydraulic Motors. (Du Bois.) . . 8vo, 5 00
Wilson's Manual of Irrigation Engineering . . . . . . Small 8vo, 4 00
Wolff's Windmill as a Prime Mover 8vo, 3 00
Wood's Turbines 8vo, 2 50
" Elements of Analytical Mechanics 8vo, 3 00

MATERIALS OF ENGINEERING.

Baker's Treatise on Masonry Construction 8vo, 5 00
Black's United States Public Works . . . . . . Oblong 4to, 5 00
Bovey's Strength of Materials and Theory of Structures 8vo, 7 50
Burr's Elasticity and Resistance of the Materials of Engineering 8vo, 5 00
Byrne's Highway Construction 8vo, 5 00
" Inspection of the Materials and Workmanship Employed in Construction 16mo, 3 00
Church's Mechanics of Engineering 8vo, 6 00
Du Bois's Mechanics of Engineering. Vol. I . . . . . . Small 4to, 7 50
Johnson's Materials of Construction . . . . . . Large 8vo, 6 00
Keep's Cast Iron 8vo, 2 50
Lanza's Applied Mechanics 8vo, 7 50
Martens's Handbook on Testing Materials. (Henning.) 2 v., 8vo, 7 50
Merrill's Stones for Building and Decoration 8vo, 5 00
Merriman's Text-book on the Mechanics of Materials 8vo, 4 00
Merriman's Strength of Materials 12mo, 1 00
Metcalf's Steel. A Manual for Steel-users 12mo, 2 00
Patton's Practical Treatise on Foundations 8vo, 5 00
Rockwell's Roads and Pavements in France 12mo, 1 25
Smith's Wire: Its Use and Manufacture . . . . . . Small 4to, 3 00
" Materials of Machines 12mo, 1 00
Snow's Principal Species of Wood: Their Characteristic Properties. (*In preparation.*)
Spalding's Hydraulic Cement 12mo, 2 00
" Text-book on Roads and Pavements 12mo, 2 00
Thurston's Materials of Engineering . . . . . . 3 Parts, 8vo, 8 00
Part I.—Non-metallic Materials of Engineering and Metallurgy 8vo, 2 00
Part II.—Iron and Steel 8vo, 3 50
Part III.—A Treatise on Brasses, Bronzes and Other Alloys and Their Constituents 8vo, 2 50
Thurston's Text-book of the Materials of Construction 8vo, 5 00
Tillson's Street Pavements and Paving Materials 8vo, 4 00
Waddell's De Pontibus. (A Pocket-book for Bridge Engineers.) 16mo, morocco, 3 00
" Specifications for Steel Bridges 12mo, 1 25
Wood's Treatise on the Resistance of Materials, and an Appendix on the Preservation of Timber 8vo, 2 00
" Elements of Analytical Mechanics 8vo, 3 00

RAILWAY ENGINEERING.

Andrews's Handbook for Street Railway Engineers. 3x5 in. mor., 1 25
Berg's Buildings and Structures of American Railroads...4to, 5 00
Brooks's Handbook of Street Railroad Location..16mo, morocco, 1 50
Butts's Civil Engineer's Field-book............16mo, morocco, 2 50
Crandall's Transition Curve....................16mo, morocco, 1 50
" Railway and Other Earthwork Tables........8vo, 1 50
Dawson's Electric Railways and Tramways. Small 4to, half mor., 12 50
" "Engineering" and Electric Traction Pocket-book. 16mo, morocco, 4 00
Dredge's History of the Pennsylvania Railroad: (1879.). Paper, 5 00
* Drinker's Tunneling, Explosive Compounds, and Rock Drills. 4to, half morocco, 25 00
Fisher's Table of Cubic Yards....................Cardboard, 25
Godwin's Railroad Engineers' Field-book and Explorers' Guide. 16mo, morocco, 2 50
Howard's Transition Curve Field-book.........16mo, morocco, 1 50
Hudson's Tables for Calculating the Cubic Contents of Excavations and Embankments.........................8vo, 1 00
Nagle's Field Manual for Railroad Engineers....16mo, morocco, 3 00
Philbrick's Field Manual for Engineers........16mo, morocco, 3 00
Pratt and Alden's Street-railway Road-bed...............8vo, 2 00
Searles's Field Engineering...................16mo, morocco, 3 00
" Railroad Spiral....................16mo, morocco, 1 50
Taylor's Prismoidal Formulæ and Earthwork.............8vo, 1 50
* Trautwine's Method of Calculating the Cubic Contents of Excavations and Embankments by the Aid of Diagrams 8vo, 2 00
* " The Field Practice of Laying Out Circular Curves for Railroads..................12mo, morocco, 2 50
* " Cross-section Sheet.......................Paper, 25
Webb's Railroad Construction..............................8vo, 4 00
Wellington's Economic Theory of the Location of Railways.. Small 8vo, 5 00

DRAWING.

Barr's Kinematics of Machinery...........................8vo, 2 50
* Bartlett's Mechanical Drawing..........................8vo, 3 00
Coolidge's Manual of Drawing..................8vo, paper, 1 00
Durley's Elementary Text-book of the Kinematics of Machines. (*In preparation.*)
Hill's Text-book on Shades and Shadows, and Perspective..8vo, 2 00
Jones's Machine Design:
Part I.—Kinematics of Machinery.......................8vo, 1 50
Part II.—Form, Strength and Proportions of Parts......8vo, 3 00
MacCord's Elements of Descriptive Geometry.............8vo, 3 00
" Kinematics; or, Practical Mechanism..........8vo, 5 00
" Mechanical Drawing............................4to, 4 00
" Velocity Diagrams..............................8vo, 1 50
* Mahan's Descriptive Geometry and Stone-cutting.......8vo, 1 50
Mahan's Industrial Drawing. (Thompson.)...............8vo, 3 50
Reed's Topographical Drawing and Sketching.............4to, 5 00
Reid's Course in Mechanical Drawing.....................8vo, 2 00
" Text-book of Mechanical Drawing and Elementary Machine Design...............................8vo, 3 00
Robinson's Principles of Mechanism......................8vo, 3 00

Smith's Manual of Topographical Drawing. (McMillan.) .8vo, 2 50

Warren's Elements of Plane and Solid Free-hand Geometrical Drawing 12mo, 1 00

" Drafting Instruments and Operations....12mo, 1 25

" Manual of Elementary Projection Drawing....12mo, 1 50

" Manual of Elementary Problems in the Linear Perspective of Form and Shadow....12mo, 1 00

" Plane Problems in Elementary Geometry....12mo, 1 25

" Primary Geometry....12mo, 75

" Elements of Descriptive Geometry, Shadows, and Perspective 8vo, 3 50

" General Problems of Shades and Shadows....8vo, 3 00

" Elements of Machine Construction and Drawing..8vo, 7 50

" Problems, Theorems, and Examples in Descriptive Geometry 8vo, 2 50

Weisbach's Kinematics and the Power of Transmission. (Herrmann and Klein.) 8vo, 5 00

Whelpley's Practical Instruction in the Art of Letter Engraving 12mo, 2 00

Wilson's Topographic Surveying....8vo, 3 50

Wilson's Free-hand Perspective....8vo, 2 50

Woolf's Elementary Course in Descriptive Geometry..Large 8vo, 3 00

ELECTRICITY AND PHYSICS.

Anthony and Brackett's Text-book of Physics. (Magie.) Small 8vo, 3 00

Anthony's Lecture-notes on the Theory of Electrical Measurments 12mo, 1 00

Benjamin's History of Electricity....8vo, 3 00

Benjamin's Voltaic Cell....8vo, 3 00

Classen's Qantitative Chemical Analysis by Electrolysis. Herrick and Boltwood.)....8vo, 3 00

Crehore and Squier's Polarizing Photo-chronograph....8vo, 3 00

Dawson's Electric Railways and Tramways..Small 4to, half mor., 12 50

Dawson's "Engineering" and Electric Traction Pocket-book. 16mo, morocco, 4 00

Flather's Dynamometers, and the Measurement of Power..12mo, 3 00

Gilbert's De Magnete. (Mottelay.)....8vo, 2 50

Holman's Precision of Measurements....8vo, 2 00

" Telescopic Mirror-scale Method, Adjustments, and Tests ....Large 8vo, 75

Landauer's Spectrum Analysis. (Tingle.)....8vo, 3 00

Le Chatelier's High-temperature Measurements. (Boudouard—Burgess.) 12mo, 3 00

Löb's Electrolysis and Electrosynthesis of Organic Compounds. (Lorenz.) 12mo, 1 00

Lyons's Treatise on Electromagnetic Phenomena....8vo, 6 00

* Michie. Elements of Wave Motion Relating to Sound and Light 8vo, 4 00

Niaudet's Elementary Treatise on Electric Batteries (Fishback.) 12mo, 2 50

* Parshall and Hobart's Electric Generators..Small 4to, half mor., 10 00

Ryan, Norris, and Hoxie's Electrical Machinery. (*In preparation.*)

Thurston's Stationary Steam-engines....8vo, 2 50

* Tillman. Elementary Lessons in Heat....8vo, 1 50

Tory and Pitcher. Manual of Laboratory Physics..Small 8vo, 2 00

LAW.

* Davis. Elements of Law..8vo, 2 50
* " Treatise on the Military Law of United States..8vo, 7 00
* Sheep, 7 50
Manual for Courts-martial....................16mo, morocco, 1 50
Wait's Engineering and Architectural Jurisprudence......8vo, 6 00
Sheep, 6 50
" Law of Operations Preliminary to Construction in Engineering and Architecture........................8vo, 5 00
Sheep, 5 50
" Law of Contracts..................................8vo, 3 00
Winthrop's Abridgment of Military Law..............12mo, 2 50

MANUFACTURES.

Beaumont's Woollen and Worsted Cloth Manufacture....12mo, 1 50
Bernadou's Smokeless Powder—Nitro-cellulose and Theory of the Cellulose Molecule..............................12mo, 2 50
Bolland's Iron Founder.............................12mo, cloth, 2 50
" "The Iron Founder" Supplement..............12mo, 2 50
" Encyclopedia of Founding and Dictionary of Foundry Terms Used in the Practice of Moulding....12mo, 3 00
Eissler's Modern High Explosives..........................8vo, 4 00
Effront's Enzymes and their Applications. (Prescott.)...8vo, 3 00
Fitzgerald's Boston Machinist............................18mo, 1 00
Ford's Boiler Making for Boiler Makers.................18mo, 1 00
Hopkins's Oil-chemists' Handbook.........................8vo, 3 00
Keep's Cast Iron..8vo 2 50
Leach's The Inspection and Analysis of Food with Special Reference to State Control. (*In preparation.*)
Metcalf's Steel. A Manual for Steel-users...............12mo, 2 00
Metcalf's Cost of Manufactures—And the Administration of Workshops, Public and Private................8vo, 5 00
Meyer's Modern Locomotive Construction...................4to, 10 00
* Reisig's Guide to Piece-dyeing............................8vo, 25 00
Smith's Press-working of Metals...........................8vo, 3 00
" Wire: Its Use and Manufacture............Small 4to, 3 00
Spalding's Hydraulic Cement..............................12mo, 2 00
Spencer's Handbook for Chemists of Beet-sugar Houses.
16mo, morocco, 3 00
" Handbook for Sugar Manufacturers and their Chemists..............................16mo, morocco, 2 00
Thurston's Manual of Steam-boilers, their Designs, Construction and Operation......................................8vo, 5 00
Walke's Lectures on Explosives............................8vo, 4 00
West's American Foundry Practice.......................12mo, 2 50
" Moulder's Text-book............................12mo, 2 50
Wiechmann's Sugar Analysis......................Small 8vo, 2 50
Wolff's Windmill as a Prime Mover.......................8vo, 3 00
Woodbury's Fire Protection of Mills......................8vo, 2 50

MATHEMATICS.

Baker's Elliptic Functions....................................8vo, 1 50
* Bass's Elements of Differential Calculus...............12mo, 4 00
Briggs's Elements of Plane Analytic Geometry..........12mo, 1 00
Chapman's Elementary Course in Theory of Equations...12mo, 1 50
Compton's Manual of Logarithmic Computations........12mo, 1 50

Davis's Introduction to the Logic of Algebra..............8vo,	1 50
***Dickson's College Algebra**......................Large 12mo,	1 50
Halsted's Elements of Geometry.........................8vo,	1 75
" Elementary Synthetic Geometry..............8vo,	1 50
* Johnson's Three-place Logarithmic Tables: Vest-pocket size, pap.,	15
100 copies for	5 00
* Mounted on heavy cardboard, 8 × 10 inches,	25
10 copies for	2 00
" Elementary Treatise on the Integral Calculus. Small 8vo,	1 50
" Curve Tracing in Cartesian Co-ordinates......12mo,	1 00
" Treatise on Ordinary and Partial Differential Equations ....Small 8vo,	3 50
" Theory of Errors and the Method of Least Squares 12mo,	1 50
* " Theoretical Mechanics.........................12mo,	3 00
Laplace's Philosophical Essay on Probabilities. (Truscott and Emory.).......................................12mo,	2 00
* Ludlow and Bass. Elements of Trigonometry and Logarithmic and Other Tables..............................8vo,	3 00
" Trigonometry. Tables published separately..Each,	2 00
Merriman and Woodward. Higher Mathematics..........8vo,	5 00
Merriman's Method of Least Squares...................8vo,	2 00
Rice and Johnson's Elementary Treatise on the Differential Calculus ....Small 8vo,	3 00
" Differential and Integral Calculus. 2 vols. in one.....................Small 8vo,	2 50
Wood's Elements of Co-ordinate Geometry...............8vo,	2 00
" Trigometry: Analytical, Plane, and Spherical....12mo,	1 00

MECHANICAL ENGINEERING.

MATERIALS OF ENGINEERING, STEAM ENGINES AND BOILERS.

Baldwin's Steam Heating for Buildings.................12mo,	2 50
Barr's Kinematics of Machinery..........................8vo,	2 50
* Bartlett's Mechanical Drawing..........................8vo,	3 00
Benjamin's Wrinkles and Recipes........................12mo,	2 00
Carpenter's Experimental Engineering....................8vo,	6 00
" Heating and Ventilating Buildings...........8vo,	4 00
Clerk's Gas and Oil Engine.......................Small 8vo,	4 00
Coolidge's Manual of Drawing..................8vo, paper,	1 00
Cromwell's Treatise on Toothed Gearing..................12mo,	1 50
" Treatise on Belts and Pulleys..............12mo,	1 50
Durley's Elementary Text-book of the Kinematics of Machines. (*In preparation.*)	
Flather's Dynamometers, and the Measurement of Power ..12mo,	3 00
" Rope Driving.................................12mo,	2 00
Gill's Gas an Fuel Analysis for Engineers.............12mo,	1 25
Hall's Car Lubrication..................................12mo,	1 00
Jones's Machine Design:	
Part I.—Kinematics of Machinery........................8vo,	1 50
Part II.—Form, Strength and Proportions of Parts......8vo,	3 00
Kent's Mechanical Engineers' Pocket-book....16mo, morocco,	5 00
Kerr's Power and Power Transmission....................8vo,	2 00

MacCord's Kinematics; or, Practical Mechanism..........8vo, 5 00
" Mechanical Drawing..........................4to, 4 00
" Velocity Diagrams..........................8vo, 1 50
Mahan's Industrial Drawing. (Thompson.)...............8vo, 3 50
Poole's Calorific Power of Fuels..........................8vo, 3 00
Reid's Course in Mechanical Drawing....................8vo, 2 00
" Text-book of Mechanical Drawing and Elementary Machine Design.................................8vo, 3 00
Richards's Compressed Air............................12mo, 1 50
Robinson's Principles of Mechanism.....................8vo, 3 00
Smith's Press-working of Metals..........................8vo, 3 00
Thurston's Treatise on Friction and Lost Work in Machinery and Mill Work..........................8vo, 3 00
" Animal as a Machine and Prime Motor and the Laws of Energetics......................12mo, 1 00
Warren's Elements of Machine Construction and Drawing..8vo, 7 50
Weisbach's Kinematics and the Power of Transmission. (Herrmann—Klein.) 8vo, 5 00
" Machinery of Transmission and Governors. (Herrmann—Klein.) 8vo, 5 00
" Hydraulics and Hydraulic Motors. (Du Bois.).8vo, 5 00
Wolff's Windmill as a Prime Mover.......................8vo, 3 00
Wood's Turbines..8vo, 2 50

MATERIALS OF ENGINEERING.

Bovey's Strength of Materials and Theory of Structures..8vo, 7 50
Burr's Elasticity and Resistance of the Materials of Engineering 8vo, 5 00
Church's Mechanics of Engineering.......................8vo, 6 00
Johnson's Materials of Construction..............Large 8vo, 6 00
Keep's Cast Iron...8vo, 2 50
Lanza's Applied Mechanics................................8vo, 7 50
Martens's Handbook on Testing Materials. (Henning.)....8vo, 7 50
Merriman's Text-book on the Mechanics of Materials....8vo, 4 00
" Strength of Materials......................12mo, 1 00
Metcalf's Steel. A Manual for Steel-users..............12mo, 2 00
Smith's Wire: Its Use and Manufacture............Small 4to, 3 00
" Materials of Machines............................12mo, 1 00
Thurston's Materials of Engineering.............3 vols., 8vo, 8 00
Part II.—Iron and Steel....................................8vo, 3 50
Part III.—A Treatise on Brasses, Bronzes and Other Alloys and their Constituents.............................8vo, 2 50
Thurston's Text-book of the Materials of Construction....8vo, 5 00
Wood's Treatise on the Resistance of Materials and an Appendix on the Preservation of Timber..........8vo, 2 00
" Elements of Analytical Mechanics................8vo, 3 00

STEAM ENGINES AND BOILERS.

Carnot's Reflections on the Motive Power of Heat. (Thurston.) 12mo, 1 50
Dawson's "Engineering" and Electric Traction Pocket-book. 16mo, morocco, 4 00
Ford's Boiler Making for Boiler Makers................18mo, 1 00
Goss's Locomotive Sparks.................................8vo, 2 00
Hemenway's Indicator Practice and Steam-engine Economy. 12mo, 2 00
Hutton's Mechanical Engineering of Power Plants........8vo, 5 00
" Heat and Heat-engines.........................8vo, 5 00

Kent's Steam-boiler Economy....8vo, 4 00
Kneass's Practice and Theory of the Injector....8vo, 1 50
MacCord's Slide-valves....8vo, 2 00
Meyer's Modern Locomotive Construction....4to, 10 00
Peabody's Manual of the Steam-engine Indicator....12mo, 1 50
" Tables of the Properties of Saturated Steam and Other Vapors....8vo, 1 00
" Thermodynamics of the Steam-engine and Other Heat-engines....8vo, 5 00
" Valve-gears for Steam-engines....8vo, 2 50
Peabody and Miller. Steam-boilers....8vo, 4 00
Pray's Twenty Years with the Indicator....Large 8vo, 2 50
Pupin's Thermodynamics of Reversible Cycles in Gases and Saturated Vapors. (Osterberg.)....12mo, 1 25
Reagan's Locomotive Mechanism and Engineering....12mo, 2 00
Rontgen's Principles of Thermodynamics. (Du Bois.)....8vo, 5 00
Sinclair's Locomotive Engine Running and Management..12mo, 2 00
Smart's Handbook of Engineering Laboratory Practice..12mo, 2 50
Snow's Steam-boiler Practice....8vo, 3 00
Spangler's Valve-gears....8vo, 2 50
" Notes on Thermodynamics....12mo, 1 00
Thurston's Handy Tables....8vo, 1 50
" Manual of the Steam-engine....2 vols., 8vo, 10 00
Part I.—History, Structure, and Theory....8vo, 6 00
Part II.—Design, Construction, and Operation....8vo, 6 00
Thurston's Handbook of Engine and Boiler Trials, and the Use of the Indicator and the Prony Brake....8vo, 5 00
" Stationary Steam-engines....8vo, 2 50
" Steam-boiler Explosions in Theory and in Practice....12mo, 1 50
" Manual of Steam-boilers, Their Designs, Construction, and Operation....8vo, 5 00
Weisbach's Heat, Steam, and Steam-engines. (Du Bois.)..8vo, 5 00
Whitham's Steam-engine Design....8vo, 5 00
Wilson's Treatise on Steam-boilers. (Flather.)....16mo, 2 50
Wood's Thermodynamics, Heat Motors, and Refrigerating Machines....8vo, 4 00

MECHANICS AND MACHINERY.

Barr's Kinematics of Machinery....8vo, 2 50
Bovey's Strength of Materials and Theory of Structures..8vo, 7 50
Chordal.—Extracts from Letters....12mo, 2 00
Church's Mechanics of Engineering....8vo, 6 00
" Notes and Examples in Mechanics....8vo, 2 00
Compton's First Lessons in Metal-working....12mo, 1 50
Compton and De Groodt. The Speed Lathe....12mo, 1 50
Cromwell's Treatise on Toothed Gearing....12mo, 1 50
" Treatise on Belts and Pulleys....12mo, 1 50
Dana's Text-book of Elementary Mechanics for the Use of Colleges and Schools....12mo, 1 50
Dingey's Machinery Pattern Making....12mo, 2 00
Dredge's Record of the Transportation Exhibits Building of the World's Columbian Exposition of 1893....4to, half mor., 5 00
Du Bois's Elementary Principles of Mechanics:
Vol. I.—Kinematics....8vo, 3 50
Vol. II.—Statics....8vo, 4 00
Vol. III.—Kinetics....8vo, 3 50
Du Bois's Mechanics of Engineering. Vol. I....Small 4to, 7 50
" " " " " Vol.II....Small 4to, 10 00

Durley's Elementary Text-book of the Kinematics of Machines. (*In preparation.*)
Fitzgerald's Boston Machinist....16mo, 1 00
Flather's Dynamometers, and the Measurement of Power. 12mo, 3 00
" Rope Driving....12mo, 2 00
Goss's Locomotive Sparks....8vo, 2 00
Hall's Car Lubrication....12mo, 1 00
Holly's Art of Saw Filing....18mo, 75
* Johnson's Theoretical Mechanics....12mo, 3 00
Johnson's Short Course in Statics by Graphic and Algebraic Methods. (*In preparation.*)
Jones's Machine Design:
Part I.—Kinematics of Machinery....8vo, 1 50
Part II.—Form, Strength and Proportions of Parts....8vo, 3 00
Kerr's Power and Power Transmission....8vo, 2 00
Lanza's Applied Mechanics....8vo, 7 50
MacCord's Kinematics; or, Practical Mechanism....8vo, 5 00
" Velocity Diagrams....8vo, 1 50
Merriman's Text-book on the Mechanics of Materials....8vo, 4 00
* Michie's Elements of Analytical Mechanics....8vo, 4 00
Reagan's Locomotive Mechanism and Engineering....12mo, 2 00
Reid's Course in Mechanical Drawing....8vo, 2 00
" Text-book of Mechanical Drawing and Elementary Machine Design....8vo, 3 00
Richards's Compressed Air....12mo, 1 50
Robinson's Principles of Mechanism....8vo, 3 00
Ryan, Norris, and Hoxie's Electrical Machinery. (*In preparation.*)
Sinclair's Locomotive-engine Running and Management. . 12mo, 2 00
Smith's Press-working of Metals....8vo, 3 00
" Materials of Machines....12mo, 1 00
Thurston's Treatise on Friction and Lost Work in Machinery and Mill Work....8vo, 3 00
" Animal as a Machine and Prime Motor, and the Laws of Energetics....12mo, 1 00
Warren's Elements of Machine Construction and Drawing. . 8vo, 7 50
Weisbach's Kinematics and the Power of Transmission. (Herrman—Klein.) 8vo, 5 00
" Machinery of Transmission and Governors. (Herrman—Klein.) 8vo, 5 00
Wood's Elements of Analytical Mechanics....8vo, 3 00
" Principles of Elementary Mechanics....12mo, 1 25
" Turbines 8vo, 2 50
The World's Columbian Exposition of 1893....4to, 1 00

METALLURGY.

Egleston's Metallurgy of Silver, Gold, and Mercury:
Vol. I.—Silver....8vo, 7 50
Vol. II.—Gold and Mercury....8vo, 7 50
** Iles's Lead-smelting....12mo, 2 50
Keep's Cast Iron....8vo, 2 50
Kunhardt's Practice of Ore Dressing in Europe....8vo, 1 50
Le Chatelier's High-temperature Measurements. (Boudouard—Burgess.) 12mo, 3 00
Metcalf's Steel. A Manual for Steel-users....12mo, 2 00
Smith's Materials of Machines....12mo, 1 00
Thurston's Materials of Engineering. In Three Parts....8vo, 8 00
Part II.—Iron and Steel....8vo, 3 50
Part III.—A Treatise on Brasses, Bronzes and Other Alloys and Their Constituents....8vo, 2 50

MINERALOGY.

Barringer's Description of Minerals of Commercial Value. Oblong, morocco, 2 50
Boyd's Resources of Southwest Virginia................8vo, 3 00
" Map of Southwest Virginia........Pocket-book form, 2 00
Brush's Manual of Determinative Mineralogy. (Penfield.).8vo, 4 00
Chester's Catalogue of Minerals....................8vo, paper, 1 00
Cloth, 1 25
" Dictionary of the Names of Minerals............8vo, 3 50
Dana's System of Mineralogy.... ......Large 8vo, half leather, 12 50
" First Appendix to Dana's New "System of Mineralogy." Large 8vo, 1 00
" Text-book of Mineralogy........................8vo, 4 00
" Minerals and How to Study Them...............12mo, 1 50
" Catalogue of American Localities of Minerals.Large 8vo, 1 00
" Manual of Mineralogy and Petrography..........12mo, 2 00
Egleston's Catalogue of Minerals and Synonyms..........8vo, 2 50
Hussak's The Determination of Rock-forming Minerals. (Smith.) ....Small 8vo, 2 00
* Penfield's Notes on Determinative Mineralogy and Record of Mineral Tests............................8vo, paper, 50
Rosenbusch's Microscopical Physiography of the Rock-making Minerals. (Idding's.)............................8vo, 5 00
* Tillman's Text-book of Important Minerals and Rocks..8vo, 2 00
Williams's Manual of Lithology..........................8vo, 3 00

MINING.

Beard's Ventilation of Mines..............................12mo, 2 50
Boyd's Resources of Southwest Virginia................8vo, 3 00
" Map of Southwest Virginia........Pocket-book form, 2 00
* Drinker's Tunneling, Explosive Compounds, and Rock Drills..............................4to, half morocco, 25 00
Eissler's Modern High Explosives........................8vo, 4 00
Fowler's Sewage Works Analyses.......................12mo, 2 00
Goodyear's Coal-mines of the Western Coast of the United States 12mo, 2 50
Ihlseng's Manual of Mining..............................8vo, 4 00
** Iles's Lead-smelting..................................12mo, 2 50
Kunhardt's Practice of Ore Dressing in Europe............8vo, 1 50
O'Driscoll's Notes on the Treatment of Gold Ores..........8vo, 2 00
Sawyer's Accidents in Mines...............................8vo, 7 00
Walke's Lectures on Explosives...........................8vo, 4 00
Wilson's Cyanide Processes..............................12mo, 1 50
Wilson's Chlorination Process...........................12mo, 1 50
Wilson's Hydraulic and Placer Mining..................12mo, 2 00
Wilson's Treatise on Practical and Theoretical Mine Ventilation 12mo, 1 25

SANITARY SCIENCE.

Folwell's Sewerage. (Designing, Construction and Maintenance.) 8vo, 3 00
" Water-supply Engineering.......................8vo, 4 00
Fuertes's Water and Public Health............ ....12mo, 1 50
" Water-filtration Works........................12mo, 2 50

Gerhard's Guide to Sanitary House-inspection..........16mo, 1 00
Goodrich's Economical Disposal of Towns' Refuse...Demy 8vo, 3 50
Hazen's Filtration of Public Water-supplies..............8vo, 3 00
Kiersted's Sewage Disposal...............................12mo, 1 25
Leach's The Inspection and Analysis of Food with Special Reference to State Control. (*In preparation.*)
Mason's Water-supply. (Considered Principally from a Sanitary Standpoint. 3d Edition, Rewritten.....8vo, 4 00
" Examination of Water. (Chemical and Bacteriological.) 12mo, 1 25
Merriman's Elements of Sanitary Engineering............8vo, 2 00
Nichols's Water-supply. (Considered Mainly from a Chemical and Sanitary Standpoint.) (1883.) 8vo, 2 50
Ogden's Sewer Design......................................12mo, 2 00
* Price's Handbook on Sanitation.........................12mo, 1 50
Richards's Cost of Food. A Study in Dietaries...........12mo, 1 00
Richards and Woodman's Air, Water, and Food from a Sanitary Standpoint....................................8vo, 2 00
Richards's Cost of Living as Modified by Sanitary Science.12mo, 1 00
* Richards and Williams's The Dietary Computer..........8vo, 1 50
Rideal's Sewage and Bacterial Purification of Sewage......8vo, 3 50
Turneaure and Russell's Public Water-supplies............8vo, 5 00
Whipple's Microscopy of Drinking-water...................8vo, 3 50
Woodhull's Notes on Military Hygiene.................16mo, 1 50

MISCELLANEOUS.

Barker's Deep-sea Soundings..............................8vo, 2 00
Emmons's Geological Guide-book of the Rocky Mountain Excursion of the International Congress of Geologists. Large 8vo, 1 50
Ferrel's Popular Treatise on the Winds..................8vo, 4 00
Haines's American Railway Management...............12mo, 2 50
Mott's Composition, Digestibility, and Nutritive Value of Food. Mounted chart, 1 25
" Fallacy of the Present Theory of Sound..........16mo, 1 00
Ricketts's History of Rensselaer Polytechnic Institute, 1824–1894..Small 8vo, 3 00
Rotherham's Emphasised New Testament...........Large 8vo, 2 00
" Critical Emphasised New Testament.......12mo, 1 50
Steel's Treatise on the Diseases of the Dog...............8vo, 3 50
Totten's Important Question in Metrology................8vo, 2 50
The World's Columbian Exposition of 1893..............4to, 1 00
Worcester and Atkinson. Small Hospitals, Establishment and Maintenance, and Suggestions for Hospital Architecture, with Plans for a Small Hospital.................12mo, 1 25

HEBREW AND CHALDEE TEXT-BOOKS.

Green's Grammar of the Hebrew Language.............8vo, 3 00
" Elementary Hebrew Grammar.................12mo, 1 25
" Hebrew Chrestomathy.........................8vo, 2 00
Gesenius's Hebrew and Chaldee Lexicon to the Old Testament Scriptures. (Tregelles.)........Small 4to, half morocco, 5 00
Letteris's Hebrew Bible......................................8vo, 2 25

Printed in Dunstable, United Kingdom

91220119R00218